Crowdsourcing Geographic Knowledge

Daniel Sui • Sarah Elwood • Michael Goodchild
Editors

Crowdsourcing Geographic Knowledge

Volunteered Geographic Information (VGI)
in Theory and Practice

Editors
Daniel Sui
Department of Geography
Centre for Urban & Regional Analysis
The Ohio State University
Columbus, OH, USA

Sarah Elwood
Department of Geography
University of Washington
Seattle, WA, USA

Michael Goodchild
Department of Geography
College of Letters & Science
University of California
Santa Barbara, CA, USA

ISBN 978-94-007-4586-5   ISBN 978-94-007-4587-2 (eBook)
DOI 10.1007/978-94-007-4587-2
Springer Dordrecht Heidelberg New York London

Library of Congress Control Number: 2012945073

© Springer Science+Business Media Dordrecht 2013
This work is subject to copyright. All rights are reserved by the Publisher, whether the whole or part of the material is concerned, specifically the rights of translation, reprinting, reuse of illustrations, recitation, broadcasting, reproduction on microfilms or in any other physical way, and transmission or information storage and retrieval, electronic adaptation, computer software, or by similar or dissimilar methodology now known or hereafter developed. Exempted from this legal reservation are brief excerpts in connection with reviews or scholarly analysis or material supplied specifically for the purpose of being entered and executed on a computer system, for exclusive use by the purchaser of the work. Duplication of this publication or parts thereof is permitted only under the provisions of the Copyright Law of the Publisher's location, in its current version, and permission for use must always be obtained from Springer. Permissions for use may be obtained through RightsLink at the Copyright Clearance Center. Violations are liable to prosecution under the respective Copyright Law.
The use of general descriptive names, registered names, trademarks, service marks, etc. in this publication does not imply, even in the absence of a specific statement, that such names are exempt from the relevant protective laws and regulations and therefore free for general use.
While the advice and information in this book are believed to be true and accurate at the date of publication, neither the authors nor the editors nor the publisher can accept any legal responsibility for any errors or omissions that may be made. The publisher makes no warranty, express or implied, with respect to the material contained herein.

Printed on acid-free paper

Springer is part of Springer Science+Business Media (www.springer.com)

This book is dedicated to the countless volunteers who have contributed time and effort to mapping the world in unprecedented detail, and to making the results freely available to all.

Acknowledgement

The editors would like to gratefully acknowledge the research support from Nick Crane and Wenqin Chen at Ohio State, Agnieszka Leszczysnki and Matt Wilson at UW, and Alan Glennon and Linna Li at UCSB for their assistance for this edited volume.

Contents

1 **Volunteered Geographic Information, the Exaflood, and the Growing Digital Divide** 1
Daniel Sui, Michael Goodchild, and Sarah Elwood

Part I Public Participation and Citizen Science

2 **Understanding the Value of VGI** 15
Rob Feick and Stéphane Roche

3 **To Volunteer or to Contribute Locational Information? Towards Truth in Labeling for Crowdsourced Geographic Information** 31
Francis Harvey

4 **Metadata Squared: Enhancing Its Usability for Volunteered Geographic Information and the GeoWeb** 43
Barbara S. Poore and Eric B. Wolf

5 **Situating the Adoption of VGI by Government** 65
Peter A. Johnson and Renee E. Sieber

6 **When Web 2.0 Meets Public Participation GIS (PPGIS): VGI and Spaces of Participatory Mapping in China** 83
Wen Lin

7 **Citizen Science and Volunteered Geographic Information: Overview and Typology of Participation** 105
Muki Haklay

Part II Geographic Knowledge Production and Place Inference

8 **Volunteered Geographic Information and Computational Geography: New Perspectives** 125
Bin Jiang

9 **The Evolution of Geo-Crowdsourcing: Bringing Volunteered Geographic Information to the Third Dimension** 139
Marcus Goetz and Alexander Zipf

10 **From Volunteered Geographic Information to Volunteered Geographic Services** 161
Jim Thatcher

11 **The Geographic Nature of Wikipedia Authorship** 175
Darren Hardy

12 **Inferring Thematic Places from Spatially Referenced Natural Language Descriptions** 201
Benjamin Adams and Grant McKenzie

13 **"I Don't Come from Anywhere": Exploring the Role of the Geoweb and Volunteered Geographic Information in Rediscovering a Sense of Place in a Dispersed Aboriginal Community** 223
Jon Corbett

Part III Emerging Applications and New Challenges

14 **Potential Contributions and Challenges of VGI for Conventional Topographic Base-Mapping Programs** 245
David J. Coleman

15 **"We Know Who You Are and We Know Where You Live": A Research Agenda for Web Demographics** 265
T. Edwin Chow

16 **Volunteered Geographic Information, Actor-Network Theory, and Severe-Storm Reports** 287
Mark H. Palmer and Scott Kraushaar

17 **VGI as a Compilation Tool for Navigation Map Databases** 307
Michael W. Dobson

18 **VGI and Public Health: Possibilities and Pitfalls** 329
Christopher Goranson, Sayone Thihalolipavan,
and Nicolás di Tada

19	**VGI in Education: From K-12 to Graduate Studies**	341
	Thomas Bartoschek and Carsten Keßler	
20	**Prospects for VGI Research and the Emerging Fourth Paradigm**	361
	Sarah Elwood, Michael F. Goodchild, and Daniel Sui	

Biographies of the Editors and Contributors ... 377

Index ... 387

Chapter 1
Volunteered Geographic Information, the Exaflood, and the Growing Digital Divide

Daniel Sui, Michael Goodchild, and Sarah Elwood

Abstract The phenomenon of volunteered geographic information is part of a profound transformation on how geographic data, information, and knowledge are produced and circulated. This chapter begins by situating this transition within the broader context of an "exaflood" of digital data growth. It considers the implications of VGI and the exaflood for further time-space compression and new forms and degrees of digital inequality. We then give a synoptic overview of the content of this edited collection and its three-part structure: VGI, public participation, and citizen science; geographic knowledge production and place inference; and emerging applications and new challenges. We conclude this chapter by discussing the renewed importance of geography and the role of crowdsourcing for geographic knowledge production.

1.1 Introduction

The past 5 years have witnessed a profound transformation of how geographic data, information, and, more broadly, knowledge have been produced and disseminated due to the phenomenal growth of a plethora of related technologies loosely known

---

D. Sui (✉)
Department of Geography, The Ohio State University, Columbus, OH, USA
e-mail: sui.10@osu.edu

M. Goodchild
Department of Geography, University of California at Santa Barbara,
Santa Barbara, CA, USA
e-mail: good@geog.ucsb.edu

S. Elwood
Department of Geography, University of Washington, Seattle, WA, USA
e-mail: selwood@u.washington.edu

as Web 2.0, cloud computing, and cyberinfrastructure. Although different lexicons have surfaced to describe this new trend by different communities, ranging from crowdsourcing to user-generated content, from Geoweb to the semantic Web, from volunteered geographic information to neogeography, PostGIS, citizen science, and eScience, the general idea coalesces around the use of the Internet to create, share, and analyze geographic information via multiple computing devices/platforms (traditional desktops, iPads, or smart phones).

Ever since the term volunteered geographic information (VGI) officially appeared in the literature (Goodchild 2007), there have been meetings and workshops devoted to the topic, including (to our knowledge) the 2007 NCGIA VGI workshop,[1] the AutoCarto 2008 workshop,[2] the USGS 2010 VGI workshop,[3] the GIScience 2010 VGI workshop,[4] and the 2011 VGI Pre-Conference at AAG.[5] Scholarly literature has also grown significantly, as evidenced by several special issues devoted exclusively to the theme of VGI in *GeoJournal* (Elwood 2008a, b), *Journal of Location-Based Services* special VGI issue (Rana and Joliveau 2009), and *Geomatica* (Feick and Roche 2010). In addition to these special issues devoted to VGI, research related to VGI has also been reported by an interdisciplinary group of researchers (Bennett 2010; Hall et al. 2010; Newman et al. 2010; Newsam 2010; Ramm and Topf 2010; Warf and Sui 2010; Kessler 2011; Obe and Hsu 2011; Roche et al. 2011).

The goal of this edited volume is to take stock of recent advances in VGI research, with particular emphasis on the role of VGI as crowdsourced data for geographic knowledge production. By doing so, we plan not only to present a state-of-the-art view of VGI as a research area but also to discuss the prospects and directions of VGI research in the near future. More than half of the chapters in this volume were based upon papers originally presented during the pre-AAG conference we organized on "Volunteered Geographic Information (VGI): Research progress and new developments"[5] on April 11, 2011, in Seattle, Washington. We also solicited additional contributions to cover topics not adequately addressed at the Seattle conference but crucial for future VGI research.

The rest of this introductory chapter is organized as follows. We first situate the phenomenon of VGI in the broader context of the big-data wave, also known as the exaflood. We then discuss the increasing digital divide and uneven practices of VGI across the world, followed by a synoptic overview of other chapters in this book. We end this chapter by discussing the role of crowdsourcing in geographic knowledge production and the evolving role of GIScience and geography in the era of big data in achieving a better understanding of the world.

---

[1] http://www.ncgia.ucsb.edu/projects/vgi (accessed February 16, 2012).
[2] http://mapcontext.com/autocarto/web/AutoCarto2008.html (accessed February 16, 2012).
[3] http://cegis.usgs.gov/vgi (accessed February 16, 2012).
[4] http://www.ornl.gov/sci/gist/workshops/agenda.shtml (accessed February 16, 2012).
[5] http://vgi.spatial.ucsb.edu (accessed February 16, 2012).

1.2 VGI and the Exaflood of Big Data

Until recently, the geospatial community has had a rather narrow definition of what is considered geographic data or information, often heavily influenced by the legacy of traditional cartography. But rapid advances in a plethora of technologies – GPS, smart phones, sensor networks, cloud computing, etc., especially all of the technologies loosely called Web 2.0 – have radically transformed how geographic data are collected, stored, disseminated, analyzed, visualized, and used. This trend is best reflected in Google's mantra that "Google Maps = Google in Maps" (Ron 2008). The insertion of an "in" between Google and Maps perhaps signifies one of the most fundamental changes in the history of human mapping efforts. Nowadays, users can search though Google Maps not only for traditional spatial/map information but also for almost any kind of digital information (such as Wikipedia entries, Flickr photos, YouTube videos, and Facebook/Twitter postings) as long as it is geotagged. Furthermore, in contrast to the traditional top-down authoritative process of geographic data production by government agencies, citizens have played an increasingly important role in producing geographic data of all kinds through a bottom-up crowdsourcing process. As a result, we now have massive amounts of geocoded data growing on a daily basis from genetic to global levels covering almost everything we can think of on or near the Earth's surface, on the average of 1 exabyte per day (Swanson 2007). For the first time in human history, we now have the capability to keep track of where everything is in real time.

Due to the ubiquity of information-sensing mobile devices, aerial sensory technologies (remote sensing), software logs, cameras, RFID (radio-frequency identification) readers, wireless sensor networks, and other types of data-gathering devices, 1–5 exabytes (1 exabyte = 10^{18} bytes) of data are created daily and 90% of the data in the world today were created within the past 2 years (MacIve 2010). The amount of data humanity creates is doubling every 2 years; 2010 is the first year that we reached 1 zettabyte (10^{21} bytes).[6] 2011 alone generated approximately 1.8 zettabytes of data. The explosive growth of big data is rapidly transforming all aspects of governments, businesses, education, and science. By 2020, the volume of the world's data will increase by 50 times from today's volume (Gantz and Reinsel 2011). We will need 75 times more IT-related infrastructure in general and ten times more servers to handle the new data. Metaphors of data storage have evolved from bank to warehouse, to portal, and now to cloud. Data storage cost has dropped dramatically during the past two decades. Between 2005 and 2011 alone, costs of storage dropped by 5/6. Not surprisingly, how to deal with the new reality of big data is on the top of the agenda of government, industry, and multiple disciplines in the academy (IWGDD 2009; CORDIS 2010; Manyika et al. 2011).

---

[6] We are aware of the inconsistencies in the estimated volume of data available so far, but we found remarkable similarities in the magnitude and range of digital data volumes. In the chapter, we relied on data primarily from EMC[2]: http://www.emc.com/leadership/programs/digital-universe.htm (accessed February 16, 2012).

Although it is a challenging task to estimate the precise volume of geospatial data out there, we can safely say geospatial data is becoming an important part of the big-data torrent. Geospatial information in general and VGI in particular should be understood in the context of big data, and indeed, crowdsourcing, the Internet of things, and big data are rapidly converging in the domain of geospatial technologies (Ball 2011). Of course, due to rapid technological advances, what is considered as big vs. small is a moving target. In the McKinsey report (Manyika et al. 2011), "personal location data" has been singled out as one of the five primary big-data streams. With approximately 600 billion transactions per day, various mobile devices are creating approximately one petabyte (10^{15} bytes) of data per year globally. Personal location data alone is a $100-billion business for service providers and $700 billion to end users (Manyika et al. 2011). The other four streams of big data identified by the McKinsey Institute – health care, public-sector administration, retail, and manufacturing – also have a significant amount of data either geocoded or geotagged. So geospatial data are not only an important component of big data but are actually, to a large extent, big data themselves. For the geospatial community, big data presents not only bigger opportunities for the business community (Francica 2011; Killpack 2011) but also new challenges for the scientific and scholarly communities to conduct ground-breaking studies related to people (at both individual and collective levels) and environment (from local to global scale) (Elkus 2011; Meek 2011; Hayes 2012).

In fact, the geospatial community was tackling big-data issues even before "big data" became trend (Miller 2010). From very early on, geospatial technologies were at the forefront of big-data challenges, primarily due to the large volumes of raster (remote-sensing imagery) and vector (detailed property surveys) data that need to be stored and managed. Back in 1997 when Microsoft Research initiated a pilot project to demonstrate database scalability, they used aerial imagery as the primary data (Ball 2011). The TerraServer Microsoft developed then is still in use and functional today and set the standard and protocol for today's other remote-sensing image serving sites such as OpenTopography.org (LiDAR data). Furthermore, to implement Al Gore's (1999) vision of a "digital earth" requires big data. Although the concept of digital earth did not evolve quite as Gore envisioned during the past decade, the growing popularity of Google Earth, Microsoft's Virtual Earth (now Bing Maps), and NASA's World Wind is an indication that geospatial and mapping tools are crucial for users to navigate through the big-data torrent.

1.3 VGI in Shrinking and Divided World

As a concomitant growth of this ever-expanding digital universe filled with big data, the world (people, made objects and things, and environment) is increasingly being recorded, referenced, and connected by vast digital networks. Geographers, along with scholars in multiple other disciplines, have noted the acceleration of our temporal experience and the reduction of the role of distance for quite some time, as evidenced by the scholarly literature on time-space compression, time-space distanciation, or

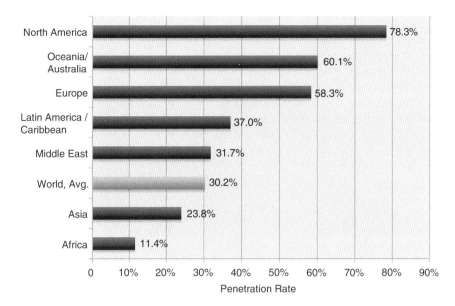

Fig. 1.1 World internet penetration rate by geographic regions – 2011 (http://www.internetworldstats.com/stats.htm)

space-time convergence (Warf 2008). The growing popularity of social media on the global scene has pushed time-space compression to new levels.

Using a more popular term, the world is rapidly becoming smaller as a result of space-time convergence. When social psychologist Stanley Milgram (1967) did his experiment back in the 1960s on how many meaningful steps are needed to connect two strangers on the surface of the Earth, Milgram and his team (Travers and Milgram 1969) concluded then that on average it takes six steps to make a meaningful connection for two randomly selected individuals, later popularized as six degrees of separation by American playwright Paul Guare (in his 1990 play "Six Degrees of Separation") and the game Six Degrees of Kevin Bacon (to connect people with Hollywood stars). In December 2011, Facebook and Yahoo! conducted a new analysis using the massive amount of data harvested from social media, and they concluded that six degrees of separation had been reduced to 4.7 by the end of 2011, largely as a result of people being increasingly connected on-line.[7]

Almost paradoxically, as some parts of the world are flooded by big data and people are increasingly connected in a shrinking world, we must also be keenly aware that this world remains a deeply divided one – both physically and digitally (Fig. 1.1). While a large majority of people in North America and Europe have access to the Internet (with Internet penetration rates at 78.3% and 58.3%, respectively, by the end of 2011), two-thirds of humanity do not have access to the rapidly expanding digital world; the world average Internet penetration rate is 30.2% with Asia (23.8%)

---

[7] http://www.physorg.com/news/2011-11-degrees.html (accessed February 16, 2012).

and Africa (11.4%) trailing at the bottom.[8] The geographical distribution of new digital data stored in 2010 reflects both the digital divide and uneven development levels across the globe, with the developed world or global north (North America and Europe) having 10–70 times more data than the developing world or global south (Africa, Latin America, and Asia) (Manyika et al. 2011). A third of humanity (about two billion people) still lives on under $2 a day.[9] We should also be mindful that sometimes simply having access to gadgets themselves is not enough. Many iPhone users in the developed world have enjoyed using one of multiple versions of restroom locators (e.g., have2p), but for a country like India, where there are more cell phones than toilets, simply having have2p installed on one's iPhone would not help much in rural areas due to the severe lack of sanitary infrastructure.[10]

In the context of geographic information (and to some extent other types of data as well), the biggest irony remains that Murphy's law is still at work – information is usually the least available where it is most needed. We have witnessed this paradox unfolding painfully in front of our eyes in the Darfur crisis in northern Sudan (2006), the aftermath of the Haiti earthquake (2010), and the BP explosion in the Gulf of Mexico (2011). Undoubtedly, how to deal with big data in a shrinking and divided world will be a major challenge for GIS and geography in the years ahead. The strengths, weaknesses, opportunities, and threats (SWOT) of VGI for improving the spatial data infrastructure (SDI) are quite different in the two global contexts of north and south (Genovese and Roche 2010). Furthermore, as Gilbert and Masucci (2011) show so clearly in their recent work on uneven information and communication geographies, we must move away from the traditional, linear conceptualization of a digital divide, concerned primarily with physical access to computers and the Internet. Instead, we must consider the multiple divides within cyberspace (or digital apartheid) by taking into account the hybrid, scattered, ordered, and individualized nature of cyberspaces (Graham 2011). Indeed, multiple hidden social and political factors are at play for determining what is or is not available on-line (Engler and Hall 2007). Internet censorship (Warf 2011; MacKinnon 2012), power laws (or the so-called 80/20 rule) (Shirkey 2006), homophile tendencies in human interactions (de Laat 2010; Merrifield 2011), and fears of colonial and imperial dominance (Bryan 2010) are also important factors to consider for the complex patterns of digital divide and uneven practices of VGI at multiple scales on the global scene.

1.4 Overview of Chapters in This Book

This book is organized into 20 chapters. Chapter 1 by the editors situates VGI in the broader context of big data and the growing global digital divide. The substantive chapters (2–19) are grouped in three parts.

---

[8] http://www.internetworldstats.com/stats2.htm (accessed February 16, 2012).

[9] http://givewell.org/international/technical/additional/Standard-of-Living (accessed February 16, 2012).

[10] http://www.globalpost.com/dispatch/india/100507/mobile-phones-toilets-sanitation-health (accessed February 16, 2012).

The six chapters in Part I focus on VGI, public participation, and citizen science. In Chap. 2, Rob Feick and Stéphane Roche extend our conceptualization of the value of VGI and geographic information (GI) more generally. The authors recognize that the proliferation of VGI has complicated our assessments of GI's value. They review these complications in their specificity and propose several new metaphors, such as unexpected discoveries, Debord's "dérives," or Lego blocks, which might be used as a guide for future valuation of VGI. Chapter 3 by Francis Harvey questions whether "volunteered" completely captures the character of crowdsourced data. The author suggests that crowdsourced data can be disaggregated into categories of "volunteered" and "contributed." The distinction between CGI and VGI is argued to be important for assessing particular crowdsourced data's fitness for use and for identifying biases or inaccuracies. In Chap. 4, Barbara Poore and Eric Wolf track the changing discourse on geospatial metadata and – through two case studies – point to ongoing transformations in popular and academic engagement with metadata in the Geoweb. The authors suggest that we are in the midst of a shift and indeed *promote* a shift, from traditional unidirectional construction of metadata to more interactive user-friendly production of metadata. Chapter 5 by Peter Johnson and Renee Sieber contextualizes VGI somewhat differently, focusing on the vicissitudes of its adoption by government for interaction with citizens. Through reflection on their work with government agencies in Québec, the authors identify the different ways that VGI has been incorporated and discuss key obstacles to and constraints on further incorporation. In Chap. 6, Wen Lin examines the politics of citizen participation and processes of subjectification that are now emerging from the encounter of Web 2.0 and public participation GIS (PPGIS). The author works with reference to a case study in China, with three specific examples of VGI mapping drawn from ethnographic fieldwork. Out of a meeting of Web 2.0 and PPGIS have emerged transformations in Chinese citizenship and new spaces of citizen participation. Finally, as a contribution to the disaggregation of the umbrella term "VGI," and also as a challenge to the exclusivity of a professionalized science that would marginalize voluntary (nonprofessional) practitioners, Chap. 7 by Muki Haklay discusses the specificity, historical trajectory, social context, power relations within, and promise of *citizen science*.

The six chapters in Part II concentrate on geographic knowledge production and place inference. With an interest in the opportunities generated by the flood of publicly available VGI, Chap. 8 by Bin Jiang contextualizes computational geography and reviews recent work in the field to demonstrate the promise of research that engages with big data. Jiang's discussion of topological thinking also clarifies the challenge that contemporary computational geography presents to conventional views of space. In Chap. 9, Marcus Goetz and Alexander Zipf attend to the transformation of VGI from its early basis in two-dimensional geographical information to its contemporary inclusion of three-dimensional (3D) data. Through specific attention to OpenStreetMap, Goetz and Zipf emphasize emerging applications of 3D data for city modeling and building modeling. Chapter 10 by Jim Thatcher engages with "volunteered geographic services" (VGS), the term he uses to describe discrete actions made possible through spatially aware mobile devices like smart phones. Thatcher suggests that VGS pushes beyond the limits of VGI by linking users through time and space and facilitating coordination of actions on the ground. Thatcher reviews possible uses of VGS in crisis response through the example of the PSUMobile.org. Chapter 11 by Darren Hardy examines the geography of VGI authorship, focusing specifically on the case of Wikipedia and its geographic articles. Exemplifying the analysis of big data, the author describes a study of 32 million contributions to those articles over 7 years. Contradicting assertions of the Internet's placelessness, Hardy finds that authorship of Wikipedia articles demonstrates distance decay. In Chap. 12, Benjamin Adams and Grant McKenzie draw together geographical insights on sense of place and techniques of computational representation, specifically latent Dirichlet allocation (LDA). The authors discuss topic modeling with VGI on travel blogs with an eye to identifying places with topics, calculating similarity between places, and evaluating changes in sense of place using computational methods. Chapter 13 is also concerned with the social relations of scientific practice: Jon Corbett writes of VGI in the context of a collaborative mapping project. Discussing his work with an aboriginal community, Corbett indicates that participatory mapping can cultivate a sense of place but that such collaborative projects demand reflexivity on the part of researchers.

The six chapters in Part III cover emerging applications and new challenges. In Chap. 14, David J. Coleman engages with underlying assumptions about VGI through consideration of conventional digital topographic mapping programs. It is argued that the updating and maintenance of maps cannot rely on VGI alone, even if VGI does indeed represent an important alternative and complementary source for data, which must be given further attention. Chapter 15 by T. Edwin Chow situates VGI in the field of Web demographics. Within this field are a whole host of Web-based systems that acquire, sort, and utilize personal data. At issue for Chow is how the field of Web demographics complicates generalizations about VGI – for instance, about the degree of voluntarism attributable to donors of geographical and personal information or the accuracy of such information. In Chap. 16, Mark H. Palmer and Scott Kraushaar employ actor-network theory (ANT) to describe a storm-tracking network that relies to a great extent on VGI. ANT presents itself as especially useful for their analysis by facilitating consideration of co-constitutive relations between society and technology. Specific to this case, ANT provides adequate flexibility for narration of both the centralized and decentralized processes on which storm reporting relies. Chapter 17 by Michael W. Dobson examines the gathering and compilation of VGI for mapping databases that, in some cases, also rely on traditional map database compilation techniques. The author reviews the promise and pitfalls of compilation systems and considers how the latter – pitfalls – might be overcome. In Chap. 18, Christopher Goranson, Sayone Thihalolipavan, and Nicolás di Tada consider the potential utility and possible pitfalls of recent advances in bringing together VGI and (public) health research. The updatability and time sensitivity of VGI are cited as central to the potential contributions. The authors also acknowledge that the use of VGI introduces new challenges – ethical and practical – particularly with regard to privacy. Chapter 19 by Thomas Bartoschek and Carsten Keßler deals with the heretofore largely neglected role of VGI in education and how it does or may transform curricula at a variety of levels, from primary education to graduate studies. The authors discuss how VGI has been introduced into classrooms and – through analysis of survey data – examine the motivations for and impediments to continued use of different VGI platforms.

In the last chapter (Chap. 20), we discuss the prospects for VGI research and its implications for GIScience and geography in the context of the fourth paradigm – data-intensive scientific inquiry.

1.5 Crowdsourcing Geographic Knowledge: From the Death of Distance to the Revenge of Geography

Back in 1995, The Economist magazine made (in retrospect) a premature announcement of the death of distance (Cairncross 1995) – the idea that distance (by implication location and more broadly geography) plays a less important role in the operation of an increasingly globally connected society in the age of rapid advances and innovations in information and communications technologies. However, it took less than 10 years before The Economist published another cover story on the theme of the revenge of geography (The Economist 2003). It turned out that, in an increasingly connected world, the precise role of distance in many societal functions may have changed, but location, and more broadly geography, has assumed a more crucial role in economic and business activities as well as in social and cultural affairs. More than ever, wireline and wireless technologies have bound the virtual and physical worlds closer (Gordon and de Souza e Silva 2011).

To us, the revenge of geography not only suggests the growing importance of location and geocoding or geotagging in the ocean of big data but also the heightened sense of and deeper appreciation for the growing divide and uneven development of an increasingly interconnected world (Hecht and Moxley 2009; Warf 2010). Situating VGI in the context of big data is only the first step to realize VGI's much broader potential and impacts down the road. Furthermore, VGI must be placed also in the context of crowdsourcing geographic knowledge about the world. Gould (1999) anticipated the arrival of a spatial century and further argued that "there is a geographer in most people (p. 314)." Multiple new technological advances during the past two decades have indeed unleashed the potential of a geographer within everybody. The phenomenon of VGI that emerged during the first decade of the twenty-first century is one of the many manifestations of a spatial century. VGI represents an unprecedented shift in the content, characteristics, and modes of geographic information creation, sharing, dissemination, and use. To us, this is the essence of the revenge of geography in the age of Web 2.0.

Big data obviously demands big machines (in terms of both speed and storage) for us to succeed in the number crunching needed to make use of them. But more importantly, big data also demands big ideas to address the world's big problems effectively. With the support of new cyberinfrastructure, new creative partnerships among government agencies, NGOs, industry and businesses, the academy, and citizens can be formed. It is gratifying to read the stories about Water Hackathon (waterhackathon.org). The World Bank has sponsored over 2,500 projects like this in more than 30,000 locations all over the world, and geospatial technologies have played crucial roles in all of these projects. Most recently, the World Bank has partnered with Google to make Google Map Maker's global mapping platform available in over 150 countries and 60 different languages, which has enabled citizen cartographers to help those in dire need.[11] We all should do no less. In the chapters that follow in the book, the reader will experience an interdisciplinary perspective on how we can rely on VGI to engage in a new mode of geographic knowledge production through crowdsourcing for a more efficient, more equitable, and sustainable world – to us that will be the most gratifying result of the revenge of geography.

References

Ball, M. (2011). How do crowdsourcing, the internet of things and big data converge on geospatial technology? http://www.vector1media.com/spatialsustain/how-do-crowdsourcing-the-internet-of-things-and-big-data-converge-on-geospatial-technology.html. Accessed January 22, 2012.
Bennett, J. (2010). *OpenStreetMap*. Birmingham: Packt Publishing Ltd.
Bryan, J. (2010). Force multipliers: Geography, militarism, and the Bowman expeditions. *Political Geography, 29*(8), 414–416.
Cairncross, F. (1995). The death of distance. *Economist, 336*(7934), 5–6 (30 September).
CORDIS (2010). *Riding the wave: How Europe can gain from the rising tide of scientific data.* Final report of the high level expert group on scientific data. http://cordis.europa.eu/fp7/ict/e-infrastructure/docs/hlg-sdi-report.pdf
de Laat, P. B. (2010). How can contributors to open-source communities be trusted? On the assumption, inference, and substitution of trust. *Ethics of Information Technology, 12*(4), 327–341.
Elkus, A. (2011). Hurricane Irene: GIS, social media, and big data shine. http://ctovision.com/2011/08/hurricane-irene-gis-social-media-and-big-data-shine. Accessed January 11, 2012
Elwood, S. (2008a). Volunteered geographic information: Key questions, concepts and methods to guide emerging research and practice. *GeoJournal, 72*(3/4), 133–135.
Elwood, S. (2008b). Volunteered geographic information: Future research directions motivated by critical, participatory, and feminist GIS. *GeoJournal, 72*, 173–183.
Engler, N. J., & Hall, G. B. (2007). The Internet, spatial data globalization, and data use: The case of Tibet. *The Information Society, 23*, 345–359.
Feick, R. D., & Roche, S. (2010). Introduction (to special issue on VGI). *Geomatica, 64*(1), 5–6.
Francica, J. (2011). Big data and why you should care. http://apb.directionsmag.com/entry/big-data-and-why-you-should-care/167326. Accessed January 21, 2012.
Gantz, J., & Reinsel, D. (2011). Extracting value from chaos. http://www.emc.com/collateral/analyst-reports/idc-extracting-value-from-chaos-ar.pdf. Accessed January 21, 2012.
Genovese, E., & Roche, S. (2010). Potential of VGI as a resource for SDIs in the North/South context. *Geomatica, 64*(4), 439–450.
Gilbert, M., & Masucci, M. (2011). *Information and communication technology geographies: Strategies for bridging the digital divide*. Vancouver: Praxis (e) Press – University of British Columbia.

---

[11] http://www.nytimes.com/2012/01/14/opinion/empowering-citizen-cartographers.html?_r=2 (accessed February 16, 2012).

Goodchild, M. F. (2007). Citizens as sensors: The world of volunteered geography. *GeoJournal, 69*(4), 211–221.

Gordon, E., & de Souza e Silva, A. (2011). *Net locality: Why location matters in a networked world*. New York: Wiley-Blackwell.

Gore, A. (1999). The digital earth: Understanding our planet in the 21st century. portal.opengeospatial.org/files/?artifact_id=6210. Accessed February 16, 2012.

Gould, P. (1999). *Becoming a geographer*. Syracuse: Syracuse University Press.

Graham, M. (2011). Time machines and virtual portals: The spatialities of the digital divide. *Progress in Development Studies, 11*(3), 211–227.

Hall, B. G., Chipeniuk, R., Feick, R. D., Leahy, M. G., & Deparday, V. (2010). Community-based production of geographic information using open source software and Web 2.0. *International Journal of Geographical Information Science, 24*(5), 761–781.

Hayes, C. (2012). Geospatial and big data: The challenge of leveraging constantly evolving Information. Presentation during 2012 Defense Geospatial Intelligence (DGI), London, January 24, 2012.

Hecht, B., & Moxley, E. (2009). Terabytes of Tobler: Evaluating the first law in a massive, domain-neutral representation of world knowledge. In *COSIT'09 Proceedings of the 9th International Conference on Spatial Information Theory* (pp. 88–105). Berlin: Springer.

Interagency Working Group on Digital Data (IWGDD) (2009). Harnessing the power of digital data for science and society. http://www.nitrd.gov/About/Harnessing_Power_Web.pdf

Kessler, F. (2011). Volunteered geographic information: A bicycling enthusiast perspective. *Cartography and Geographic Information Science, 38*(3), 258–268.

Killpack, C. (2011). Big data, bigger opportunity. http://www.geospatialworld.net/images/magazines/gw-april11–18–26%20Cover%20Story.pdf. Accessed January 31, 2012.

MacIve, K. (2010). Google chief Eric Schmidt on the data explosion. http://www.i-cio.com/features/august-2010/eric-schmidt-exabytes-of-data. Accessed February 16, 2012.

MacKinnon, R. (2012). *Consent of the networked: The worldwide struggle for Internet freedom*. New York: Basic Books.

Manyika, J., Chui, M., Brown, B., Bughin, J., Dobbs, R., Roxburgh, C., & Byers, A. H. (2011). Big data: The next frontier for innovation, competition, and productivity. http://www.mckinsey.com/Insights/MGI/Research/Technology_and_Innovation/Big_data_The_next_frontier_for_innovation

Meek, D. (2011). YouTube and social movements: A phenomenological analysis of participation, events, and cyberplace. In *Antipode*. Epub ahead of print. Accessed January 4, 2012. doi: 10.1111/j.1467–8330.2011.00942.x

Merrifield, A. (2011). Crowd politics: Or, 'Here Comes Everybuddy'. *New Left Review, 71*, 103–114.

Milgram, S. (1967). The small world problem. *Psychology Today, 2*, 60–67.

Miller, H. J. (2010). The data avalanche is here: Shouldn't we be digging? *Journal of Regional Science, 50*, 181–201.

Newman, G., Zimmerman, D., Crall, A., Laituri, M., Graham, J., & Stapel, L. (2010). User-friendly web mapping: Lessons from a citizen science website. *International Journal of Geographical Information Science, 24*(12), 1851–1869.

Newsam, S. (2010). Crowdsourcing what is where: Community-contributed photos as volunteered geographic information. *IEEE Multimedia, 17*(4), 36–45.

Obe, R., & Hsu, L. (2011). *PostGIS in action*. Stamford: Manning Publications.

Ramm, F., & Topf, J. (2010). *OpenStreetMap: Using and enhancing the free map of the world*. Cambridge: UIT Cambridge Ltd.

Rana, S., & Joliveau, T. (2009). Neogeography: An extension of mainstream geography for everyone made by everyone? *Journal of Location Based Services, 3*(2), 75–81.

Roche, S., Propeck-Zimmermann, E., & Mericskay, B. (2011). GeoWeb and crisis management: Issues and perspectives of volunteered geographic information. *GeoJournal,* Epub ahead of print. Accessed January 4, 2012. doi: 10.1007/s10708–011–9423–9.

Ron, L. (2008). Google maps = Google on maps. http://blip.tv/oreilly-where-20-conference/lior-ron-google-maps-google-on-maps-975838

Shirkey, C. (2006). Power laws, weblogs, and inequality. In J. Dean, J. W. Anderson, & G. Lovink (Eds.), *Reformatting politics* (pp. 35–42). New York: Routledge.
Swanson, B. (2007). The coming exaflood. *Wall Street Journal* (January 20). http://www.discovery.org/scripts/viewDB/index.php?command=view&id=3869. Accessed March 1, 2010.
The Economist (2003). The revenge of geography. http://www.economist.com/node/1620794
Travers, J., & Milgram, S. (1969). An experimental study of the small world problem. *Sociometry, 32*, 425–434.
Warf, B. (2008). *Time-space compression: Historical geographies*. London: Routledge.
Warf, B. (2010). Uneven geographies of the African Internet: Growth, change, and implications. *African Geographical Review, 29*(2), 41–66.
Warf, B. (2011). Geographies of global Internet censorship. *GeoJournal, 76*(1), 1–23.
Warf, B., & Sui, D. (2010). From GIS to neogeography: Ontological implications and theories of truth. *Annals of GIScience, 26*(4), 197–209.

Part I
Public Participation and Citizen Science

Chapter 2
Understanding the Value of VGI

Rob Feick and Stéphane Roche

Abstract Growing investments of time, money and other resources in the production of geographic information (GI) in concert with the increasingly widespread use of GI throughout society are often accompanied by statements that reference the economic, cultural and social value of GI. Despite considerable effort over the past decade, our capacity to quantify the value of GI or even understand how value should be conceptualized remains limited. The recent emergence of volunteered geographic information (VGI) has introduced several new facets to the challenge of understanding the value of (V)GI. This chapter examines how VGI use and production are challenging our understanding of how GI and VGI alike are valued. Following a review of the traditional approaches to valuing GI, the chapter explores the distinctive characteristics of VGI use and production that introduce new dimensions to value. More specifically, the chapter proposes several metaphors (serendipitous and unexpected discovery, Debord's 'Dérives' metaphor, Lego block theory) that can be used to conceptualize VGI value and the potential to adapt the 'fitness-for-use' concept to guide user assessments of VGI value in practice.

2.1 Introduction

There is growing evidence that many decision makers in the public, private and non-governmental sectors recognize that geographic information (GI) is, or can be, of considerable value (Crompvoets et al. 2010). Definitions of value vary considerably;

R. Feick (✉)
School of Planning, University of Waterloo, Waterloo, ON, Canada
e-mail: rdfeick@uwaterloo.ca

S. Roche
Département des sciences géomatiques, Université Laval, Québec, QC, Canada
e-mail: Stephane.Roche@scg.ulaval.ca

however, the term value is broadly understood to capture, at least in a nominal sense, the concepts of importance or worth (Longhorn and Blakemore 2008). Public expenditures on spatial data infrastructures, mapping agencies and spatial data sets and a proliferation of private investments in GI-centric products and services (e.g. GM's OnStar, Oracle Spatial) are some of the most tangible examples of the economic importance of GI. Similarly, concerns related to the social value of GI (Roche et al. 2003) and particularly equity imbalances associated with differential access to GI provided much of the initial impetus for public participation GIS (PPGIS) research and practice.

While there may be widespread recognition that GI has significant economic and social value, our understanding of how to quantify or even conceptualize the value of GI remains limited (Crompvoets et al. 2010; Longhorn and Blakemore 2008). The rapid emergence of user-generated, or volunteered, geographic information (VGI) as a complementary model for spatial data use and production has introduced several new facets to the challenge of understanding the value of (V)GI (Goodchild 2007). Some of these changes, such as advertising-supported GI business models and cost savings that firms realize through user-generated and crowdsourced data, are relatively straightforward to conceptualize and monetize as they can be seen to some extent as variants of past practices (Rana and Joliveau 2009). In contrast, the sociopolitical dimensions of VGI value are both more interesting and more problematic to consider in light of traditional valuing approaches. For example, questions of who benefits, and to what extent, from the evolving processes for creating and using VGI become less tractable as one considers that the nature of social relations across the Geoweb is often transient and issue and place specific (Elwood 2010). Similarly, conceptualizing the value of the representations, processes and outcomes associated with VGI is all the more challenging given that the boundaries of what constitutes 'geographic information' are substantially more fluid and opaque in the Web 2.0 realm than in the preceding era (Haklay et al. 2008).

This chapter explores these dimensions of VGI value and more generally examines how the phenomenon of VGI use and production is challenging our understanding of how GI and VGI alike are valued. A foundation for this discussion is built in the next two sections which (a) discuss how the concept of value has been defined and subsequently applied to geographic information and (b) review briefly concepts and approaches integral to valuing traditional or authoritative GI. Next, the discussion turns to the specific characteristics of VGI which require that traditional valuing approaches be reconsidered. The chapter concludes by suggesting several metaphors that can be used to conceptualize VGI value and the potential to adapt the 'fitness-for-use' concept to guide user assessments of VGI value in practice.

2.2 Defining Value and the Value of Geographic Information

Although the term 'value' permeates both popular discourse and academic research, the task of defining exactly what the word means has proven to be somewhat elusive. Rodriguez (2005) noted that a broad distinction can be made between philosophical conceptualizations of value and those that are rooted in concepts of economic worth. From a philosophical perspective, value is used most frequently to refer to individuals' core beliefs and is linked to concepts of morals, ethics and behaviour, and judging whether individuals' or societal choices can be deemed to be right or wrong. This conceptualization is central to discussions relating to the ethical use of information technology as illustrated in standards of behaviour for the GI profession (e.g. GIS Certification Institute's Code of Ethics) and ongoing debates concerning the social impact of GI technology, particularly but not exclusively in terms of privacy and surveillance (see, e.g. Perkins and Dodge 2009; Pickles 1995; Elwood and Leszczynski 2011). While this research is clearly of interest to the broader GIScience community, this chapter focuses on the perspective where value is seen as an indicator of the worth or significance of a specific good or service.

Economic theory has historically associated value with concepts such as utility, benefit and willingness to pay and has used it to explore issues as diverse as consumer behaviour, local patterns of government expenditures and the economic worth of protecting endangered species (Musgrave 1939; Tiebout 1956; Richardson and Loomis 2009). Despite its central presence in economics theory and continued efforts to view value through a common monetary lens, economists have long grappled with the multifaceted nature of the concept. The value of private or near-private goods (i.e. excludable, rivalrous), for example, can be approximated reasonably well as a function of purchase price paid (exchange value) and/or the sum of production, distribution and implementation costs (Genovese et al. 2009a). In contrast, the value of goods and services that display characteristics of common property or public goods (e.g. national parks, clean air and ocean fish stocks) cannot be monetized as readily since markets fail when potential users cannot be excluded by pricing or exclusive ownership mechanisms. Some goods and services (e.g. health care, telecommunications systems) are recognized for being of value to both individuals and to society as a whole, albeit to different degrees. Others may be considered of worth for cultural or political reasons (e.g. religious sites, historic structures). Even in the cases where there is widespread agreement on their importance, quantifying these dimensions of value is still particularly difficult. Geographic information typically has characteristics of both private and public goods or services (e.g. cadastral parcels, land cover, road networks) which makes it problematic to express through monetary metrics and more generally contributes to difficulty understanding its value.

Irrespective of whether value is articulated through economic metrics such as money or qualitative comparisons (e.g. quality of life indices), it is apparent that value is both relative and dynamic. Through personal experience, we know that the value of particular goods or services varies with individuals' preferences, the use that good is applied to, available alternatives and the situation in which it is consumed, among other factors. For example, a person may value a tourist map of an unfamiliar city quite highly during their travels and simply discard that map when they return home, while another person with firsthand knowledge of the city may not see any value in the tourist map.

The contingent nature of value introduces complexities into assessing the value of geographic information. First, information can be seen to possess primarily latent value that is only realized when it is applied to an appropriate problem or cognitive task (Longhorn and Blakemore 2008). In this way, the value of an information product (e.g. stream network) is catalysed by some uses (e.g. hydrological modelling) and not others (e.g. transportation planning) and by selected users. Second, information is an intangible product that often resembles a *de facto* public good since use by one person does not preclude others' use; however, use restrictions owing to license agreements, copyright and confidentiality concerns often mean that it can approximate a *de jure* private good. In these cases, the value of geographic data produced through traditional, authoritative processes can be quantified as a function of inputs (e.g. labour, time, software, distribution costs) and profit. Third, many of the benefits associated with GI use are intangible or qualitative in nature. Investments in GIS, spatial data infrastructures (SDIs) and associated technologies, for example, are justified not only in terms of cost savings or economic spinoffs but also with reference to important yet difficult to quantify benefits such as improved decision making, customer service, public safety or social utility value (Roche and Raveleau 2009; Obermeyer 2006).

2.3 Approaches to Defining the Value of Authoritative GI

As mentioned by Genovese et al. (2009a), assessing the value of digital information products services and infrastructures in general, and GI in particular, is challenging for several reasons. First, since no consensus exists about how GI itself should be defined (is it a product, a process, a public good?), defining and assessing the value of GI are complicated. Moreover, even if the quantitative part of the value, including monetary costs to create, update, market and distribute a particular data set, can be calculated with reasonable certainty, the qualitative dimension of GI value is still difficult to estimate. Longhorn and Blakemore (2008) also note that the value of a given data set will often vary between individuals and/or organizations because of differences in their needs, resources and objectives. In this way, they refer to the extrinsic characteristics of GI and consider its value relative to its capacity to meet users' needs and expectations. From this fitness-for-use approach, the economic value of GI can be expressed in terms of its exchange value or the amount that a purchaser is willing to pay for it. More commonly, GI displays a mixture of public and private good characteristics. These multistage processes of modifying GI from their original form to create newly derived products are particularly important when assessing the value of GI to an economy.

Nevertheless and primarily because of the needs of official GI producers (i.e. government agencies, private data creators/resellers), several research efforts over the past 15 years have been devoted to the question of GI value and developing economic evaluation methods in particular (see reviews in Longhorn and Blakemore 2008; Genovese et al. 2009a). Most of this work has centred on developing quantitative economic methods and models to calculate the economic value of GI through approaches such as nonlinear prices (Krek and Frank 2000), cost-benefit analysis (CBA) (Halsing et al. 2004), return on investment (ROI) (Craglia and Nowak 2006) and avoided costs (Didier 1990). One of the main motivations and expectations from the authoritative and primarily public sector GI producers' point of view was to define the economic benefits of GI production associated with government operations. Several national studies on this issue of quantifying the net benefits of state investment in GI production to regional and national economies have been completed for countries such as Australia (ACIL Tasman 2008), the United States (GITA 2007) and Canada (Natural Resources Canada 2006). Trying to provide rationale for cost recovery or revenue generation is another motivation for governments given recent periods of budgetary restrictions. On the private sector side, GI producers have consistently tried to identify and delimit value with reference to potential GI markets and the need to maximize profits. The recent extension of spatial data infrastructures (SDIs) to regional scales (e.g. Quebec, Canada (PGGQ 2004); Catalonia, Spain (Garcia Almirall 2008)) has generated more interest in understanding the economic value of SDI investments on regional and local economies.

Methods developed for products and services other than information, including value chain and market segmentation methods, have been applied in recent years to estimate the value of GI for various fields and uses (Krek 2006; Genovese et al. 2009b, 2010). Longhorn and Blakemore (2008, p. 38) define the value chain as 'the set of value-adding activities an organization performs in creating and distributing goods and services, including direct activities such as production and sales, and indirect activities such as managing human resources and providing finance'. Value is created step by step along the chain; thus, pricing in a value chain serves to determine the way in which the value created for the end user is distributed among the contributors (Genovese et al. 2010). The sum of all such margins, at the end of the chain, equals the total value added (Krek and Frank 2000). While the value chain is considered as one of the most suitable approaches to assess GI value (Krek and Frank 2000; Longhorn and Blakemore 2008; Genovese et al. 2009b), it is also one of the most complex due to the number of variables that have to be taken into account to characterize production and dissemination of GI. Consequently, an operational value chain method dedicated to GI still remains to be developed.

Understanding the qualitative value of authoritative GI has proven to be more complex than attempting to measure value through monetary approaches such as cost/benefit or return on investment. Usually measures of social benefit have been expressed in terms of outcomes such as improved service delivery, better informed decision making and enhanced capacity for disadvantaged groups to capitalize on publicly available GI and spatial information technology. Craglia and Novak (2006) identified three main types of social-political benefits associated with authoritative GI use:

1. Benefits to citizens through greater access to information and more transparent and accountable governance, improved empowerment and participation, customer/citizen goodwill and quality of life
2. Benefits to government that arise from improved collaboration with other stakeholders within and outside government, greater political legitimacy, improved

decision making, enhanced service delivery (e.g. health services) and better management and planning of land use change, environmental issues and sustainable development

3. Benefits to business related to increased innovation and knowledge, new business opportunities and applications, and job creation

Defining the value of authoritative GI across economic, social and political dimensions remains a real challenge. Even when there is an agreement on the definition of value, the concept is inherently complex as it encompasses many variables and theoretical and practical positions that relate to the intrinsic characteristics of data, characteristics of the end user (e.g. novice to expert continuum) and variability in the requirements of different fields and use contexts (e.g. contrasting demands for accuracy, precision and currency for cartographic uses vs. scientific analyses). Most of the methods already developed usually define value from a particular user group's perspective (e.g. company, individual), although sometimes there are attempts to extend this to the broader society through methods such as cost/benefit, multiplier effects, ROI and value chains.

2.4 What Is Different About Valuing VGI?

There are some signs that the approaches outlined above for understanding the value of authoritative GI can be applied, at least to some extent, to VGI. For example, some firms have effectively outsourced aspects of data maintenance and creation by encouraging their end users to submit reports of data errors and omissions. Google Maps' 'report a problem', Garmin's 'report a map problem' and TomTom's 'Map Share' applications are some of the more widely known examples of private authoritative GI being augmented and enhanced through VGI. Quantifying the economic value of integrating customers' VGI into firm's operations is difficult since this information is privately held; however, some approximation could be made with reference to avoided costs and, potentially, greater customer allegiance associated with GI services that are updated on an ongoing basis. The maturation of VGI projects such as OpenStreetMap has also had recognizable economic spinoffs as firms have built new business models based on enhancing a royalty-free data source (e.g. CloudMade) or developing new Geoweb applications and services that rely on 'free' data and often open-source software (e.g. JOSm, Potlatch, CloudMade or even Ushahidi).

Evidence of the social value of VGI is most apparent within the areas of crisis mapping, humanitarian aid and disaster relief. Goodchild and Glendon (2010) and others (e.g. Roche et al. 2011) document well the contributions that citizens' location-encoded media (e.g. SMS messages, geotagged photos and movies) can make to relief efforts. The Ushahidi-Haiti platform represents well the social value of such an application for crisis management. Operational only 3 days after the 2010 earthquake, it received over 3,000 testimonies of relief needs in 2 weeks with more than half being posted via SMS. Basic mapping needs were accentuated by the earthquake as well, given that there had been no recent mapping by the government and that the national Haitian map agency was totally destroyed by the earthquake. Following the earthquake, mapping campaigns grouped under the initiative Drawing Together were carried out by hundreds of Internet volunteers all around the world. The work achieved by these 'tech volunteers' was subsequently continued by many other existing initiatives such as OpenStreetMap, which quickly developed an OSM-based collaborative mapping platform specifically dedicated to Haiti (Roche et al. 2011). Clearly, it is not possible in cases such as this to assign quantitative values to information that can reduce suffering or possibly save lives. Moreover, given that VGI may be the only current and relevant form of GI in some crisis situations, even the possibility of estimating value through comparisons to private or public sector alternatives is removed.

Personal capacity building and more precisely spatial enablement refer to another field where the social value of VGI figures prominently. From an individual's point of view, 'spatial enablement' as defined by Williamson et al. (2010) refers to a person's ability to use any geospatial information and location technologies as a means to activate their spatial skills and then improve their spatiality. Lussault (2007) defines spatiality as the way people (or groups of people) interact with space and other individuals (or groups) on, in and through space. Hence, in the current social context, individual spatiality should not be considered detached from possibilities offered by information-enabled mobility (info-mobility) and real-time geo-communication (location-based communication). Therefore, a spatially enabled citizen is characterized by their ability to express, formalize, equip (technologically and cognitively) and of course consciously – or unconsciously – activate and efficiently use their spatial skills. We argue that being involved in VGI activities is actually one of the more preeminent means for individuals to develop their spatial enablement. This improvement is another characteristic of the social value of VGI. Improving spatial enablement, for a group of people, or even an organization or the society itself, is part of geosocial added value.

The high profile examples mentioned above illustrate that it is possible to at least approximate the relative value of certain types of VGI. However, understanding how or if this value can be measured, what dimensions of value are most pertinent and to whom the benefits and possible detriments (e.g. privacy encroachments) of VGI accrue is more problematic. This is due in part to the fact that, more than its authoritative counterparts, VGI's value is a function of both the intrinsic characteristics of the data that citizen volunteers create and the socio-technological processes through which these data are produced and used. This duality of product and process is captured succinctly in Goodchild's (2007) original definition of VGI which refers to '…the widespread engagement of large numbers of private citizens, often with little in the way of formal qualifications, in the creation of geographic information'. As a first step toward understanding these issues, we identify next several key characteristics of VGI that cause the approaches developed to value authoritative GI to be only partially suitable within a volunteer-based and collaborative production context. For convenience, these characteristics are presented through two broad and overlapping dimensions that focus on the value of VGI relative to (a) its intrinsic characteristics as data and (b) its user and production processes.

2.4.1 VGI Data Characteristics

Spatial data sets created through expert-led processes are often evaluated in light of their intrinsic characteristics such as resolution, currency, level of positional accuracy as well as the professional reputation of their private or public sector authors. Many of these properties can be examined through well-known procedures (e.g. root-mean-square error), compared to published standards and conflated to produce measures of a spatial data set's overall quality (Burrough and McDonnell 1998; Haklay 2010). The spatial data quality literature has long recognized, through the concept of 'fitness for use', that quality, like value, is not absolute because a given data set has different degrees of suitability for specific purposes and users' demands (Chrisman 1983; Devillers and Jeansoulin 2006). Fitness for use does not translate directly to exchange value largely because it is not feasible to charge people different prices based on their willingness to pay. However, in most instances there is a positive correlation between overall data quality and price when demand is aggregated (Genovese et al. 2009a; Longhorn and Blakemore 2008).

Compared to authoritative GI, there are several limitations to assessing the value of VGI data sets as a function of their properties. Given the absence of market forces and professional standards within the VGI context, documentation and measurement of these properties are likely to be more variable for VGI than comparable GI produced by experts. Individuals who create VGI for personal or very limited group use, for example, have little incentive to document their data and in this respect are similar to many experts who have 'working quality' data for internal use. In contrast, mature VGI projects such as OpenStreetMap can be well documented and allow inspection of edits on a record-by-record basis (OpenStreetMap 2011).

The heterogeneous nature of VGI introduces further challenges into the assessment of its value. As documented throughout this book, VGI ranges from data that are experiential and largely personal in nature (e.g. geotagged vacation photos) through passively contributed information concerning personal activity spaces (e.g. credit card transactions, cellular phone tracking) to what might be considered quasi-scientific data (e.g. locations of animal sightings, amateur weather station readings). In general, it is reasonable to assume that the closer a specific VGI resource resembles authoritative GI and is focused on quantifiable and undisputed events, phenomena or 'facts', the more applicable traditional metrics of data quality and worth based on data set properties are. Other dimensions of VGI heterogeneity introduce uncertainty into value appraisals. For example, since VGI is driven by amateurs' interests, it is often more diverse in thematic focus than is typical with authoritative GI (Goodchild 2008). In some respects, novel forms of VGI such as geographically referenced text messages require established understandings of the nature of spatial data, their potential uses and, ultimately, their value to be reconsidered and extended (Sui 2008). Finally, the internal heterogeneity of VGI data sets that are assembled through the collaborative efforts of many volunteers complicates value assessments. In particular, the standard approaches for evaluating data accuracy and completeness referred to earlier were designed to apply to GI that was produced entirely by a single author or entity. Notwithstanding the self-policing nature of communities engaged in user-generated content, data quality can vary substantially between contributors to a given data set.

2.4.2 Use and Production Processes

From the perspective of value, three of the most apparent ways that VGI use and production processes differ from their authoritative counterparts are (a) spatial data use and production have been transformed from niche activities involving experts to processes that engage large numbers of amateurs with varying interests and abilities, (b) the distinctions between spatial data users and producers are blurred as individuals participate in both roles at different times (i.e. 'produsers' as defined by Bruns (2008), and more precisely in the GI field by Budhathoki et al. (2008) and Coleman et al. (2009)) and (c) data use and production are loosely organized, if at all, and are not constrained by market forces or the same regulatory and standards as authoritative GI.

In the context of this chapter, the net effect of these processes is greater uncertainty and variability concerning how the value of VGI should be conceptualized and operationalized. On the positive side of the ledger, since VGI production is driven by personal interests and motivation, it often results in data that otherwise would not be generated by private firms or government agencies (Goodchild 2008). In some instances, these data may only be of value to their author; however, in other cases, rich data sets comprised of local and/or experiential knowledge are created (Hall et al. 2010). This type of citizen-led process of spatial data production and use has clear social value. First, it can foster a culture of bottom-up participation where citizens can have direct control over how their viewpoints are represented within a digital mapping environment with less expert oversight and control than is typical with more structured PPGIS cases (Roche 2011). Second, participation in collaborative VGI builds individual and community capacity through improved social networks and technical skill sets. Third, the potential to mobilize hundreds, if not thousands, of loosely organized contributors over short periods of time and collaboratively construct data through distributed means is often far more responsive to emerging needs than governments or even private firms could be on their own, as Goodchild and Glendon (2010) demonstrate for crisis management purposes. The so-called Arab Spring, rooted in Tunisia and rapidly disseminated to most of North African and Middle East countries in early 2011, as well as the more recent Occupy Wall Street social movements, has clearly shown how powerful and effective a large mobilization throughout social networks can be. User-generated location-based content has played, and still plays, a key role in these important spatial and social disseminations of ideas and thoughts. While not easy to economically assess, the value of VGI has been dramatically high in terms of social and political changes and also in terms of social cohesion.

The organic and volunteer-led processes of VGI use and production clearly do not have only positive impacts on its economic and sociopolitical value. For example, concerns have been expressed about the exploitative ('geoslavery') effects of relying on 'free' labour for spatial data production (Dobson and Fisher 2003; Obermeyer 2008). Others have noted the potential for personal privacy and confidentiality to be violated as few limitations exist regarding what can be posted, shared and/or collected through passive monitoring of individuals' behaviour (e.g. mining personal activity spaces through their cellular phones) within a largely unregulated Geoweb environment. This issue is, for instance, clearly underlined by the 22 legal complaints the law student Max Schrems has filed against Facebook during the last 2 years (Yahoo! 2011). Lack of professional oversight and control can also raise doubts about the quality of VGI and its usefulness for problem solving especially in comparison to authoritative GI (Grira et al. 2010). Some of this uncertainty relates to the value of individuals' contributions to collaborative VGI resources as volunteers' technical skill sets, objectivity and motivations for creating and sharing VGI can vary substantially (Coleman et al. 2009; Budhathoki et al. 2008; Flanagin and Metzger 2008). What makes this particularly challenging in the VGI context is that this uncertainty applies not only to entire data sets but also to individual features or records within a data set that has multiple authors. Finally, the responsiveness of VGI production processes to changes in volunteers' interests and emerging events can be somewhat of a double-edged sword as maintaining a cadre of active contributors over longer periods of time can be difficult (Haklay 2010). This challenge is not new to most volunteer-centred activities in society. However, the value of VGI as a reliable source of spatial data is diminished if there is uncertainty concerning the longevity of volunteers' efforts.

2.5 From Value Chain to Lego Blocks: VGI as Extensible and Reusable Data Components

This chapter aimed to explore the specific dimensions of VGI value and more generally to examine how the phenomenon of VGI use and production is taxing our understanding of how GI and VGI alike are valued. We have argued that the specific characteristics of VGI (heterogeneous, time sensitive, responsiveness, geosocial based, etc.) require traditional valuing approaches to be reconsidered. Rather than exploring in detail how traditional approaches to GI valuing can be adapted to the VGI context, in this section we aim to initiate a discussion within the field by proposing that one of the more important dimensions of VGI value is its potential to foster innovation and learning. In this case, innovation and learning are driven by users' idiosyncratic and diverse needs, backgrounds and objectives and result from both direct content generation and collaborative efforts where others' data are extended or repurposed. Previous sections of this chapter highlighted that there is value in VGI outputs (data sets and content) and value in their underlying socio-technological processes (learning, networking, etc.). Indeed VGI should contribute to geo-literacy by providing extended opportunities to move learning outside the formal educational institution. Geo-enabled volunteers could then 'record observations, combine them with observations of others, and analyze them for geospatial patterns' (Edelson 2011).

The value chain concept provides a good metaphor for conceptualizing the industrial nature of authoritative GI production and the distinctions between GI use and production processes. While aspects of the value chain can apply to more structured VGI production (e.g. OSM, Christmas Bird Count), it is clear that it is not as appropriate for many other types of actively contributed VGI where users have more latitude to customize their efforts to suit their specific skills and needs. Instead, we suggest that other metaphors that recognize explicitly the organic and sometimes contradictory nature of innovation are more applicable to the VGI milieu. Raymond's (1999) metaphor of the cathedral and the bazaar, for example, is used widely to contrast proprietary and open-source software. Proprietary software is developed in a highly structured and for-profit environment where teams of experts are responsible for assessing needs and developing and marketing monolithic software packages. Due to the complexity of proprietary software, lack of access to source code and marketing realities, fundamental change to proprietary software is almost exclusively controlled by the firm's experts. In contrast, open-source software is created in an open and often chaotic manner by individuals and a changing coalition of users. Innovation in this case is not controlled by experts but instead is generated through multiple ongoing threads of independent and shared community development.

The cathedral and the bazaar metaphor does apply well to some aspects of VGI use and production. However, Debord's (1958) 'Dérives' metaphor is richer and more appropriate in the ways that it captures the continual interplay between serendipitous innovation and social learning in VGI production and use. Debord describes the 'Dérives' or drift as a way of wandering in an urban environment without specific aims other than to discover and become immersed in a narrative network of experiences and lives. Just as a psychogeographer explores urban space, a VGI producer often explores, uses and produces new VGI without necessarily having a predefined plan, timetable or schedule. Navigation in VGI processes allows producers to encounter unexpected data sources, discover anticipated phenomenon and then increase the potential value of VGI as processes and as data sets. In this way, at least some aspects of VGI value are indeed clearly linked to the concept of serendipity. Serendipity is an evolving concept that is extensively used in the Web 2.0 literature to refer to unexpected discoveries that are made randomly and through intelligence in a process that initially targets a different object than those that are discovered.

We do not propose that the serendipitous dimension of VGI value is based solely on unanticipated discoveries that result from Dérives-like explorations of existing VGI. Serendipity, for example, is constrained by the richness of the environment (i.e. what is available to discover) and a producer's inductive capacity (Bourcier and Van Andel 2008). More importantly, while the Dérives concept can explain how innovation may be fostered in individuals, it does not adequately account for the synergistic effects that collaborative use and repurposing of VGI can have on innovation. At the risk of exceeding the limit of metaphors for a single chapter, it is possible to also view innovation in the VGI context in much the same way as children create their own toys from a container of Lego building blocks. One child may select components (blocks, windows, etc.) to build a house, while another may create a car. A third child may be inspired by the efforts of the others and help the first child to finish the house or, after seizing possession, repurpose and modify a playmate's 'output' according to his or her own preferences and abilities. In this light, different VGI data sets represent data 'blocks' that can be extended or reused within collaborative Web 2.0 environments, through joint data construction and mashup-based compilations.

To varying degrees, these metaphors can help us to conceptualize the linkages between innovation in VGI processes and the value associated with individual and social learning processes. However, translating what may be an interesting metaphor into an operational approach for establishing VGI value remains a substantive challenge. One approach that appears promising is to harness the collaborative capacity of the Web 2.0 model to allow users to develop community-based evaluations of VGI value. For example, Grira et al. (2010) adapted the fitness-for-use concept to explore how uncertainty regarding the quality of a VGI data set can be addressed in terms of users' assessments of its suitability for various tasks. We suggest that a parallel adaptation of fitness for use could be made to provide mass appraisals of the value that users ascribe to specific VGI data sets and their associated processes. These user-based assessments of VGI could play the same role as users' evaluations of restaurants and vacation destinations do on sites such as Yelp! and TripAdvisor. In this case, even simple user notes (e.g. 'This data set of streams is complete in my county, but the adjacent watershed is poorly covered …') or ratings could provide the basis for community learning, innovation and communication and could facilitate identification of what types of uses and users a data set is valued for. More importantly, this form of mass value assessments may provide interesting insights into the ways that individuals and groups can use VGI to reshape and redefine how places are represented and understood (Graham and Zook 2011).

2.6 Summary and Conclusion

This chapter sought to investigate how the growing VGI phenomena and its associated use and production patterns in business, government and the broader public are altering how we understand the value of geographic information. Following a discussion of how the concept of value has been conceptualized in general and the challenges inherent to valuing information, several of the traditional methods for valuing GI were reviewed. The contingent nature of value was seen to be particularly important in the case of geographic information, given that users, applications and localized needs can vary substantially. This general argument for a context-sensitive value of GI was seen to be more appropriate in the context of VGI in light of its unique characteristics. As a data source, the lack of expert oversight, the absence of professional standards and the inherent heterogeneity of VGI across thematic, media and spatial dimensions were identified as key contributors to the complexity of valuing VGI data. Further complexity in valuing VGI was seen to be rooted in the socio-technological processes from which VGI is created by individuals and groups who may not otherwise interact or converse. As one way to manage this complexity, the appropriateness of several metaphors (e.g. serendipity, Dérives, Lego block) was explored.

We would like to see future research engage with and extend this Lego block concept toward a more formal conceptualization, and we would like to end this chapter by offering a few directions that might be taken. Since this theory is mainly based on the idea that every VGI data set should be considered as an extensible and reusable data component (i.e. a basic Lego block), building a stronger conceptual framework should be based on an ontology for VGI and an evaluative framework. To be efficiently built, this ontology needs to be fed with observations from various field works, from different VGI 'produse' contexts and from VGI built for different purposes. As Grira et al. (2010) demonstrated in the context of spatial data uncertainty, there is good potential to develop an operational evaluation framework for VGI valuing that draws upon user communities' assessments. Given the heterogeneity of the VGI phenomenon, we suggest that this framework be tested first in some of the more established VGI contexts (e.g. OpenStreetMap, Wikimapia). Moreover, and following the earlier discussion of serendipity and Dérives, we suggest that grounded theory approaches may be the most appropriate means of identifying the main 'laws' of this Lego block 'theory' and evaluating its merits for VGI valuing.

References

ACIL Tasman (2008). *The value of spatial information: The impact of modern spatial information technologies on the Australian economy*. Report prepared for the CRC for Spatial Information and ANZLIC, the Spatial Information Council, Australia. http://www.anzlic.org.au/Publications/Industy/Downloads_GetFile.aspx?id=251. Accessed January 2, 2012.

Bourcier, D., & Van Andel, P. (Eds.). (2008). *La Sérendipité: Le hasard heureux*. Paris: Hermann.

Bruns, A. (2008). *Blogs, Wikipedia, second life, and beyond. From production to produsage*. New York: Peter Lang.

Budhathoki, N. R., Bruce, B., & Nedovic-Budic, Z. (2008). Reconceptualizing the role of the user of spatial data infrastructure. *GeoJournal, 72*, 149–160.

Burrough, P. A., & McDonnell, R. A. (1998). *Principles of geographic information systems* (2nd ed.). New York: Oxford University Press.

Chrisman, N.R. (1983). The role of quality information in the long term functioning of a Geographical Information System. *Proceedings of the International Symposium on Automated Cartography (Auto Carto 6), Ottawa, Canada*, pp. 303–321.

Coleman, D. J., Georgiadou, Y., & Labonté, J. (2009). Volunteered geographic information: The nature and motivation of produsers. *International Journal of Spatial Data Infrastructure Research, 4*, 332–358.

Craglia, M., & Nowak, J. (2006). *Report of international workshop on spatial data infrastructures: Cost-benefit/return on investment: Assessing the impacts of spatial data infrastructures, European Commission, Directorate General Joint Research Centre* (Technical report). Ispra: Institute for Environment and Sustainability.

Crompvoets, J., de Man, E., & Macharis, C. (2010). Value of spatial data: Networked performance beyond economic rhetoric. *International Journal of Spatial Data Infrastructures Research, 5*, 96–119.

Debord, G. E. (1958). Théorie de la dérive, *Internationnale situationniste*, n.2, décembre. http://debordiana.chez.com/francais/is2.htm#theorie. Accessed January 2, 2012.

Devillers, R., & Jeansoulin, R. (Eds.). (2006). *Fundamentals of spatial data quality*. London: ISTE.

Didier, M. (1990). *Utilité et valeur de l'information géographique*. Paris: Presses Universitaires de France.

Dobson, J. E., & Fisher, P. F. (2003 Spring). Geoslavery. *IEEE Technology and Society Magazine*, 47–52.

Edelson, D. C. (2011). "GeoLearning": Tricorders – The next tool for geographic learning? *ArcNews*, Winter 2010/2011. http://www.esri.com/news/arcnews/winter1011articles/tricorders.html. Accessed January 2, 2012.

Elwood, S. (2010). Geographic information science: Emerging research on the societal implications of the geospatial web. *Progress in Human Geography, 34*(3), 349–357.

Elwood, S., & Leszczynski, A. (2011). Privacy reconsidered: New representations, data practices, and the geoweb. *Geoforum, 42*(1), 6–15.

Flanagin, A. J., & Metzger, M. J. (2008). The credibility of volunteered geographic information. *GeoJournal, 72*, 137–148.

Garcia Almirall, P., Bergadà, M. M., & Queraltó Ros, P. (2008). The socio-economic impact of the spatial data infrastructure of Catalonia, European Commission, EUR 23300 EN. Accessed January 2, 2012.

Genovese, E., Cotteret, G., Roche, S., Caron, C., & Feick, R. (2009a). Evaluating the socio-economic impact of geographic information: A classification of the literature. *International Journal of Spatial Data Infrastructure Research, 4*, 218–238.

Genovese, E., Roche, S., & Caron, C. (2009b). The value chain approach to evaluate the economic impact of geographic information: Towards a new visual tool. In B. van Loenen, J. W. J. Besemer, & J. A. Zevenberger (Eds.), *SDI convergence: Research, emerging trends, and critical assessment* (pp. 175–187). http://www.gsdi.org/gsdiconf/gsdi11/SDICnvrgncBook.pdf. Accessed January 2, 2012.

Genovese, E., Roche, S., Caron, C., & Feick, R. (2010). The ecoGeo cookbook for the assessment of geographic information value. *International Journal of Spatial Data Infrastructure Research, 5*, 120–144.

GITA (2007). Building a business case for shared geospatial data and services: A practitioners guide to financial and strategic analysis for a multi-participant program. http://www.fgdc.gov/policyandplanning/50states/roiworkbook.pdf. Accessed January 2, 2012.

Goodchild, M. F. (2007). Citizens as voluntary sensors: Spatial data infrastructure in the world of Web 2.0. *International Journal of Spatial Data Infrastructures Research, 2*, 24–32.

Goodchild, M. F. (2008). Commentary: Whither VGI? *GeoJournal, 72*, 239–244.

Goodchild, M. F., & Glennon, J. A. (2010). Crowdsourcing geographic information for disaster response: A research frontier. *International Journal of Digital Earth, 3*(3), 231–241.

Graham, M., & Zook, M. (2011). Visualizing global cyberscapes: Mapping user-generated placemarks. *Journal of Urban Technology, 18*(1), 115–132.

Grira, J., Bédard, Y., & Roche, S. (2010). Spatial data uncertainty in the VGI world: Going from consumer to producer. *Geomatica, 64*(1), 61–71.

Haklay, M. (2010). How good is volunteered geographic information? A comparative study of OpenStreetMap and ordnance survey datasets. *Environment and Planning B, 37*, 682–703.

Haklay, M., Singleton, A., & Parker, C. (2008). Web mapping 2.0: The neogeography of the GeoWeb. *Geography Compass, 2*(6), 2011–2039.

Hall, B., Chipeniuk, R., Feick, R., Leahy, M., & Deparday, V. (2010). Community-based production of geographic information using open source software and Web 2.0. *International Journal of Geographic Information Science, 24*(5), 761–781.

Halsing, D., Theissen, K., & Bernknopf, R. (2004). A cost-benefit analysis of the National Map. Circular 1271, U.S. Department of the Interior, U.S. Geological Survey, Reston, Virginia. http://pubs.usgs.gov/circ/2004/1271. Accessed January 2, 2012.

Krek, A. (2006). Geographic information as an economic good. In M. Campagna (Ed.), *GIS for sustainable development*. Boca Raton: Taylor and Francis.

Krek, A., & Frank, A. U. (2000). The production of geographic information – The value tree. *Geo-Informations-Systeme – Journal for Spatial Information and Decision Making 13*(3), 10–12. ftp://ftp.geoinfo.tuwien.ac.at/krek/3226_value-tree.pdf. Accessed January 2, 2012.

Longhorn, R., & Blakemore, M. (2008). *Geographic information: Value, pricing, production and consumption*. Boca Raton: CRC Press.

Lussault, M. (2007). *L'homme spatial: la construction sociale de l'espace humain*. Paris: Seuil.

Musgrave, R. A. (1939). The voluntary exchange theory of public economy. *Quarterly Journal of Economics, 53*(2), 213–237.

Natural Resources Canada (2006). *Résultats du recensement 2004 de l'industrie géomatique* (Technical report). Sherbrooke: Natural Resources Canada.

Obermeyer, N. (2006). Measuring the benefits and costs of GIS. In P. Longely, M. Goodchild, D. Maguire, & D. Rhind (Eds.), *Geographical information systems: Principles, techniques, management and applications* (2nd ed., pp. 601–610). Hoboken: Wiley.

Obermeyer, N. (2008). Thoughts on "Volunteered (Geo)Slavery". http://www.ncgia.ucsb.edu/projects/vgi/docs/position/Obermeyer_Paper.pdf. Accessed January 2, 2012.

OpenStreetMap (2011). OpenStreetMap Changesets. http://www.openstreetmap.org/browse/changesets. Accessed January 2, 2012.

Perkins, C., & Dodge, M. (2009). Satellite imagery and the spectacle of secret spaces. *Geoforum, 40*, 546–560.

Pickles, J. (1995). *Ground truth: The social implications of geographic information systems*. New York: The Guilford Press.

Plan géomatique du gouvernement du Québec (PGGQ). (2004). *Profil financier de la géomatique des ministères et des organismes* (Technical report). Ministère des Ressources naturelles et de la Faunes, Québec, 23.

Rana, S., & Joliveau, T. (2009). NeoGeography: An extension of mainstream geography for everyone made by everyone? *Journal of Location Based Services, 3*(2), 75–81.

Raymond, E. S. (1999). *The cathedral and the bazaar: Musings on Linux and open source by an accidental revolutionary*. Cambridge, MA: O'Reilly.

Richardson, L., & Loomis, J. (2009). The total economic value of threatened, endangered and rare species: An updated meta-analysis. *Ecological Economics, 68*, 1535–1548.

Roche, S. (2011). De la cartographie participative aux WikiSIG. In O. Walser, L. Thévoz, F. Joerin, M. Schuler, S. Joost, B. Debarbieux, & H. Dao (Eds.), *Les SIG au service du développement territorial* (pp. 117–129). Lausanne: Presses polytechniques et universitaires romandes.

Roche, S., & Raveleau, B. (2009). Social use and adoption models of GIS. In S. Roche & C. Caron (Eds.), *Organizational facets of GIS* (pp. 115–144). London: ISTE Ltd/John Wiley.

Roche, S., Sureau, K., & Caron, C. (2003). How to improve the social-utility value of geographic information technologies for the French local governments? A Delphi study. *Environment and Planning B: Planning and Design, 30*(3), 429–447.

Roche, S., Propeck-Zimmerman, E., & Mericskay, B. (2011). GeoWeb and risk management: Issues and perspectives of volunteered geographic information. *GeoJournal*. doi:10.1007/s10708-011-9423-9.

Rodriguez, P. O. (2005). *Cadre théorique pour l'évaluation des infrastructures d'information geospatial*. PhD thesis, Département des Sciences Géomatiques, Faculté de Foresterie et de Géomatique, Laval University, Québec.

Sui, D. (2008). The wikification of GIS and its consequences: Or Angelina Jolie's new tattoo and the future of GIS. *Computers, Environment and Urban Systems, 32*, 1–5.

Tiebout, C. M. (1956). A pure theory of local expenditures. *Journal of Political Economy, 64*(5), 416–424.

Williamson, I., Rajabifard, A., & Holland, P. (2010). Spatially enabled society. *Proceedings of the FIG Congress 2010, "Facing the Challenges – Building the Capacity", Sydney*. http://www.fig.net/pub/fig2010/papers/inv03%5Cinv03_williamson_rajabifard_et_al_4134.pdf. Accessed January 2, 2012.

Yahoo! (2011). Austrian student takes on Facebook. http://news.yahoo.com/austrian-student-takes-facebook-074701796.html. Accessed December 7, 2011.

Chapter 3
To Volunteer or to Contribute Locational Information? Towards Truth in Labeling for Crowdsourced Geographic Information

Francis Harvey

Abstract Geographers, planners, and others increasingly refer to crowdsourced data in geography as volunteered geographic information (VGI). But is volunteered the right adjective to use for all types of crowdsourced geographic information? This chapter examines this question by making the following distinction along an ethical line for crowdsourced data collection: data collected following an "opt-in" agreement is volunteered; data collected under an "opt-out" provision is contributed (CGI). Opt-in agreements provide some clarity and control in the collection and intended reuse of collected data. Opt-out agreements are, in comparison, very open-ended and begin with few, if any, possibilities to control data collection. The chapter suggests that distinguishing *contributed* crowdsourced data from *volunteered* crowdsourced data is important to start to understand the nature of sources of crowdsourced data of any provenance and to help begin to identify possible biases. In the concluding discussion, this chapter argues that the simple distinction between CGI and VGI is valuable for assessments of data's fitness for use. Following the truth-in-labeling principle known for food products, differentiating between CGI and VGI is also helpful to identify cases where lax approaches or even malfeasance leads to inaccurate or biased crowdsourced data.

3.1 Introduction

In many discussions, crowdsourced geographic information is unambiguously referred to as volunteered geographic information (VGI; Goodchild 2007). This chapter considers what the term *volunteered* indicates and how distinguishing volunteered from contributed along ethical lines signals important differences in the processes of acquisition and the uses of crowdsourced data. The issue has many dimensions. Most people

---

F. Harvey (✉)
Department of Geography, University of Minnesota,
Minneapolis, MN, USA
e-mail: fharvey@umn.edu; francis.harvey@gmail.com

Table 3.1 Simplified indicators of differences between clarity and control of crowdsourced data collection following volunteered (opt-in) and contributed (opt-out) approaches

	Volunteered (opt-in)		Contributed (opt-out)	
	Collection	Reuse	Collection	Reuse
Clarity	+	?	?	?
Control	+	?	–	–

Explanations: +/– indicates possibility or absence, ? indicates ambiguous possibilities

Table 3.2 Summary of key distinctions between volunteered geographic information and contributed geographic information

Opt-in (volunteered)	Opt-out (contributed)
Clarity and specifics	Vagueness and generalities
Control over data collection	Uncontrolled data collection
Limited control over data reuse	No control over data reuse

would agree that data is volunteered when people freely choose to collect data. When collecting data is part of automated, open-ended, or uncontrollable processes, volunteered seems an inappropriate term. When data collection is part of a selective activity or the looseness of the term crowdsourcing is used to mask inaccurate data collection, biases, or even outright malfeasance, differentiating modes of collection and participation can be of great value in assessing data quality. This chapter sets out to show that a simple distinction between volunteered geographic information (VGI) and contributed geographic information (CGI), following the difference between the clarity and control afforded by, respectively, opt-in and opt-out provisions, provides valuable guidance for assessing crowdsourced data. Both OpenStreetMap (OSM) data and Geocaching offer examples of *opt-in* volunteered geographic information. Examples of *opt-out* crowdsourced, or contributed, data include cell phone tracking and RFID-equipped transport cards. This chapter concludes with the suggestion that people working with crowdsourced data distinguish VGI from CGI to help provide information about the data's provenance, or origins. The distinction between VGI and CGI helps to clarify the underlying approach to data collection, use, and reuse potential and helps to address users' questions about the origin and quality of crowdsourced geographic information. The focus is on the importance of informing other creators and users about key characteristics of crowdsourced geographic information to help ensure clarity and make better assessments of data quality.

3.2 Volunteering or Contributing: An Important Distinction for Crowdsourced Data

Crowdsourced geographic information are now ubiquitous parts of the information society (Dobson and Fisher 2003; Goodchild 2007). Examples include the detail and amount of data collected by most smart phone users without their knowledge and without any ability to control the collection. In 2010, a German Green party member, Malte Spitz, filed a suit in court and received detailed records of his mobile phone use for a 7-month period from his service provider. The detail in the locational data astonished him. The 35,831 individual records of information transferred from his smart phone for this period, either together or in pieces, can be used to create a profile of his activities that shows when and where he was during this period: what streets he walked down, when he took a train or a plane, where he worked, where he likes to go for a beer, and even where and when he slept (Biermann 2011). Detailed data is being constantly collected for smart phone users, unless they turn off their phone or disable location services.

Many examples have recently proliferated of how crowdsourced locational data is collected, often without our knowledge and frequently without information about reuses (Liptak 2011; National Research Council 2007; Acohido 2011). Some people have reported that the locations of their cell phone usage were unknowingly made available. Apple, Google, and Microsoft all faced embarrassing situations within the past 2 years (2010–2011) when software engineers discovered the wealth of location data being recorded and transferred to their mobile devices' applications and service providers, often without the user's knowledge. Indeed, worse yet, even choosing to disable location services did not stop sending location data, making turning off the phone the only way to protect location information. The technical and privacy issues are beyond the scope of this paper, but the hubbub that came with these recent cases of location information provided by users points to the importance and social perceptions of location privacy. It also points to complicated relationships between data collection and reuses (reuses, rather than simply uses to stress the many multiple uses of crowdsourced geographic information). While location privacy and surveillance issues are part of larger questions about mobile computing use and abuse, crowdsourced geographic information already can play an important role in the ability of companies and government agencies to know and predict people's activities.

The distinction between crowdsourced geographic information that is volunteered and contributed is a key distinction proposed in this chapter. Drawing on that distinction, the concept of truth in labeling, following pragmatic ethics, helps explain the provenance of crowdsourced geographic information, assess its fitness for use, and determine if lax standards and even malfeasance diminish the data's accuracy (see also the chapter from J. Dobson (Chap. 17) in this book). The example of Mr. Spitz's mobile phone data can provide an initial illustration of the distinction before a longer review of the underlying concepts in the following section. Assuming that he had signed a contract giving his provider the right to collect the data, as most people using smart phones regularly do, did he voluntarily agree to the collection of his locational data? What did the provider have to do to access the data for technical, marketing, or other purposes? Could other companies purchase the raw, aggregated, or anonymized data of Mr. Spitz? Making the distinction between volunteered and contributed can help in assessing not only this data but any data that is labeled crowdsourced data. Crowdsourced data collected with user control is volunteered, whereas crowdsourced data collected with no or limited user control is contributed.

That he had to go to court to access his own data is perhaps a surprising twist to Mr. Spitz's story, but in this case perhaps personal access was limited by the contract. In any case, the necessity of that step is further evidence that this data was contributed. Stipulations in contracts that limit access to locational data recorded about individuals are not unusual. Often we click and accept many pages of contract, which include information about the collection of location data and its reuse, on our way to exciting uses of technology applications and hardware. In many cases, and an increasingly due to concerns about privacy and the use of mobile devices, the choices to provide location data are clearer cut.

The distinction between geographic information that has been volunteered and contributed offers a valuable touchstone for considering crowdsourced data and later use. A straightforward distinction can be made between data we have chosen to collect, such as a geotagged photo on a shared website or social networking service, and data that is collected automatically, for example, stationary air pollution monitors or other sensor webs. Geographic information collection and reuse involving devices used by people are very complex when considered in detail, but the fundamental distinction between volunteered and contributed data offers a pragmatic way to engage key choices. Pictures we choose to upload to a social networking site that allows us control over access seem a clear-cut case of volunteered information. If the same images are used by the same site to advertise this functionality without knowledge or approval of the person making the pictures, the example involves the contribution of information. More complex examples arise when access and reuse are controlled by the person who made the pictures, but the company operating the site uses geographic information about the location of the pictures to profile users and sell aggregated data to mobile advertisers. The fundamental difference in terms of collection is between overtly choosing a course of action to collect geographic information on the one hand and on the other committing actions in which the collection of geographic information occurs without control and is possibly designed to be beyond the user's direct ability to influence. Distinctions also need to be drawn between the ways in which people participate in controlling access and use of data and derivatives. These distinction can be measured first in the possibility to gain *clarity* about the purposes for collecting and reusing the data and second in abilities to *control* collection and reuse.

Volunteered geographic information, following these differences, is crowdsourced information with clarity about purposes and abilities to control collection and reuse. Contributed geographic information, or CGI, refers to geographic information that has been collected without the immediate knowledge and explicit decision of a person using mobile technology that records location. Volunteered geographic information, or VGI, refers to geographic information collected with the knowledge and explicit decision of a person. Some examples include geographic information collected in the course of using a cell phone (CGI), geographic information collected in the operation of a car navigation system (also CGI), geographic information collected through Geocaching (VGI), and geographic information created during an OpenStreetMap (OSM) data party (also VGI).

The distinction between volunteered and contributed geographic information opens up possibilities for considering choices and other issues including the provenance of crowd-based geographic information, the better assessment of reuse potential, considerations of location privacy, and liability concerns

3.3 Ethical and Legal Issues

This section describes how the difference between volunteered and contributed reflects ethical distinctions that also find their expression in laws and principles of the law. Mr. Spitz's story and other examples point to important ethical choices we make when assessing the clarity of our contributions to crowdsourced data and our potential to control its collection and reuse. Related to privacy issues, the consequences for developing applications and implementing applications can be of broad use and substantial significance. While many ethical approaches are brought to bear on privacy issues, pragmatic ethics offers a very helpful framework to engage the practical consequences of our actions (Kwame 2008; Critchley 2007; Harvey 2012). A principle of pragmatic ethics in terms of crowdsourced data is that as creators we bear responsibilities for what we make possible and as users we have the responsibility to understand what we do. This principle holds relevance for considering the origin (provenance) of crowdsourced data. Finding satisfactory clarity and accepting possibilities of control inform our decisions to provide crowdsourced data and our agreement with the collection and reuse of our contributed data. As there are no universal values that hold for all people at all times, ethical issues and peoples' decisions reflect a vast range of choices. Laws and legal principles in theory reflect this heterogeneity and strive to delineate a framework that regulates through a minimum of restriction and a maximum of clarification to support people making adequate decisions.

To make this distinction between volunteered and contributed geographic information based on clarity and control, some additional considerations of ethical and legal issues offer important context. Starting with online discussions following revelations about storage of personal locational data storage on Apple's iPhone mobile devices, the so-called Locationgate offers interesting perspectives on different ethical stances towards privacy and the legal context of liability (Pogue 2011). Some people participating in these discussions stated they had no problem with constant and uncontrolled collection of location data, while others commented that this was frighteningly close to situations where governments have secretly collected information about citizens. Some commentators suggested we simply move on into this new era and leave privacy as an outdated concept behind, while others were adamant that the underlying protection of human rights to privacy be assured. A similar range of views on the ethical, legal, privacy, and liability issues connected to crowdsourced geographic information would be found among those people with economic stakes in geographic information or a strong professional or personal interest. The attitudes of the public at large may range from fear or concern to acceptance and even support, as revealed by larger studies of public perceptions of information technologies (Pew Research Center 2010). In contrast, many practitioners are far more likely to ascertain challenges, even dilemmas, in contradictory positions held between various professional groups, academics, and social groups (Blakemore and Longhorn 2004). Indeed, legal studies point to the need for a fundamental reworking of outdated laws to reflect the possibilities of the new technologies (Samuelson, n.d.). In the current void, liability frequently surfaces as the concept for concern among data providers and what contracts seek to limit (Onsrud 1995, 1997).

In the changing, complex, and even contradictory legal landscape related to mobile technologies, pragmatic ethical concepts may provide a good guide for deliberation around choices and issues related to protection of individual privacy as well as for clarifying the origins of crowdsourced geographic information. As valuable as principles are, the complexity of situations points to the need to translate concepts into actions or the creation of policies to guide actions. To explain clarity- and control-related choices connected to crowdsourced geographic information collection and use, established distinctions between opt-out and opt-in principles in agreeing to use devices and applications offer a very sound approach to translating ethical concepts into actions and uses of geographic information (Elwood and Leszczynski 2011).

Central differences between opt-in and opt-out provisions lie first in the choice a potential user has to control the service (opt-in) or to accept the service and all its terms and conditions unconditionally (opt-out). Opt-in provisions afford more flexibility and control, for example, the possibility of using some location service functions while disabling others. Under opt-out provisions, a potential user faces a starker choice between using a service or a device and rejecting the service or device entirely.

Most software and applications that make use of personal location information have opt-out provisions, for example, the mid-2011 Apple privacy policy describes how to opt out of aspects of their interest-based provision of advertisements to end users.[1] The open-ended iPhone software license agreement[2] explains that the use of location-based services indicates agreement and consent to Apple and its partners to collect, maintain, process, and use the customer's location data, subject to withdrawal at any time. However, third-party applications and services use different terms and privacy policies.

For the collection of contributed geographic information, many agreements require the creator or user of the CGI to agree before use to accept all terms or otherwise not use the device or application. For example, users of Google's Map Maker application can find this explanation of terms and privacy as part of the opt-out policy for any submissions they make:

By submitting User Submissions to the Service, you give Google a perpetual, irrevocable, worldwide, royalty-free, and nonexclusive license to reproduce, adapt, modify,

---

[1] http://www.apple.com/privacy

[2] http://images.apple.com/legal/sla/docs/iphone.pdf

translate, publish, publicly perform, publicly display, distribute, and create derivative works of the User Submission. You confirm and warrant to Google that you own or have all of the necessary rights or permissions to grant this license. You also grant to end users of Google services the right to access and use, including the right to edit, the User Submissions as permitted under the applicable Google terms of service.[3]

Twitter's location service provides an example of opt-in provision (Trapani 2009). To use its location services, users have to explicitly choose to utilize the location service; by default it is unavailable. In the case of crowdsourced geographic information, data collected by OpenStreetMap (OSM) mapping parties offers an example of volunteered geographic information, compared to the geographic information contributed through Google's Map Maker. The legal differences reflect abilities to clarify and control collection and reuse of crowdsourced geographic information.

While there are many complexities in every example of crowdsourced data that impact the scope of opt-in or opt-out provisions, the legal distinction between opt-in (volunteered) and opt-out (contributed) licenses also touches on a fundamental ethical understanding about volunteering (Cloke et al. 2007). Many people maintain that volunteering comes with an explicit understanding of what we are contributing, how we are contributing, and future possibilities of the collected information's uses. This widely held attitude towards volunteering to crowdsource information in many cultures reflects a pragmatic understanding of the control a person has over all volunteered actions. The ease of data reuse necessitates extending this pragmatic understanding to include recognition of possible future reuses.

Opting-in agreements clarify to collectors of crowdsourced information the specifics of how the data they agree to provide is collected and indicates the possibilities for reuse. This agrees with widely accepted principles of volunteering. Opt-out provisions may be clear but often are totalizing in scope; their acceptance involves the loss of control and influence over the collection and uses of information. This difference speaks to the practical importance of distinguishing crowdsourced information collected by people actively deciding to collect data, and having some measure of control, from crowdsourced data collected automatically or with a clear abrogation of possibilities to influence the collection and also reuse of the data.

3.4 Truth in Labeling

Clarification of the nature of crowdsourced geographic information can begin with the distinction between contributed and volunteered geographic information (CGI and VGI). The different labels help indicate the provenance. Provenance is a term widely used in e-Science and by librarians and information professionals to refer to attributes of the source of information that allow for assessment of the materials (Simmhan et al. 2005; Moreau et al. 2008; Cheney et al. 2009). It helps distinguish

---

[3] http://www.google.com/mapmaker/mapfiles/s/terms_mapmaker.html

the fitness for use and opens possibilities to assess lax standards in data collection and processing and possible malfeasance. Assessing data provenance is a fundamental issue that surfaced in the late 1980s and early 1990s in developing geographic information metadata (Lanter 1991; Smith 1996; Bowker 2000; Harvey 2002; West and Hess 2002). How we share data when we lack first-hand knowledge of the measurements and processes used to collect and prepare the data is complicated (Goodchild 1992; Goodchild 1995), but users will want to avoid this complexity and still, however, know they can trust the data (Harvey 2003). With the development of the Internet, the ways we can account for complexities remain limited (Chrisman 1994; West and Hess 2002; Tosta 1999). In this sense, the truth-in-labeling concept refers to the scientific empirical approach to assuring that information about the origin, processing, and current status of the data are made available and ensuring that data can be assessed whenever needed to determine its quality (Chrisman 1999).

Truth in labeling is simply a principle, in relationship to the provenance of crowdsourced geographic information, to record and make accessible characteristics of the collection, processing, and reuses of information to existing and potential users. As part of provenance, the distinction between CGI and VGI is an important truth-in-labeling indicator for potential users about quality and potential biases.

Objections to this approach may arise due to the cost and time it will take to collect and make information about geographic information provenance available. These objections speak to important practical issues but should be balanced by the consideration of how long it will take users to assess the suitability of the data when they have to meticulously determine data quality in relationship to prospective uses. Considering the many uses of each geographic information source, the multiplier for the cost and time of each group determining quality can make this a more considerable undertaking that far outstrips the costs and time of the creator to provide provenance information. Certainly, distinguishing VGI from CGI will not add much effort or time to the preparation of metadata.

Other objections may arise in the complexities and lack of a sharp line between CGI and VGI. For instance, a corporate-organized or even corporate-supported OSM mapping party would be for some an example of CGI, even if the mappers volunteered. In their eyes, the corporate influence constitutes a bias that can be both explicit and implicit. The CGI/VGI distinction may not be in accord with every person's ideology or perceptions, but it remains an accessible way to distinguish the type of collection and to clarify other questions about provenance.

3.5 Towards Truth in Labeling for Crowdsourced Geographic Information

Knowing if crowdsourced data is CGI or VGI can be very helpful in assessing crowdsourced geographic information for continued or new reuses. A scenario where a local activist group draws on OSM data to show the location of squats, only to find out that the OSM data has been volunteered by people working for the police, shows the possibility of manipulation and the importance of knowing not only who collected the data but also the provenance of the data. A scenario where crowdsourced data for neighborhood shops and mobile food stands has been collected by the merchants or by people working for the agents offers another example where knowing the nature of the data's origin is important. Distinguishing CGI from VGI cannot answer all questions about provenance and fitness for use, but following the principle of truth in labeling is a start to distinguish the primary relationship of people involved in collecting data to their role in data collection.

The main argument of this chapter is that truth in labeling and making the distinction for crowdsourced geographic information between VGI and CGI support needs to determine provenance, assess data quality, and support the assessment of possible use of lax standards and malfeasance. The determination of different levels of clarity and control relies on distinctions in terms of data collection and data reuses. The complexity of crowdsourced data is multifaceted. Uploads of photographs with locational information to Flickr are simply not the same as taking part in an OpenStreetMap mapping party. Nor is logging in to a Foursquare account to find recommended local restaurants the same as receiving unsolicited SMS advertisements with coupons for nearby food trucks, based on one's cell phone provider selling location data to marketing companies.

Clarity and control over the information individuals provide are larger issues than crowdsourced geographic information. They are part of mobile technology uses involving crowdsourced data and discussions about location privacy. Since it is impossible to exhaustively determine the reuses of geographic information in advance, it seems very relevant to ensure that future users are able to assess and qualify data, starting with a distinction between geographic information that has been contributed (CGI) and geographic information that has been volunteered (VGI).

Distinguishing CGI from VGI is more than academic. As part of evident and growing concerns about privacy, the distinction helps geographic information collectors and users distinguish on the fly a key parameter of data quality (Duval et al. 2002; Kim 1999; Tsou 2002) without overwhelming them with the details necessary for thorough consideration. The CGI and VGI distinction signals a fundamental difference that analysis of the other provenance information – about the who, what, when, where, why, and how of the geographic information in question – should serve to clarify.

Truth in labeling is a principle for clearly distinguishing CGI and VGI during creation and distribution of crowdsourced data. While the underlying position and attribute accuracy issues are complex, usage of the terms allows for the easy differentiation of geographic information collected *with* direct human control, and broad understanding of possible reuses, from geographic information collected *without* direct human control and with limited or even no knowledge of reuses. Adoption of this distinction would be an important first step in clarifying the origins of crowdsourced geographic information for users and further creators.

3.6 Summary and Conclusions

Volunteered geographic information is only and just that. Other types of crowd-sourced geographic information collected by automated sensors, as part of other activities or under opt-out agreements, are contributed geographic information. Distinguishing volunteered and contributed as ethical concepts is an important way to move forward with less ambiguity.

In particular, this is related to privacy and liability. Dan Sui has expressed the centrality of privacy and liability issues among the concerns for VGI (Sui 2008), asking important questions about how VGI will develop in the changing computer networked ecosystem. Some choices to consider now and in the future include the following: When we create geographic information, how are we taking privacy issues into consideration and clarifying both the context and modes through which geographic information was collected and could be used? When we use crowd-sourced geographic information, how are we assessing the context and modes of data collection and preparation?

The chapter by J. Dobson (Chap. 17) in this volume discusses relevant data quality problems in greater depth than this chapter. A potentially fruitful way to merge the CGI and VGI distinction lies in publishing data quality assessments and linking them to metadata in repositories as additional detail to the fundamental CGI/VGI difference in labels. As D. Colman discusses using this approach for conventional topographic base mapping in his chapter in this volume. Clarifications of data quality can help with the assessment of fitness for use and also inform potential users of scale-related differences and potential lax adherence to accuracy standards or even potential malfeasance. The decision to incur charges for using licensed data may find support if clarity of documentation and of information provenance facilitates greater clarity and control for individuals who collect the data.

Acknowledgements In addition to helpful comments and discussion at the 2011 workshop in Seattle, the comments of two anonymous reviewers and editors have been very helpful in improving this chapter.

References

Acohido, B. (2011). Privacy implications of ubiquitous digital sensors. *USA Today*, January 26, 2011, P1B.

Biermann, K. (2011). Betrayed by our own data. Die Zeit Online, March 26, 2011. http://www.zeit.de/digital/datenschutz/2011–03/data-protection-malte-spitz. Accessed August 26, 2011.

Blakemore, M., & Longhorn, R. (2004). Ethics and GIS: The practitioner's dilemma. In *AGI 2004 Conference Workshop on GIS Ethics*.

Bowker, G. C. (2000). The world of biodiversity: Data and metadata. *International Journal of Geographical Information Science, 14*(8), 739–754.

Cheney, J., Chiticariu, L., & Tan, W. C. (2009). Provenance in databases: Why, how, and where. *Foundations and Trends in Databases, 1*(4), 379–474.

Chrisman, N. R. (1994). Metadata required to determine the fitness of spatial data for use in environmental analysis. In W. K. Michener, J. W. Brunt, & S. G. Stafford (Eds.), *Environmental*

information management and analysis: Ecosystem to global scales (pp. 177–190). London: Taylor and Francis.
Chrisman, N. R. (1999). Speaking truth to power: An agenda for change. In K. Lowell & A. Jaton (Eds.), *Spatial accuracy assessment. Land information uncertainty in natural resources*. Chelsea: Ann Arbor Press.
Cloke, P., Johnsen, S., & May, J. (2007). Ethical citizenship? Volunteers and the ethics of providing services for homeless people. *Geoforum, 38*(6), 1089–1101.
Critchley, S. (2007). *Infinitely demanding: Ethics of commitment, politics of resistance*. London: Verso.
Dobson, J. E., & Fisher, P. F. (2003). Geoslavery. *IEEE Technology and Society Magazine, 22*(1), 47–52.
Duval, E., Hodgkins, W., Sutton, S., Weibel, S. L. et al. (2002). Metadata principles and practicalities. *D-Lib Magazine 8*(4). http://dlib.org/dlib/april02/weibel/04weibel.html.
Elwood, S., & Leszczynski, A. (2011). Privacy, reconsidered: New representations, data practices, and the Geoweb. *GeoJournal, 42*(1), 6–15.
Goodchild, M.F. (1992). Sharing imperfect data. Available on-line at: http://www.geog.ucsb.edu/~good/papers/228.pdf. Accessed August 26, 2011
Goodchild, M. F. (1995). Sharing imperfect data. In H. J. Onsrud & G. Rushton (Eds.), *Sharing geographic information* (pp. 413–425). New Brunswick: Rutgers University Press.
Goodchild, M. F. (2007). Citizens as voluntary sensors: Spatial data infrastructure in the world of Web2.0. *International Journal of Spatial Data Infrastructures Research, 2*, 24–32.
Harvey, F. (2002). Visualizing data quality through interactive metadata browsing in a VR environment. In P. F. Fisher & D. Unwin (Eds.), *Re-presenting GIS*. Chichester: Wiley.
Harvey, F. (2003). Developing geographic information infrastructures for local government: The role of trust. *The Canadian Geographer, 47*(1), 28–37.
Harvey, F. (2012). Practical ethics for professional geographers. In M. Solem, K. Foote, & J. Monk (Eds.), *Practicing geography: Careers for enhancing society and the environment*. Upper Saddle River: Pearson Prentice Hall.
Kim, T. J. (1999). Metadata for geo-spatial data sharing: A comparative analysis. *The Annals of Regional Science, 33*(2), 171–181.
Kwame, A. A. (2008). *Experiments in ethics*. Cambridge, MA: Harvard University Press.
Lanter, D. P. (1991). Design of a lineage meta-database for GIS. *Cartography and Geographic Information Systems, 18*(4), 255–261.
Liptak, A. (2011). Court case asks if 'Big Brother' is spelled GPS. *The New York Times*, online. http://www.nytimes.com/2011/09/11/us/11gps.html. Accessed August 26, 2011.
Moreau, L., Groth, P., Miles, S., Vazquez, J., Ibbotson, J., Jiang, S., Munroe, S., Rana, O., Schreiber, A., Tan, V., & Varga, L. (2008). The provenance of electronic data. *Communications of the ACM, 51*(4), 52–58.
National Research Council. (2007). *Putting people on the map: Protecting confidentiality with linked social-spatial data*. Washington, DC: National Academy Press.
Onsrud, H. (1995). Identifying unethical conduct in the use of GIS. *Cartography and Geographic Information Systems, 22*(1), 90–97.
Onsrud, H. (1997). Ethical issues in the use and development of GIS. *Paper read at GIS/LIS'97*.
Pew Research Center (2010). The future of online socializing. http://pewresearch.org/pubs/1652/social-relations-online-experts-predict-future. Accessed August 26, 2011.
Pogue, D. (2011). Wrapping up the Apple location brouhaha. http://pogue.blogs.nytimes.com/2011/04/28/wrapping-up-the-apple-location-brouhaha/?pagemode=print. Accessed August 26, 2011.
Samuelson, P. (n.d.). Privacy as intellectual property? http://people.ischool.berkeley.edu/~pam/papers/privasip_draft.pdf. Accessed August 26, 2011.
Simmhan, Y. L., Plale, B., & Gannon, D. (2005). A survey of data provenance in e-science. *ACM SIGMOD Record, 34*(3), 31–36.
Smith, T. R. (1996). The meta-information environment of digital libraries. *D-Lib Magazine* (July/August). http://dlib.org/dlib/july96/new/07smith.html.

Sui, D. (2008). The wikification of GIS and its consequences: Or Angelina Jolie's new tattoo and the future of GIS. *Computers, Environment & Urban Systems, 32*, 1–5.

Tosta, N. (1999). NSDI was supposed to be a verb. In B. Gittings (Ed.), *Innovations in GIS 6* (pp. 3–24). London: Taylor and Francis.

Trapani, G. (2009). Details on Twitter's imminent geolocation launch. Smarterware.org. http://smarterware.org/3419/details-on-twitters-imminent-geolocation-support-launch

Tsou, M.-H. (2002). An operational metadata framework for searching, indexing, and retrieving distributed geographic information services on the Internet. In M. J. Egenhofer, & D. M. Mark (Eds.), *Proceedings, geographic information science.* Second International Conference, GIScience 2002, Boulder, CO, USA, September 2002. New York: Springer.

West, L. A., Jr., & Hess, T. J. (2002). Metadata as a knowledge management tool: Supporting intelligent agent and end user access to spatial data. *Decision Support Systems, 32*, 247–264.

Chapter 4
Metadata Squared: Enhancing Its Usability for Volunteered Geographic Information and the GeoWeb

Barbara S. Poore and Eric B. Wolf

> *Arguably, given the rich interactivity of geographic information, usability applies not only to the systems but also to the content of those systems: the structure and portrayal of the data and metadata within them. This is where the usability industry is relatively weak, and therefore where one of the biggest research challenges lies.*
>
> (Davies et al. 2005)

> *In the earlier world dominated by paper maps the body of information described by metadata was a single map, and an intimate association existed between a map's contents and its marginalia. In the digital world, however, the concept of a data set is much more fluid.*
>
> (Goodchild 2007a)

Abstract The Internet has brought many changes to the way geographic information is created and shared. One aspect that has not changed is metadata. Static spatial data quality descriptions were standardized in the mid-1990s and cannot accommodate the current climate of data creation where nonexperts are using mobile phones and other location-based devices on a continuous basis to contribute data to Internet mapping platforms. The usability of standard geospatial metadata is being questioned by academics and neogeographers alike. This chapter analyzes current discussions of metadata to demonstrate how the media shift that is occurring has affected requirements for metadata. Two case studies of metadata use are presented—online sharing of environmental information through a regional spatial data infrastructure in the early 2000s, and new types of metadata that are being used today in OpenStreetMap, a map of the world created entirely by volunteers.

B.S. Poore (✉) • E.B. Wolf
Center of Excellence in GIScience, U.S. Geological Survey, Saint Petersburg, FL, USA
e-mail: bspoore@usgs.gov; ebwolf@usgs.gov

Changes in metadata requirements are examined for usability, the ease with which metadata supports coproduction of data by communities of users, how metadata enhances findability, and how the relationship between metadata and data has changed. We argue that traditional metadata associated with spatial data infrastructures is inadequate and suggest several research avenues to make this type of metadata more interactive and effective in the GeoWeb.

4.1 Introduction

Geospatial metadata is commonly referred to as data about data. Metadata describes the content, quality, and origins of a geospatial data set. According to the US Federal Geographic Data Committee (FGDC), which pioneered geospatial metadata standards in the 1990s, metadata was critical for the online delivery of data, allowing users to find, understand, and reuse data sets produced by others (FGDC 2000). Metadata allowed organizations to better manage their investments in geospatial data and provide information to online catalogs and clearinghouses (FGDC 2000). The metadata standards were developed when the Internet was in its infancy, but since then, the use of the Internet as a medium of communication and exchange among data producers and users has burgeoned. Metadata has taken on a more vital role in locating and managing the enormous amounts of geospatial data now available in the GeoWeb (Scharl and Tochterman 2007; Tsou 2002).

There is evidence that many who work with geospatial data sets consider metadata inconvenient, complex, and difficult to produce, creating a "metadata bottleneck" (Batcheller et al. 2009; Batcheller 2008; Tsou 2002). Although GIS professionals may acknowledge the importance of metadata, it often falls off the working agenda. The paradox of metadata is that while the costs accrue to the data-producing organization, many of the benefits accrue to the users (National Research Council 2001). Data producers have encouraged their employees to generate metadata by various means: simple fiat—thou shalt write metadata; providing specialized metadata tools; and in some cases by not allowing data to be submitted to a system-absent completed metadata. Despite these efforts, many data sets lack associated metadata (for a recent example, see Hennig et al. 2011).

If metadata is a usability issue for data producers, it is equally so for the end user. In an online discussion in 2010 about geospatial metadata in the GeoWeb (Fee 2010), a commenter points out that the usability of metadata for discovering the contents of a geospatial data file can be critically impacted by something as simple as the title of the data set:

> Improve how the metadata Title is constructed. This sounds so basic, but it's really important over time. Somehow require the user to create a human friendly Title for their data right up front, so that the metadata doesn't default to some cryptic file name. (Haddad 2010)

The producer/user disconnect has been recognized as a key reason that metadata may impact the usability of geospatial data sets (Comber et al. 2008). The Federal Geographic Data Committee was advised to improve the usability of geospatial metadata by structuring data-sharing partnerships to bring data producers and users into closer alignment (National Research Council 2001). But the usability problem does not just consist of the binary of expert producers and nonexpert users; there is a third element, technology (Moore 2010). Moreover, Internet technologies now permit users to become the producers of geospatial data (Coleman et al. 2009; Budhathoki et al. 2008). Experts understand controlled vocabularies and domains, but their solutions do not scale. Users understand local contexts and use cases, and they are more numerous than experts, but they do not necessarily understand expert vocabularies and domains. Machines can process large volumes of data and can be programmed to identify and process structured data, but they are poor at interpretation and contextual meaning.

This chapter examines how new types of metadata, spawned in the technology-mediated shift from the paper world to the online world, might lead to a more usable, interactive model for metadata. This interactive model would result from active negotiation among expert data producers, machines—defined here as structured programs or software—and data users who are empowered by software to produce data for themselves. This new model overturns the traditional view of metadata in which information about the data set is simply conveyed through a transparent communication system from expert to users (Poore and Chrisman 2006).

The impetus for examining the role of metadata in the GeoWeb resulted from a project the USGS undertook in 2010 to test whether volunteered geographic information (Goodchild 2007b) could be incorporated into *The National Map* (www.nationalmap.gov) of the USGS (Wolf et al. 2011). This ongoing project is using the database structure and the editor Potlatch 2, developed by the OpenStreetMap (OSM) community, to collect and manage geospatial data produced by volunteers. OSM is an open-source street map of the world, created and maintained entirely by volunteers (www.openstreetmap.org). To support simultaneous use by many different users and to record a complete history of all edits, the OSM database stores metadata at the level of the node, which many institutional GIS do not.

In researching the technical aspects of how elements of open and crowdsourced projects can be adapted to the needs of spatial data infrastructures (Onsrud 2007), the authors noted a recent uptick in theoretical discussions of metadata in both the formal GIScience literature and informal discussions online in what we are calling the geoblogosphere. A search of the Web of Science index shows papers on geospatial metadata increased from 1 to 2 per year to an average of 5 or more per year in 2004, perhaps reflecting an increasing interest in ontologies (e.g., Rodriguez et al. 2005). The geoblogosphere is decidedly nonacademic, but it too has been the site of prolonged discussions of metadata. A blog post by Fee (2010) "Let's Save Metadata" attracted attention from both GIS professionals and so-called neogeographers (Turner 2006) who are recasting the role of mapping on the Internet. These recent journal articles and blog discussions form the backdrop against which we evaluate how metadata is being remade in the age of the GeoWeb.

Our thesis is that there are qualitative differences between today's collaborative online mapping projects and the previous generation of multi-institutional data-sharing projects—spatial data infrastructures. We consider geographic information and metadata to be media for communication (McLuhan 1964; Sui and Goodchild 2001, 2011; Sui 2008) and explore the relationship between the new media practices being developed by the community of neogeographers and the efforts of traditional mapping endeavors such as *The National Map* to incorporate citizen contributions.

Media shifts have been profound, and yet metadata practices have not changed much since the mid-1990s. We categorize media changes that metadata might need to undergo in four areas:

- *Usability*—a quality attribute describing how easy it is for the user to interact with a program or piece of software (Nielsen and Loranger 2006).
- *Support for coproduction of data by communities of users*—the recognition that nontraditional users of geospatial data, whether called neogeographers or "citizens as sensors" (Goodchild 2007b), are producing and sharing large quantities of geospatial data online. This is related to changes that have been characterized as Web 2.0 (O'Reilly 2005). This media shift requires metadata to match the scale and dynamism of the current GeoWeb, reflecting the simultaneous edits of large volumes of data and supporting applications enabled by online sensor networks and location-based services (Pultar et al. 2010).
- Shifts in requirements for *findability*—that is, "the degree to which a particular object is easy to discover or locate" (Morville 2005: 4), applicable to both individual objects and systems.
- Altered *relationships between data and metadata*—the idea that in the disorder that is the current Internet, everything is data (Weinberger 2007); metadata and geospatial data are no longer distinguishable.

To examine changing attitudes to metadata, we draw on online sources, interviews, and case studies from information-sharing communities both before and after the so-called Web 2.0 revolution. We consider how metadata enhances the usability of geospatial data and how metadata itself may have to change to accommodate the media shift. The first case comes from a study of information sharing by a number of groups in the Pacific Northwest in the late 1990s as they constructed a regional spatial data infrastructure to help remediate environmental conditions responsible for the decline of native salmon stocks (Poore 2003). The second study is based on a content analysis of online discussions by volunteer neogeographers as they debate metadata and use an open-source mapping platform to map the streets and buildings of Haiti in the aftermath of the 2010 Haitian earthquake. Finally, we suggest alternatives to current metadata paradigms that might bring data producers, data users, and technology together into one interactive system in which users have more input into the production of metadata.

4.2 Background

The Content Standard for Geospatial Metadata (CSDGM) (FGDC 1994) was developed and promoted by the Federal Geographic Data Committee (FGDC) in the mid-1990s on the cusp of the Internet era. It was further adapted and published as a standard while retaining its essential form (International Organization for Standardization 2003). The metadata standard was primarily intended to help large organizations manage their geospatial data holdings; accommodating the end user was a lesser goal. The metadata standard codified common elements to describe geospatial data including data set identification, data quality, data set organization, spatial references, entity and attribute information, distribution constraints, and information about the metadata producer (FGDC 2000). In media transitions, new media often borrow and repurpose the forms of old media. Because the CSDGM straddled the paper and the digital eras, it cobbled together elements of two earlier media forms, the library card and the map collar.

4.2.1 The Library Model of Metadata

The card in the card catalog of a traditional library contained metadata about a physical object—the book—but the card was just a pointer to the book. It said little about the content of the book. Once the reader had located the book on the shelf, she or he had no further need for the metadata—the book *was* the content. The CSDGM and its descendants were built using a structured language based on the Standard Generalized Markup Language (International Organization for Standardization 1986), adapted from the library community (FGDC 2006; Goodchild et al. 2007). In the metadata standard, information about how the data set should be described was rigidly specified by logically constrained production rules that identified each permitted element (or field such as "keyword"), how the elements fit together, which elements were compound, which could repeat, which were required, and what expressions were permitted within each element.

This structure allowed metadata records to be parsed by machine. Metadata describing data sets housed on distributed servers were indexed and stored in a centralized registry or digital card catalog (FGDC 2006) called a clearinghouse or portal. The contents of these spatial data portals were frequently not exposed to the Internet at large to be passively crawled by spiders and indexed for full-text searching by text engines. Rather, spatial data portals became specialized for geographic information and relied on the logical structure of the metadata to facilitate precise searches. For example, a user could specify a land-use data set in Florida from 2009 using the keywords from the metadata record and receive just the data set she or he required without having to wade through a million documents from a Google search on the open Web.

A big problem for the library model is the constant need for updating and maintaining the catalog (Li et al. 2010). Furthermore, producing metadata that will meet the requirements of a complex and highly structured standard with 334 elements is hard. The "added rigor" of adhering to the production rules exacts a huge price in human labor (Shirky 2005a). Due to the persistent legacy of the library model and the complexity of the standard, metadata has typically been managed quite separately from geospatial databases. This has led to an expanding role for metadata managers or curators who extract metadata from the actual data producers and resolve data integration issues (Millerand and Bowker 2009; Schuurman 2009). This tends to distance the metadata from both the data producer and the user.

4.2.2 The Map Model of Metadata

Having found the record of a potential data set, the user will want further information about the quality of the data it describes as well as information on how to obtain it. This is where the CSGDM departs from the library model. The metadata points to the data, but once one has the metadata, one still does not have the data. Maps and the geospatial data derived from them do not yield everything a user needs to know, the former due to its status as an image, the latter because it is expressed in machine language; thus, metadata takes on an explanatory role in addition to its pointing function. Like the map collar or legend, metadata contains information about the contents and quality of the data set.

In the paper world, the map collar provides additional information on the author, location, map scale, the subject, the symbols used, etc. But even the most detailed legend cannot adequately "explain" a map (Wood et al. 2010). Most map legends obliterate the traces of the work practices that went into making the map, compressing the map information into numbers and other symbols (Latour 1999). Strictly in terms of processing data, metadata is a narrative form (first we did this, then that). In theory it can be expansive, describing work processes in detail, but in practice, like the map collar, it often fails to adequately explain the genesis of the data to an outside user. In standard FGDC metadata, narrative explanations of work practices are chopped into data elements and separated from the geospatial database as surely as though the map collar had been cut away. The severing of metadata from the data can lead to user confusion in the GeoWeb.

The library catalog as a finding aid depends on the separation of metadata from the data, but this separation induces new usability problems for the user as well as rendering the metadata incapable of reflecting the rapid change to database transactions in real time. To suit the media changes that have accompanied the GeoWeb, metadata must become interactive and embedded directly with the data, reflecting changes from data producers and users alike.

4.2.3 Interactive, Embedded Metadata in the Digital Age

Amazon.com provides a model for how metadata operates in the online world and demonstrates the three-cornered relationship of producer, user-producer, and technology. A search for a book on Amazon will result in a virtual page that contains descriptive metadata similar to that in a traditional library card catalog—typically title, author, publisher, and publication date. This descriptive metadata points not to the physical copy of the book but to information about how to purchase a physical or digital copy of the book, much as in the FGDC model. In addition, Amazon supplies professional reviews of the book, suggests additional books the user might like based on their past purchasing behavior and the behavior of other users, and allows the user to easily save a link to information about the book to a "wish list." By linking user behavior—search terms and purchasing decisions—directly to the book page, Amazon is using transactional user-centered metadata to enhance usability of the site. In addition, Amazon has supplied a means for users to interact with each other around books by having user reviews and user lists linked to the page. This ever-expanding universe of explanatory and transactional information, much of it generated by the site's users, is metadata, even though it is not formalized or authoritative. This miscellaneous explanatory information is networked on top of a unified platform that supports multiple simultaneous edits.

Amazon behaves somewhat like a physical library in that the page about the book (metadata) points to the location where the book can be found. But the book has, in many cases, also been digitized. As the instance of a physical book, the digital book can also be called metadata because it is not the "real" book in much the same way that the map is not the territory (Korzybski 1933: 58). There is no longer a distinction between data and metadata. As Goodchild (2007a, above) noted, not only is the concept of a data set fluid, so is the concept of metadata. In fact, metadata *is* spatial data (Chrisman 1994). Search provides a way into the information about the book, but the other metadata is accessed through links. Thus in the user's experience, Amazon avails itself of the model of the Internet. The way Amazon deploys metadata as a rich context of explanatory information follows the new media model described above. Metadata about the book enhances usability, is produced by communities of users, and enhances findability of related information.

4.3 Formal and Informal Discussions of Metadata

Two seemingly opposed themes—that geospatial metadata is not simple enough but at the same time not complex enough—appear in present-day metadata discussions. The former is most apparent in the neogeography community; the latter among academics. These are essentially usability issues, though they are different in kind. As Goodchild notes (2007a, quoted above), the digital age has rendered distinctions between data and metadata more slippery. In distributed online mapping systems, usability is a more complex problem than just making a map interface easier and more intuitive to use. Usability must start with the data and their metadata (see Davies et al. 2005, quoted above).

4.3.1 "Let's Save Metadata": Neogeographers

A challenge "Let's save metadata" was recently posted on a geoblog (Fee 2010). Fee's main complaint is the lack of usability in the FGDC standard, and he cites the human/machine dichotomy. Producing metadata following the standard and reading the typical metadata record are hard. Servers can use XML to talk to each other. "But servers rarely read and write metadata on their own without human interaction. Thus the reality of the situation is we poor humans have to ingest and parse metadata regularly<XML>YIKES<?XML>" (Fee 2010). An example of this is the XML

```
<?xml version="1.0" encoding="ISO - 8859 - 1"?>
< ! DOCTYPE metadata SYSTEM "http : // www . fgdc. gov/ metadata/ fgdc - std - 001 - 1998 . dtd">
<metadata>
    <idinfo>
        <citation>
            <citeinfo>
                <origin>University of Florida GeoPlan Center</origin>
                <pubdate>20101220</pubdate>
                <title>GENERALIZED LAND USE DERIVED FROM 2010 PARCELS - FLORIDA DOT DISTRICT 7</title>
                <geoform>vector digital data</geoform>
                <pubinfo>
                    <pubplace>Gainesville, FL</pubplace>
                    <publish>University of Florida GeoPlan Center</publish>
                </pubinfo>
                <othercit>FDOT District 7</othercit>
                <onlink>http ://www.fgdl.org</ onlink>
                <lworkcit>
                    <citeinfo>
                        <othercit>Source - 2010 Automated - 2010</othercit>
                    </ citeinfo>
                </ lworkcit>
                < ftname Sync="TRUE">ETAT.D7_LU_GEN_2010</ftname></ citeinfo>
        </ citation>
```

Fig. 4.1 Snippet of XML code of the metadata for a land-use map of Florida downloaded from the University of Florida GeoPlan Center

expression of a land-use data set downloaded from the Florida GeoPlan Center (Fig. 4.1). In practice, this format is rarely encountered—more frequently the metadata is rendered by the machine in a more common indented format (Fig. 4.2), but even this more approachable format demands that the reader do most of the work of decoding the questions she or he wants to ask about the data set.

What matters to users are answers to the who, what, when, where, how and why questions, but "those questions are hard to parse out of metadata" (Fee 2010). This discussion about the usability of metadata and what users really want essentially rehashes those of the mid-1990s when the metadata standard was proposed, indicating that there are unresolved usability issues that have persisted for nearly two decades (Schweitzer 1998).

Professional GIS software such as ArcGIS automates the production of metadata to some extent, although Fee rightly observes that current GIS software could do a better job at this. Furthermore, many of the people commenting on this blog online do not use professional GIS software. Many of the 72 responses can be identified with the tenets of neogeography, although there were also some responses from

**GENERALIZED LAND USE DERIVED FROM 2010
PARCELS -FLORIDA DOT DISTRICT 7**

Metadata also available as

Metadata:

- Identification_Information
- Data_Quality_Information
- Spatial_Data_Organization_Information
- Spatial_Reference_Information
- Entity_and_Attribute_Information
- Distribution_Information
- Metadata_Reference_Information

Identification_Information:

   Citation:
      Citation_Information:
         Originator: University of Florida GeoPlan Center
         Publication_Date: 20101220
         Title:
            GENERALIZED LAND USE DERIVED FROM 2010 PARCELS -FLORIDA DOT DISTRICT 7
         Geospatial_Data_Presentation_Form: vector digital data
         Publication_Information:

            Publication_Place: Gainesville, FL
            Publisher: University of Florida GeoPlan Center

         Other_Citation_Details: FDOT District 7
         Online_Linkage:<http://www.fgdl.org>
         Larger_Work_Citation:

            Citation_Information:

               Other_Citation_Details: Source - 2010 Automated - 2010

Fig. 4.2 Metadata shown in Fig. 4.2 in indented format

individuals who could be called professional GIS users from the academia, the industry, and the government.

There was a general agreement that metadata needed to be simpler to produce and make. One remark summed up many of the comments. "Unless a caveman can do it, users won't read or write meaningful metadata. And relevant metadata must be stored and travel with the data" (Entchev 2010).

This last point—the necessity to store and transmit metadata with the data—elicited a good deal of discussion on the geoblog. In our comparison of two case studies, we show the difference between the standard approach to metadata that has been developed by the professional GIS community in spatial data infrastructures, and the approach of the open-source mapping community which is experimenting with data structures that store metadata at the level of the individual data object.

In an era of too much information, findability becomes preeminent (Morville 2005). Google has accustomed us to the idea that simple keyword searching (land use, Florida) should be all that is necessary for the user to search for data, and yet these fields are buried in an overly complex metadata structure (Gould 2006a).

In fact, if one does a Google search for "land use, Florida", one comes up with several FGDC metadata records near the top of the search, based most likely on the "keyword" field. So the original theory of structured metadata was correct. It can be useful for findability. The problem in this case lies in the uneven adoption of the standard and the issue of not exposing more metadata records to the Internet at large. The need for compliance with a top-down mandate may also have doomed metadata's potential. Alternative bottom-up, user-generated taxonomies or folksonomies (Vander Wal 2007) may work better (Gould 2006b). Capturing and exploiting user tags, which are the locally generated equivalent of the metadata standard's keyword fields, could potentially produce an emergent ontology that would aid findability. This issue is discussed further below using the example of OpenStreetMap.

4.3.2 Metadata and Meaning: GIScience

There has also been an increase in academic papers on metadata since 2005. In his review of the adoption and spread of metadata standards since the mid-1990s, Goodchild (2007a) calls for user-centric rather than producer-centric metadata, emphasizing easy-to-understand measures of data quality and tools to assess fitness for users' unique purposes. This is in concert with the discussions in the neogeography community.

On the other hand, several researchers have moved in a different direction, calling for more complexity, either a different kind of metadata or further metadata extensions. These arguments largely center on the idea that metadata, as currently structured, does a poor job of capturing differing meanings, or semantics embodied in a database (Comber et al. 2008). Schuurman and Leszczynski (2006) have proposed additional formal metadata categories for data semantics to assist data interoperability, achieved through database ethnographies (Schuurman 2008). Although machine understanding of semantics, as embedded in metadata, might be desirable, extending the metadata standard will complicate an already complex structure. Gahegan et al. (2009), in their work on community-based knowledge in cyberinfrastructures, caution that ontologies alone cannot capture meaning because they ignore "use-cases, provenance data, social networks and workflows."

4.4 Metadata Top Down

Traditional metadata associated with spatial data infrastructures can be examined according to the four criteria set out at the beginning of this chapter: Is it easy to use? Does it reflect coproduction by a community of users? Does it enhance findability? What is the relationship between data and metadata?

In the late 1990s, federal and state agencies in the Pacific Northwest of the USA were building a shared, multi-organization regional data set of rivers and streams to assist recovery planning for the 22 species of salmon that had been listed as endangered or threatened in 1999 (US Department of Commerce 1999). Over 40 organizations participated in the development of a common data model for hydrography (water) data and built an online clearinghouse fashioned after the FGDC model (Poore 2003).

4.4.1 Usability

The usability of traditional metadata is affected by the compression of work practices, the modular structure of the metadata, and the separation of the metadata from the data. Metadata aims to describe the products (data) of work practices (data analysis and data production) that are mediated by technologies. These practices emerge from situated learning when communities work together on particular problems (Lave and Wenger 1991). Situated knowledge, being primarily tacit knowledge developed over the course of a project, is difficult to translate for other communities. An example of what is lost when working procedures are compressed into metadata can be seen in the records of stream databases that were kept in the Pacific Northwest Hydrography project.

A metadata record of the stream layer from the Six Rivers National Forest (1999), which covers a large area of federal land in Northern California, demonstrates the problem faced by these regional integrators. The process of producing the stream layer for the forest is described in the metadata, but the work details are of necessity condensed. The early history of this data is omitted. The ultimate source of much of the digital data on streams in the Pacific Northwest was digital data derived from the 1:24,000-scale USGS topographic maps in the early 1990s. These data, known as digital line graphs (DLGs), were generalized to 1:100,000 scale and shared with the US Forest Service and other agencies. But the data were inadequate for watershed level work; the generalizing process omitted much detail on intermittent and smaller streams, and the maps from which the data were derived were out of date.

To be useful at a local scale within the Six Rivers National Forest, the stream network had to be densified—adding back in the stream information that had been removed when the DLGs were created. The Forest relied on a then-current densification process called crenulation to delineate streams that were not included in the USGS DLGs. The metadata refers to this process and includes a reference to existing practice (Maxwell et al. 1995). Crenulation is a process of inferring the course of stream channels by tracing the folds or crenulations down a slope on a contour map. This process can be traced back to the scientific literature of the 1930s on geomorphology of stream channels. Thus, a long history of scientific discovery and insight is compressed and translated into the one word, "crenulation," that appears in the metadata record. This process of compression and translation is characteristic of the circulation of scientific knowledge (Latour 1999), but in order to understand the how the data were created, the user must dig into various scattered sources.

Eventually, the original Six Rivers water data set was integrated into a larger data set composed of all the stream layers in the Forest Service's Pacific Southwest Region (US Department of Commerce 2004). This metadata record shows further compression of data techniques and origins. A newer software-based modeling technique, based on flow accumulation, was used to densify the streams in several of the watersheds. The older ephemeral streams from the Six Rivers National Forest that had been densified by the hand crenulation method were discarded. The link backward to the Maxwell method was severed. Severed as well are the work practices of a previous community that drew on a long tradition of local observation and scientific knowledge. Does the end user of the integrated data set need to know the back story of crenulation? Perhaps not, depending on the use to which the data set is put, but the new metadata emphatically warns that there is not necessarily a link between what the data set portrays and what was directly observed in the watershed:

> IT MUST BE CLEARLY UNDERSTOOD THAT THIS DATA SET, AT THIS TIME, IS NOT INTENDED BY, NOR IS CAPABLE OF, DISPLAYING WHERE WATER IS ACTUALLY FLOWING ON THE LANDSCAPE.

4.4.2 Community

It is not so easy to reconstruct the community that built these data sets of the streams. This community is not reflected in the metadata. Nor is there any way to recover the specific history of this data set. The "Time Period of Content"—a field in the metadata—only reflects the time at which the data set was produced. It is static and does not reflect this long history of where the data or the production techniques originated.

4.4.3 Findability and the Separation of Metadata from Data

As to findability, the results are mixed. We could not locate the original Six Rivers National Forest data and the Maxwell reference through the National Forest website, but we were able to find the integrated Southwest Division metadata record discussed above. After much searching through Google and other sites, we found an Esri geodatabase through the National Hydrography Dataset (www.nhd.gov). The reference to the Maxwell process was preserved several layers deep in the geodatabase and without citation of the relevant literature. Findability and usability suffer when the history of the data is hard to reconstruct through the reorganization of the software packages and processes by which the data have been conveyed to the user.

4.4.4 Metadata Bottom Up or Metadata Squared

Digital media have spawned new practices for categorizing data. Classification practices that were adequate for the physical world, in which each unique object had a unique place, have of necessity changed (Shirky 2005a). In the digital world, an object can be in many places at once and can be "about" many different things at the same time, leading to a proliferation of information. This proliferation demands networked user-generated classifications, a bottom-up ontology, a "new order of order" (Weinberger 2007).

4.4.5 OpenStreetMap

OpenStreetMap (OSM) (http://www.openstreetmap.org) was started by Steve Coast in 2004 (Wikipedia 2010). The goal of OSM is to make an all-volunteer online map of the world that will be free of use restrictions and open to all (http://wiki.openstreetmap.org/wiki/FAQ). Anyone can edit the map, discuss the map, create tutorials and other explanatory material, freely access the data, and influence the future direction of the map. OSM volunteers are attracted to the project through engagement with the online data and mailing lists. Frequent mapping parties, held throughout the world, solidify the community by adding the face-to-face experience. A distinctive feature of the wiki software that underlies OSM is a history of all changes to the map over time. OSM can be evaluated according to the same criteria used for evaluating traditional metadata: usability, coproduction by community, findability, and an altered relationship between data and metadata.

In OSM, there is no longer any separation between map and collar, that is, the data and the metadata. This is a new type of mapping medium. The map becomes a platform or canvas on which the user is invited to draw, that is, to edit the map. Users are motivated to contribute to the map for various reasons (Budhathoki 2010), including the strong user community, but a desire to assert creativity is important (Budhathoki 2010). The system supports almost instantaneous updates, validating the mapping platform as a creative endeavor.

The data structure also gives the OSM community the ability to respond quickly to emergency situations in which better maps are needed. Volunteers from across the globe began mapping the street network in Haiti within hours of the earthquake in January 2010. Announcements on the OSM wiki, mailing lists, and social networking spread the word about the need for mappers. Face-to-face crisis camps brought together at least 700 mappers in cities around the world to map (Waters 2010). High-quality satellite images released for public use by GeoEye and DigitalGlobe were the primary vehicles, along with old CIA maps from the 1940s, that volunteers used to trace streets and buildings in the damaged areas (Silver 2010; Maron 2010). The resulting maps were used by many organizations in the response and recovery (Ball 2010). The community organized and coordinated itself entirely using the ancillary metadata that surrounds the map.

4.4.6 Metadata Types

There are two different types of metadata preserved by the OSM platform: object-level metadata and ancillary metadata; there is no overarching metadata document as in traditional geospatial data. Object-level metadata is incorporated directly into the data structure, making no distinction between data and metadata (Weinberger 2007.) The data is very simple and expressed in XML. Data elements or data primitives are nodes (a point expressed in latitude and longitude), ways (ordered interconnection of nodes), and relations (sets of nodes or ways) (http://wiki.openstreetmap.org/wiki/Data_Primitives). The metadata identifies the node, its coordinates, the user who created the node, the editing session (changeset) of which it is a part, the version number of the edit, and the date and time it was edited. Elements can have any number of user-generated tags, consisting of a key and a value. Below is an example of OSM metadata drawn form a user-created map of Port-au-Prince, Haiti, in the aftermath of the earthquake.

<node id="613826766" lat="18.5450619" lon="-72.3305089" user="samlarsen1" uid="5974" visible="true" version="2" changeset="3636891" timestamp="2010-01-16 T23:34:34Z">
<tag k="building" v="collapsed"/>
<tag k="source" v="GeoEye"/>
</node>

In this case, the node is identified as a building that is collapsed (e.g., tag k="building", v="collapsed"). The source is from a particular user, samlarsen1, who digitized the node based on the GeoEye imagery of Haiti.

Because OSM changes continually over time, sometimes quite rapidly as in the Haitian earthquake crisis, object-level metadata is necessary. Object-level metadata facilitates the communal character of map production. Changes to the map can be tracked, and the history can be "rolled back" to a previous state easily if an error is detected by another editor.

In addition to the metadata that resides directly in the data structure, there is a vast, swirling universe of ancillary data describing the map, explaining how to use it, and facilitating community discussion and debate. These data are quite diverse, consisting of computer programs, tiling schemes, IRC chat rooms, YouTube video demonstrations, tweets, e-mail discussion groups, and wiki pages. All of these are socially mediated, produced by the community, and accessible to any user. They are not unlike the Maxwell manual described in the hydrography example above—not formal metadata but associated documents that can amplify the meaning of the data by describing mapping practices. In the format of traditional metadata, these linkages are often lost. But in the GeoWeb, there is no extra expense in linking to them. These ancillary data are much like the cloud of information surrounding digital objects on Amazon.com. We refer to this as metadata squared due to the possibility of endless proliferation.

The community of users makes final decisions in OSM, unlike Wikipedia, where edits are semi-anonymous and controversial topics are supervised by a group of editors. In OSM, one must have an account to edit the map, making the user not only identifiable but also accountable to the community. Trust is placed in the adage, common among open-source computer programmers, that "given enough eyeballs, all bugs are shallow" (Raymond 1999). Questions about the map can be posted on the wiki help pages. Users vote for their favorite question and badges are awarded for participation in answering. This gamelike feature serves to build community.

Fig. 4.3 Port OpenStreetMap changeset 3654854, screenshot from www.openstreetmap.org

One can search for a particular user in the wiki. This redirects one back to the user's page in the map interface. The page for the user "samlarsen1," who edited the node discussed above, lists the areas that he has mapped and links to a history of his edits. One editing session (or changeset) took place on January 18, 2010 (Fig. 4.3), when Larsen mapped a road southwest of Port-au-Prince near Grand Goâve. This page is undeniably metadata. It gives the geographic location of the nodes the user has contributed, the editing software used (JOSM), the imagery from which the mapping was derived (Digital Eye), a list of the nodes, and the roads (ways) that the nodes contributed to.

Each of these nodes and ways has its own dynamically generated page with a graphic designed to facilitate human exploration of the data. For example, on the main page for changeset 3654854, the user can click on the graphic, bringing him or her back to the full OSM map so he or she can see the geographic context in which the nodes and ways fit.

This close coupling between wiki, map, and user information and the deployment of different media makes for a rich understanding of the data set. In addition, the user has the ability to download this changeset or any number of other changesets that might have been produced by "samlarsen1" or other users. Users can manipulate this changeset data and its associated metadata in many different ways, by user, by tag, or by geographic area. Examples of relevant code are given in the wiki. In addition, programmers who work on OSM have begun to construct various tools to manipulate user tags to provide interesting ways to visualize OSM data. For example, Tagwatch aggregates user tags three times a week and provides statistical information on which tags are being used by the community (http://tagwatch.stoecker.eu/).

A tag is a key=value pair that can be attached to the nodes, ways, relations and even changesets (http://wiki.openstreetmap.org/wiki/Tags). Tags serve a role similar to attributes in a more traditional, relational database model except that they are not constrained by a top-down schema. The content of tags is up to the user. Any tag can be used, as long as it is verifiable, although this is not strictly enforced. If a user cannot find a relevant tag, he or she can propose a new tag which is voted on by the community. This can lead to confusion, as demonstrated by the tags that emerged during the Haiti mapping. Collapsed buildings were identified in a number of different ways, as "earthquake:damage=collapsed_building", "earthquake:damage=collapsed", "building=collapsed", and several misspelled variants. Tags recommended by the community for use in future disasters reflect the most frequent usage.

As OSM matures, tag analysis with such programs as Tagwatch could be used to generate a user-centered, bottom-up ontology (Shirky 2005b). Work in the library community advances the notion that social semantics—relationships between tags generated through a social process such as OSM participation—can capture local meaning best and can be disambiguated and systematized using controlled vocabularies (Qin 2008). In the case of geospatial metadata, OSM tags on structures could be compared to a controlled database such as the USGS's Geographic Names Information System (http://geonames.usgs.gov/domestic/index.html) to generate a gazetteer that would include semantic relations as well as unofficial names for structures that mirrored local customs.

4.4.7 Evaluation

In terms of the criteria for evaluation—usability, coproduction by community, findability, and the altered relationship between data and metadata—the OSM approach to metadata seems to do well at accommodating a community of mappers who are not necessarily professionals and allowing for simultaneous, distributed updating.

The Humanitarian OpenStreetMap Team (HOT) (http://wiki.openstreetmap.org/wiki/Humanitarian_OSM_Team) in the Haitian crisis demonstrated the rapidity with which maps could be made of an area that had not been not previously well mapped, and these maps were widely used by first responders (Osborne 2010). The simplicity of the underlying data structures and the built-in support for distributed communications in OSM provided a platform which could rapidly scale for multiple simultaneous edit sessions during crisis events.

Usability: Fitness for use is one aspect of usability most familiar to professional mapping endeavors. However, there are places, such as Haiti, that are poorly mapped, if they are mapped at all. Any given map of such a place is useful when compared to having no map at all. The OSM data have the added benefit of being free from governmental or corporate license restrictions. Furthermore, crowdsourced maps tend to improve over time. In Europe, OSM data are nearing the positional accuracy of data from the national mapping agencies, many of which restrict the use of their data (Haklay 2010a). OSM data has also proven useful for commercial interests such as MapQuest.

But just because these maps can be useful, especially in areas where there are no maps, does not mean that all aspects of usability for the end user have been carefully considered. Usability refers to the interactive affordances of the system as a whole. How easy is it to find, access, edit, or understand the data? In his analysis of the completeness of two volunteered data sets—the OSM map of Haiti produced by the volunteers of the HOT and a map of Haiti produced by volunteers working within Google Map Maker (GMM)—versus a more official data set produced by the United Nations Stabilization Mission in Haiti (MINUSTAH), Haklay (2010b) found that the official data set was the most complete. Each data set contained features that the others lacked; however, the lack of metadata to explain data semantics, which differ among the maps, would make it difficult for the end user to integrate the three data sets. Even though the OSM and GMM data were produced by volunteers aligned with the neogeography community and the MINUSTAH data were produced by professionals, the OSM and Map Maker data were delivered to the user in a manner similar to traditional GIS products. That is, knowledge of the semantics of the data was left to the end user to interpret. The MINUSTAH data set contained operational geographic information such as road conditions that reflected its use for humanitarian workers in the field. In an online discussion that followed this analysis, discussants generally agreed that in situations where there are specific first-responder needs that are not being met by the metadata that accompanies generic data sets, some intermediary is needed to translate or adapt the data to these needs. Thus, metadata can directly impact the usability of the data sets and the tools.

Some users of the two mapping platforms in Haiti argued for simplicity, claiming that the proliferation of metadata squared made things more difficult:

> I believe that GMM can be a serious competition to OSM if it is simpler to use, easier to learn and thus more inviting to the casual newcomer. With GMM you have one way of mapping a simple item e.g. a bicycle track. Everybody can do it in ten minutes, no questions arise. With OSM you have two major tools, a huge load of tags, a wiki, a forum, several mailing lists, three different answers to the question, pages of contradictory documentation, plenty of old discussions and after working through all this, you realize that the question has not been resolved yet. I can see how many people would prefer the simple way offered by Google.—"Nop" on January 1, 2010 [OSM-talk] Countering Google's propaganda. (http://lists.openstreetmap.org/pipermail/talk/2010-January/046358.html)

Findability: In a paradoxical way, the proliferation of explanatory material, dueling tags, and community discussions that makes OSM useful and usable is the essence of findability in the media shift that has taken place in the GeoWeb. Morville (2005) describes the shift from theories of information retrieval, on which catalog-centered traditional metadata is based, to information browsing or foraging, in which a number of different strategies are used and the information seeker does not proceed by systematic logical steps but by pursuing leads as they emerge (Bates 2002). Information seeking relies on context, the frame of reference, environment, or setting within which information seeking is performed (Courtright 2008). So do usability and usefulness.

Community: The environment created by OSM with its links back and forth between the graphic, the textual, and the interactive differs from the spatial data infrastructure notion of neatly contained and formalized data—the catalog, the data element, and the interface. Context is community and community has a close affinity with gossip and recommendation culture of the current Internet. For people active in OSM, being able to help decide what to map may be of equal importance as the act of mapping. For the neogeographers, being able to shape the direction of the project and the map itself augurs well for a profound shift in media from the old geospatial world.

4.5 Conclusion

It has been suggested by Goodchild (2008) and Schuurman and Leszczynski (2006) that formal metadata standards need to be rethought to become more user-centric. They also propose adding new data elements to metadata for nonspatial attributes. This would afford the user better access to the context beyond technical and geometric elements and convey the tacit information that went into the making of a data set. Goodchild (2008) would augment current metadata about data geometries and lineages to better express data quality in data sets of mixed origins. These are both good suggestions, but we argue that the practices of neogeographers in the GeoWeb have shown that the metadata genie may be already out of the bottle.

The Internet has supported greater and greater interactivity, user collaboration, and the co-creation of geographic data. Metadata generated both automatically and by direct user contribution from descriptive and transactional work practices is closely coupled with the geographic data itself in these new GeoWeb systems. This proliferation of metadata, or metadata squared, facilitates finding, assessing, using, and making geospatial data.

What is the balance between centralized formalized metadata and the freewheeling metadata of an open-source community such as OSM? We argue that the neogeographers and the academics are both right. Standard metadata is not simple enough and at the same time not complex enough to give the context of data creation. These problems largely result from the media shift we have described. Many people find traditional metadata hard to manage, hard to produce, hard to use, and based on an outmoded static model of the way the Internet works. In theory, traditional metadata could be considered superior to the messy pseudo-metadata of OSM in its precise descriptions and consistent terminology; however, the difficulty lies in getting people to conform to the rigorous standards and keeping them updated. Formalized cataloging systems cannot scale at rates that match the growth of information in the GeoWeb.

Systems like OSM provide a lower cost of entry for producing and using data. The metadata is simplified and object-based, allowing for flexibility and rapid development. The ability to freely download both data and metadata supports emergent use. The proliferating ancillary material, the metadata squared, may be difficult to navigate at times, but the internal linking and the ability to switch back and forth rapidly between text and graphic modes make for an environment that can be richly rewarding for discovering meaning. The map grows and gains validity from the instantaneous feedback among users. And this growth can occur at exponential rates not realized by mapping agencies.

The OSM community might become more organized over time. Weber (2004) observes that open-source software projects can become powerful magnets that attract standards. As discussed above, there is talk on the OSM lists of developing a formal ontology of disaster-related terms. Van Exel and Dias (2011) are exploring how analysis of user behavior in OSM can serve as a proxy for trust and authority. Spatial data infrastructures could benefit by looking at the close coupling of the map and its explanatory context and could develop better systems for encouraging user feedback, as the USGS research on volunteered geographic information referred to above may show (Wolf et al. 2011; van Oort et al. 2009).

As one geoblogger put it:

> Traditional concepts like error bounds will fundamentally change because data collection is no longer happening on an annual basis, but will occur persistently from millions of globally distributed sensors. Error will be a fluid concept and not a static measure. Metadata needs to also change to be a fluid concept. The requirement for dedicated GIS metadata librarians with hundreds of metadata elements will not scale. Most importantly I think we need to stop thinking of the crowd as volunteers and amateurs. We should think of them as data collection points. This new reality is going to require innovative concepts around not only leveraging the crowd for data, but also using the crowd to ascertain the veracity of data. The crowd needs to be leveraged to verify and update metadata. (Gorman 2011)

The suggestion that people should be thought of as "collection points" along with Goodchild's (2007b) idea of "citizens as sensors" dehumanizes the relationship between people and the places they inhabit. The suggestion that the "crowd" should be treated as a field of automata dismisses the immense value of possibly capturing the individual's (or localized community's) unique perspectives on place. The key here is to allow interaction and feedback (Grira et al. 2010). A system like Tagwatch might be leveraged into an ontology based on user tags. Van Exel's work on trust and authority might provide the basis for a system of metadata verification produced by users. The ongoing work on qualitative GIS and metadata (Schuurman 2009) could lead to new avenues for enriching metadata. In short, the research frontier on how user-generated data might contribute additional meaning to the GeoWeb starts with metadata.

Acknowledgments The authors are grateful to Daniel Sui, Michael Goodchild, and Sarah Elwood for the invitation to submit this chapter to the volume on volunteered geographic information. We thank Peter Schweitzer, Martin van Exel, and two anonymous reviewers for helping us improve the structure and concepts of the paper.

References

Ball, M. (2010). What can be learned from the volunteer mapping efforts for Haiti? *Spatial Sustain*, January 31, 2010. http://vector1media.com/spatialsustain/what-can-be-learned-from-the-volunteer-mapping-efforts-for-haiti.html. Accessed July 28, 2011.

Batcheller, J. (2008). Automating geospatial metadata generation—An integrated data management and documentation approach. *Computers & Geosciences, 34*, 387–398. doi:10.1016/j.cageo.2007.04.001.

Batcheller, J., Gittings, B., & Dunfey, R. (2009). A method for automating geospatial dataset metadata. *Future Internet, 1*, 28–46.

Bates, M. (2002, September 11). Toward an integrated model of information seeking and searching. *Fourth international conference on information needs, seeking, and use in different contexts*, Lisbon, Portugal.

Budhathoki, N. (2010). Participants' motivations to contribute geographic information in an online community. Dissertation, University of Illinois Urbanna-Champaign.

Budhathoki, N., Bruce, B., & Nedovic-Budic, Z. (2008). Reconceptualizing the role of the users of spatial data infrastructure. *GeoJournal, 72*(3–4), 149–160. doi:10.1007/s10708-008-9189-x.

Chrisman, N. (1994). Metadata required to determine the fitness of spatial data for use in environmental analysis. In W. Michener, J. Brunt, & S. Stafford (Eds.), *Environmental information management and analysis: Ecosystem to global scales* (pp. 177–190). London: Taylor and Francis.

Coleman, D. J., Georgiadou, Y., & Labonte, J. (2009). Volunteered geographic information: The nature and motivation of produsers. *International Journal of Spatial Data Infrastructures, 4*, 332–358.

Comber, A., Fisher, P., & Wadsworth, R. (2008). Semantics, metadata, geographical information and users. *Transactions in GIS, 12*, 287–291. doi:10.1111/j.1467-9671.2008.01102.x.

Courtright, C. (2008). Context in information behavior research. In B. Cronin (Ed.), *Annual review of information science and technology* (pp. 273–306). Medford: Information Today, Inc.

Davies, C., Wood, L., & Fountain, L. (2005, Nov 8–10). User-centred GI: Hearing the voice of the customer. *Annual Conference of the Association for Geographic Information: AGI 05: People Places and Partnerships*, London.

Entchev, A. (2010). Comment to "Let's save metadata", February 16, 2010. http://www.spatially-adjusted.com/2010/02/15/lets-save-metadata/#comment-13108. Accessed July 27, 2011.

Federal Geographic Data Committee. (1994). *Content standards for digital geospatial metadata*. Washington, DC: Federal Geographic Data Committee.

Federal Geographic Data Committee. (2000). *Content standard for digital geospatial metadata workbook, Version 2.0*. Reston: Federal Geographic Data Committee.

Federal Geographic Data Committee. (2006). Clearinghouse concepts q&a. Federal Geographic Data Committee. http://www.fgdc.gov/dataandservices/clearinghouse_qanda. Accessed November 10, 2010.

Fee, J. (2010). Let's save metadata. http://www.spatiallyadjusted.com/2010/02/15/lets-save-metadata/. Accessed November 12, 2010.

Gahegan, M., Luo, J., Weaver, S. D., Pike, W., & Banchuen, T. (2009). Connecting GEON: Making sense of the myriad resources, researchers and concepts that comprise a geoscience infrastructure. *Computers & Geosciences, 35*, 836–854.

Goodchild, M. (2007a). Beyond metadata: Towards user-centric description of data quality. Spatial Data Quality 2007: ISSDQ. 13–15 June at Enschede, the Netherlands.

Goodchild, M. (2007b). Citizens as sensors: The world of volunteered geography. *GeoJournal, 69*(4), 211–221.

Goodchild, M. (2008, June 25–27). Spatial accuracy 2.0. In *8th international symposium on spatial accuracy assessment in natural resources and environmental sciences*, Shanghai.

Goodchild, M., Fu, P., & Rich, P. (2007). Sharing geographic information: An assessment of the geospatial One-stop. *Annals of the Association of American Geographers, 97*(2), 250–266.

Gorman, S. (2011). Statistical challenges of data at scale: Bringing back the science. http://blog.geoiq.com/2011/06/20/statistical-challenges-of-data-at-scale-bringing-back-the-science/. Accessed July 27, 2011.

Gould, M. (2006a). Meta-findability: Part 1. *GeoConnexion International Magazine, 5*(7), 36–38.

Gould, M. (2006b). Meta-findability: Part 2. *GEOconnexion International Magazine, 5*(8), 28–29.

Grira, J., Bédard, Y., & Roche, S. (2010). Spatial data uncertainty in the VGI world: Going from consumer to producer. *Geomatica, 64*(1), 61–71.

Haddad, T. C. (2010). Comment to let's save metadata. http://www.spatiallyadjusted.com/2010/02/15/lets-save-metadata/#comment-13113. Accessed July 20, 2011.

Haklay, M. (2010a). How good is volunteered geographical information? A comparative study of OpenStreetMap and Ordnance Survey datasets. *Environment and Planning B, 37*(4), 682–703.

Haklay, M. (2010b). Haiti—Further comparisons and the usability of geographic information in emergency situations. http://povesham.wordpress.com/2010/01/29/haiti-—further-comparisons-and-the-usability-of-geographic-information-in-emergency-situations/. Accessed July 11, 2011.

Hennig, S., Belgiu, G., Wallentin, K., & Hormanseder, K. (2011). User-centric SDI: Addressing users in a third-generation SDI. In *Inspire Conference 2011*, Edinburgh.

International Organization for Standardization. (1986). *ISO 8879:1986 Information processing – Text and office systems – Standard Generalized Markup Language (SGML)*. Geneva: International Organization for Standardization.

International Organization for Standardization. (2003). *ISO 19115: 2003, Geographic information–metadata*. Geneva: International Organization for Standardization.

Korzybski, A. (1933). *A non-Aristotelian system and its necessity for rigour in mathematics and physics. Science and sanity*. Laxeville: International Non-Aristotelian Library.

Latour, B. (1999). *Pandora's hope: Essays on the reality of science studies*. Cambridge, MA: Harvard University Press.

Lave, J., & Wenger, E. (1991). *Situated learning: Legitimate peripheral participation*. Cambridge: Cambridge University Press.

Li, W., Yang, C., & Yang, C. (2010). An active crawler for discovering geospatial web services and their distribution pattern – A case study of OGC web map service. *International Journal of Geographical Information Science, 24*(8), 1127–1147. doi:10.1080/13658810903514172.

Maron, M. (2010). Haiti OpenStreetMap response. http://brainoff.com/weblog/2010/01/14/1518. Accessed November 11, 2010.

Maxwell, J., Edwards, C., Jensen, M., Paustian, S., Parrott, H., & Hill, D. (1995). *A hierarchical framework of aquatic ecological units in North American (Nearctic Zone)*. St. Paul: U.S. Department of Agriculture, Forest Service.

McLuhan, M. (1964). *Understanding media: The extension of man*. London: Sphere Books.

Millerand, F., & Bowker, G. (2009). Metadata standards: Trajectories and enactment in the life of an ontology. In M. Lampland & S. Star (Eds.), *Standards and their stories* (pp. 149–166). Ithaca: Cornell University Press.

Moore, M. (2010). Cyborg metadata: Humans and machines working together to manage information – Part 1: Text. *Online Currents, 24*(3), 131–138.

Morville, P. (2005). *Ambient findability*. Sebastopol: O'Reilly.

National Research Council. (2001). *National spatial data infrastructure partnership programs: Rethinking the focus*. Washington, DC: National Academies Press.

Nielsen, J., & Loranger, H. (2006). *Prioritizing web usability*. Berkeley: New Riders Press.

Onsrud, H. (Ed.). (2007). *Research and theory in advancing spatial data infrastructure concepts*. Redlands: ESRI Press.

O'Reilly, T. (2005). What is web 2.0?: Design patterns and business models for the next generation of software. http://www.oreillynet.com/lpt/a/6228. Accessed November 15, 2010.

Osborne, C. (2010). Mapping a crisis. *Guardian Online* http://www.guardian.co.uk/open-platform/blog/mapping-a-crisis. Accessed July 27, 2011.

Poore, B. (2003). Blue lines: Water, information, and salmon in the Pacific Northwest. Dissertation, University of Washington.

Poore, B., & Chrisman, N. (2006). Order from noise: Toward a social theory of geographic information. *Annals of the Association of American Geographers, 96*(3), 508–523.

Pultar, E., Cova, M., Yuan, M., & Goodchild, M. (2010). EDGIS: A dynamic GIS based on space time points. *International Journal of Geographical Information Science, 24*(3), 329–346. doi:10.1080/13658810802644567.

Qin, J. (2008). Controlled semantics vs. social semantics: An epistemological analysis. In *Proceedings of the Tenth International ISKO Conference: Culture and identity in knowledge organization* (pp. 5–8), Montreal, August 5–8, 2008.

Raymond, E. (1999). *The Cathedral and the Bazaar*. Sebastopol: O'Reilly.

Rodriguez, M., Cruz, I., Egenhofer, M., & Levashkin, S. (Eds.). (2005). *GeoSpatial semantics. Lecture notes in computer science* (Vol. 3799). Berlin: Springer.

Scharl, A., & Tochterman, K. (2007). *The geospatial web*. London: Springer.
Schuurman, N. (2008). Database ethnographies using social science methodologies to enhance data analysis and interpretation. *Geography Compass, 2*(5), 1529–1548.
Schuurman, N. (2009). Metadata as a site for imbuing GIS with qualitative information. In M. Cope & S. Elwood (Eds.), *Qualitative GIS: A mixed media approach* (pp. 41–56). Los Angeles: Sage.
Schuurman, N., & Leszczynski, A. (2006). Ontology based metadata. *Transactions in GIS, 10*(5), 709–726.
Schweitzer, P. (1998). Easy as ABC – Putting metadata in plain language. *GIS World, 11*(9), 56–59.
Shirky, C. (2005a). Ontology is overrated: Categories, links, and tags. http://www.shirky.com/writings/ontology_overrated.html. Accessed January 28, 2009.
Shirky, C. (2005b). Folksonomies + controlled vocabularies. http://many.corante.com/archives/2005/01/07/folksonomies_controlled_vocabularies.php. Accessed October 28, 2010.
Silver, J. (2010). Data information: How visual tools can transform lives. http://www.wired.co.uk/wired-magazine/archive/2010/09/features/data-information?page=all. Accessed November 11, 2010.
Six Rivers National Forest. (1999). *Metadata for stream*. Eureka, CA: U.S. Forest Service. http://www.ncgic.gov/GIS_Data/smf/hydro/stream.metadata.html. Accessed September 10, 2001.
Sui, D. (2008). The wikification of GIS and its consequences: Or Angelina Jolie's new tattoo and the future of GIS. *Computers, Environment and Urban Systems, 32*, 1–5.
Sui, D., & Goodchild, M. (2001). GIS as media? *International Journal of Geographical Information Science, 15*(5), 387–390. doi:10.1080/13658810110038924.
Sui, D., & Goodchild, M. (2011). The convergence of GIS and social media: Challenges for GIScience. *International Journal of Geographical Information Science, 25*(11), 1737–1748.
Tsou, M. (2002). An operational metadata framework for searching, indexing, and retrieving geographic information services on the Internet. In M. Egenhofer & D. Mark (Eds.), *GIScience 2002: Lecture notes in computer science* (Vol. 2478, pp. 313–332). Berlin: Springer.
Turner, A. (2006). *Introduction to neogeography*. Sebastopol: O'Reilly Media.
U.S. Department of Commerce, National Oceanic and Atmospheric Administration. (1999). 50 CFR Part 223: Endangered and threatened species; proposed rule governing take of threatened Snake River, Central California Coast, South/Central California Coast, Lower Columbia River, Central Valley California, Middle Columbia River, and Upper Willamette River evolutionarily significant units (ESUs) of West Coast steelhead. *Federal Register, 64*(250), 73479–73506.
U.S. Department of Commerce, U.S. Forest Service, Remote Sensing Lab, Pacific Southwest Region. (2004). Metadata for NWCSTRM03_2 2004. http://www.fs.fed.us/r5/rsl/projects/gis/data/calcovs/nwcstrm03_2.html. Accessed July 27, 2011.
Van Exel, M., & Dias, E. (2011). Towards a methodology for trust stratification in VGI. *VGI Pre-Conference at AAG*, Seattle. http://vgi.spatial.ucsb.edu/sites/vgi.spatial.ucsb.edu/files/file/aag/van_Exel_abstract.pdf. Accessed July 8, 2010.
van Oort, P., Hazeu, G., Kramer, H., Bregt, A., & Rip, F. (2009). Social networks in spatial data infrastructures. *GeoJournal, 75*(1), 105–118.
Vander Wal, T. (2007). Folksonomy: Coinage and definition. http://www.vanderwal.net/folksonomy.htm. Accessed November 15, 2010.
Waters, T. (2010). The OpenStreetMap project and Haiti earthquake case study. http://www.slideshare.net/chippy/openstreetmap-case-study-haiti-crisis-response. Accessed November 10, 2010.
Weber, S. (2004). *The Success of open source*. Cambridge: Harvard University Press.
Weinberger, D. (2007). *Everything is miscellaneous: The power of the new digital disorder*. New York: Times Books.
Wikipedia. (2010). OpenStreetMap. http://en.wikipedia.org/wiki/OpenStreetMap. Accessed November 15, 2010.
Wolf, E., Matthews, G., McNinch, K., & Poore, B. (2011). *OpenStreetMap collaborative prototype, phase one* (Open-file report of 2011–1136). Reston: U.S. Geological Survey. http://pubs.usgs.gov/of/2011/1136/. Accessed December 12, 2011.
Wood, D., Fels, J., & Krygier, J. (2010). *Rethinking the power of maps*. New York: Guilford.

Chapter 5
Situating the Adoption of VGI by Government

Peter A. Johnson and Renee E. Sieber

Abstract Governments have long been active online, providing services and information to citizens. With the development of Web 2.0 technology, many governments are considering how they can better engage with and accept citizen input online, particularly through the gathering and use of volunteered geographic information (VGI). Though there are several benefits to governments accepting VGI, the process of adopting VGI as a support to decision-making is not without challenge. We identify three areas of challenge to the adoption of VGI by government; these are the costs of VGI, the challenges for governments to accept non-expert data of questionable accuracy and formality, and the jurisdictional issues in VGI. We then identify three ways that governments can situate themselves to accept VGI—by formalizing the VGI collection process, through encouraging collaboration between levels of government, and by investigating the participatory potential of VGI.

5.1 Introduction

Western-style democratic governments at all levels are often interested in connecting with citizens through the use of Internet-based communications technologies, such as Web 2.0. Creating this new online relationship between governments and citizens can support greater transparency, efficiency, and effectiveness of government services (Brewer 2006; Dovey and Eggers 2008; Saebo et al. 2008). This also can increase the level of citizen participation in decision-making. Numerous technologies have emerged

---

P.A. Johnson (✉)
Department of Geography and Environmental Management, University of Waterloo, Waterloo, ON, Canada
e-mail: pa2johns@uwaterloo.ca

R.E. Sieber
Department of Geography, McGill University, Montreal, QC, Canada
e-mail: renee.sieber@mcgill.ca

to support these types of e-governance initiatives; these platforms can be used to address place-based aspects where governments impact everyday life (Drummond and French 2008). Applications range from easing the daily commute through accessing crowdsourced traffic reports to facilitating discussion on land use development scenarios and to identifying service provision locations. The Geospatial Web 2.0 (or Geoweb) is a set of geospatially enabled online tools and data that can be used to support these types of initiatives (Ganapati 2010; Rouse et al. 2007). The Geoweb serves as a conduit for volunteered geographic information (VGI) sourced from or contributed by citizens as part of a one-way 'government-to-citizen' (G2C) or two-way 'citizen-to-government-to-citizen' (C2G2C) process. By requesting citizen contributions of VGI, a government can potentially create a two-way conversation with citizens that demonstrates their responsiveness to specific concerns.

Two reasons drive the collection and use of VGI by governments and government agencies. First is the potential for citizens, whether they reside inside or outside a given jurisdiction, to act as sensors of their environment (Goodchild 2007). The general trend in downsizing governments at the provincial/state level, driven by neoliberalization, has reduced the resources available to support municipal level decision-making (Dovey and Eggers 2008; Johnson and Sieber 2011a). In North America and Europe, municipalities are being asked to take on increased planning and land management responsibilities but without a corresponding increase in resources or staff support. This creates an opportunity, for better or worse, for the use VGI as a type of 'contracting out' of data collection tasks (Newman et al. 2010), creating a government spatial data infrastructure that is dependent on volunteer effort. This approach to VGI use by governments treats citizens as a distributed set of sensors (Goodchild 2007) to be networked together to supply decision-makers with rich sources of data. Citizens, specifically local residents, are supposedly closer to the phenomena, can identify changes, and report those changes more quickly than government employees reliant on infrequently collected data. Citizens as reference is ambiguous hold a valuable local knowledge of place, and considering that pride of place is a prime motivator of citizens who contribute geospatial information, they are more likely to volunteer that information in digital form (Budhathoki et al. 2010; Elwood 2008). Tulloch (2008) provides a case study of the citizen-based verification of official government-collected data on vernal pools. This task was conducted by a group of citizen scientists, contributing information via a custom-made Geoweb site. This example shows how citizen volunteer efforts can be incorporated into a government process as a way of both saving the government money and utilizing the knowledge of citizens to support decision-making and management.

Second, VGI can be valuable to governments as a form of citizen participation. As opposed to the citizens-as-sensors view, this treats the process of VGI usage as an opportunity for citizens-as-partners to co-produce social, economic, and environmental goals, with the mission of strengthening civil society. For governments, an increased focus on the process of VGI collection and two-way communication, rather than the unidirectional sensor relationship, can support essential participatory components of democratic governance, particularly in reinforcing the transparency and responsiveness of a government to its electorate (Dovey and Eggers 2008;

Ganapati 2011). Similar to how town hall assemblies or letter writing to representatives is a method of sharing citizen perspective with elected officials, VGI holds promise to act as a digital and geospatially referenced conduit to connect elector with both elected official and specific government departments. This strengthening of C2G2C linkages includes the potential for constituents to form a power base for government employee initiatives or support for official policies.

Despite the identification of reasons why governments would adopt VGI and the Geoweb (Ganapati 2011; Johnson and Sieber 2011a), these motivations alone do not determine whether a government will proactively solicit information or integrate content into policy. Shifting government motivations to participate in concrete actions relies on explicating the organizational and cultural challenges, technological issues, and issues involving the scaled and interconnected nature of governance. As governments adopt different formal and informal processes and tools for gathering, evaluating, and incorporating this type of data into decision-making, critical reflection on the relationship between VGI as a form of citizen participation and the needs and constraints of government is required.

Past research emphasizes the opportunity for VGI and the Geoweb to realize the promise of public participation geographic information systems (PPGIS) (Miller 2007; Sieber 2006), providing a conduit for citizens to share their local knowledge with decision-makers, effecting change, establishing two-way communication, or even circumventing traditional pathways of public participation. To overcome these challenges, VGI researchers can draw on critical GIS studies (Crampton 2009; Schuurman 2000; Sieber 2004, 2006). For example, we are warned of the continued slippage of privacy in a Web 2.0 world (Elwood and Leszczynski 2011; Zook and Graham 2007), and how citizen participation in online deliberation may not match the high levels of participation that are often seen in other online activities, such as social networking and gaming (Chadwick 2009). Echoing critiques of GIS (e.g. Pickles 1995), government adoption of VGI could represent a strategy by government of co-optation or distraction from other more effective forms of citizen engagement. These concerns highlight the need for a better understanding of how VGI and government can integrate and what factors, both of VGI and governments as organizations, affect this adoption. This research differs from the majority of the research around government adoption of VGI, which emphasizes data handling (e.g. evaluating accuracy, understanding citizen motivations to contribute as a mechanism to perpetuate information flows) largely to the exclusion of the democratic process by which governments must act—hence, our use in the chapter title of the word, 'situating', which expresses a social and critical practice.

This chapter draws on our experience developing and implementing VGI applications for government partners in the Canadian province of Quebec. The context of our reflections makes these recommendations particularly relevant to other similar western-style democracies, particularly the United Kingdom, Australia and New Zealand, Western Europe, and the United States. Though there are likely to be differences in VGI adoption in government between municipal and state/provincial governments within one country, or between countries with different political traditions, there are comparative lessons to be drawn from this work that can guide

VGI adoption across multiple contexts. We have been co-developing Geoweb and VGI collection platforms in the rural area of Acton, a municipal regional area, similar to a county but with greater jurisdictional authority (Municipal Regional County of Acton or MRC Acton), approximately 1 h east of Montreal, Quebec. The development is being coordinated by five provincial agencies; the hope is that this will form the basis for further Geoweb developments throughout the province. The specific goals of this project are to engage citizens within community economic development and environmental management using Geoweb applications and to co-develop a sustainable software platform for sharing with other jurisdictions. We implemented projects to create user-generated maps of local economic assets to support regional marketing efforts and to gather citizen reports of riverbank erosion as a component of municipal remediation efforts. Through these two projects in particular, we have been engaged with governments attempting to adopt VGI collection and their associated reactions to the challenges that this adoption creates. These adoption challenges and organizational constraints are the focus of this chapter.

The government agencies with which we work have been initially enthusiastic about the potential of VGI, but this dampens closer to deployment, as they raise questions about the fitness of VGI within government. Across many different types of government, we have repeatedly seen a resistance to the acceptance and use of VGI. Based on these experiences, we have identified a series of broad constraints to the adoption of VGI by governments. These constraints include both the motivational factors that lead government to consider the use of VGI and how these play out on the landscape of organizational structure to impede adoption and use. This means examining government adoption through two models of citizenry: citizens as sensors and citizens as partners. We frame a discussion of ways that governments can situate themselves amongst the constraints so they can adopt VGI.

5.2 The Practice of VGI in Government

The concept of users contributing geographically related information online as participation in governance is not new. In an overview of the use of geographic information systems (GIS) to increase citizen engagement, Ganapati (2010) outlines four broad thematic areas where governments can use the Geoweb for e-government applications: citizen-oriented transit information, citizen relationship management, citizen-volunteered geographic information, and citizen participation in planning and decision-making. Each of these application areas builds on identified roles for information and communications technology (ICT) for e-governance and PPGIS research, providing services to citizens and increasing participation in governance, in many cases through the government acceptance of VGI. The first two of these areas are based on data provision from government to citizens as a form of improved service. Whether this in the form of real-time transit tracking, or by making government data more freely available, these types of initiatives show how governments can improve services and become more transparent, though these types of information provision activities do not equate with public participation or involvement.

Many examples of e-government activities focus on the prosaic functioning of governance-service provision and strategic planning issues. The most prominent usage of VGI has been in acute responses to a natural or human-generated crisis. Many of these occurred exogenously to government, in a sense responding to the failures of governments to act swiftly to identify hotspots and distribute aid. The earliest example is the use of Google Maps to encourage volunteers to contribute information about Hurricane Katrina impacts in 2005 (Miller 2007). This information was used to support official government rescue efforts. Similar uses of VGI have been seen in responses to wildfires (De Longueville et al. 2010; Goodchild and Glennon 2010) and earthquakes (Zook et al. 2010). Given that each of these examples was developed externally to government organizations, yet has as an outcome the mobilization of decision-making and demand for government action, it raises questions about the ability and desire for governments to directly accept and act upon VGI. The utility of VGI for many decision-making tasks has been identified, but the process through which VGI can be adopted by government remains to be negotiated.

5.3 Adoption of VGI in Government

In addition to learning from existing e-government examples, governments will likely look to their experiences with GIS when they decide to adopt VGI. Although both are geospatial and rooted to place, adoption of the Geoweb, that is, the underlying platform upon which VGI is added, differs from GIS implementation in government. One significant area of difference is in the locus of development. GIS implementation steps, such as the customization of the software, acquisition of framework data sets, and the purchase of hardware, are conducted under a mandate of a government agency or department. Even if GIS forms a part of a multi-agency activity, governments—here, we refer largely to municipalities or regional agencies—have considerable control over purchases, staff, and data. By contrast, many Geoweb platforms operate outside of existing government mandates and processes. The hardware and software stack are now hosted in the Web-based 'cloud' and often reliant on a software-as-a-service (SaaS) model of distribution. VGI generation is external to the organization compared to the internal data of the organization typically fed into a GIS. This allows citizens to circumvent government, making the VGI adoption process different than that of other types of technologies within government, such as GIS (Budic 1994; Goelman 2005), planning support systems (PSS), and spatial decision support systems (SDSS) (Geertman 2006; Vonk et al. 2007). GIS, PSS, and SDSS adoption within an organization often starts with software being purchased or developed to accomplish a set series of tasks (Nedovic-Budic 1998).

Technology adoption research has focused on the bridging of a perceived gap between technology developer, tool, and user, where addressing identified constraints is within the sphere of influence of either developer or organization (Johnson and

Sieber 2011b; te Brömmelstroet and Bertolini 2008; te Brömmelstroet and Schrijnen 2010). VGI demonstrates the need to negotiate adoption in a more fluid fashion, one that is not simply focused on a developer meeting the needs of the user. In a governance context, these roles are shared by citizens, community organizations, non-governmental organizations (NGOs), universities, information technology (IT) companies, and multiple levels of government. For example, the citizen can fill roles as the developer of technology, VGI contributor, and the user of contributed data sets. Citizens can build their own mash-ups and develop their own mobile apps. Citizens produce (the subject of VGI), they consume (traditional GIS), and they create value-added, derived information as well. An IT company may develop tools but is also the user, accessing citizen VGI for marketing purposes. Governments now develop Geoweb applications to gather VGI for specific purposes, using third-party platforms such as Twitter and Google Maps, though they have little control, ownership, or input into these platforms. With these competing priorities, Geoweb applications have multiple objectives, which may be only peripherally related to those of government. From a developer perspective, the tools to collect VGI are not simply refined, retracted, or revised to better meet the needs of government; instead, they evolve in response to corporate and user-community preferences. Government adoption of VGI and its underlying Geoweb platform operates through a more interconnected and complicated set of pathways compared to traditional types of geospatial technology, where software is provided by one developer and data is shared internally or sourced from other government agencies (Harvey 2003; Onsrud and Pinto 1991).

We define broad characteristics of VGI that challenge the technology adoption processes in government, generate new organizational constraints, or reinforce existing constraints that can impede adoption. These are the costs of VGI, the challenges for governments to accept non-expert data of questionable accuracy and formality, and the jurisdictional issues in VGI.

5.3.1 The Costs of VGI

Any new technology will introduce additional resource costs for government. For VGI, these costs can be both financial, for software and services, and human resource costs, for training to negotiate the VGI learning curve and additional staff to support the VGI gathering process. Each of these areas of resource cost will differ depending on the type of government organization. In general, for government agencies with larger budgets or an existing GIS division, there is likely greater capacity to absorb the costs of gathering and using VGI. These agencies have existing spatial data that can be combined or refined with VGI. They likely have staff trained in system administration, computer server maintenance, and ability to build computer applications. Geoweb platform development requires a shift in domain knowledge that more resembles this type of system administration as opposed to spatial analysis. For local and regional governments, there still may be many good reasons to use VGI, yet the infrastructure required to support its gathering may not exist or be otherwise inadequate. This is a similar dynamic as is found with many types of IT and GIS adoption in municipal planning organizations, where availability of financial and human resources represents a significant contributor to adoption (Al-Kodmany 2000; Carver et al. 2001).

The resource implications of the Geoweb framework used to collect VGI are generally ignored in academic literature, with the prevailing view being that these tools are easy to deploy and lightweight (Haklay et al. 2008; Hudson-Smith et al. 2009; Turner 2006). Though the financial cost of access to many platforms, such as Google Maps (http://code.google.com/apis/maps/index.html) and Open Layers (http://openlayers.org/) may be low or free, the skills cost to develop anything more than a basic solution can be prohibitive, requiring advanced computer programming skills. Add to this the cost of maintaining and eventual refreshing of a site, and there may be a substantial resource cost for governments. Even governments that are active on free social media sites, such as Twitter and Facebook, may find that they require additional training and resource expenditure to gather, respond to, and analyze contributed information. The human resources cost of VGI is also based on the modification of existing workflow and organizational process to accept this new form of input. Depending on the specific government process into which VGI is to be incorporated, this could require a chain of employees to adapt their workflow. For example, with a government service municipal request system, citizens submit a request for service to repair issues such as broken street lights or potholes. Without a robust integration into a municipal workflow including dispatchers, workers, and a response to citizens once work is completed, such a system may add considerably to workload.

In MRC Acton, the costs of VGI have been encountered in several ways. First, resource costs have largely focused on human resources, as staff negotiate the learning curve for understanding and using VGI. Despite the presence and frequent use of traditional desktop GIS in MRC Acton, we found that this did not adequately prepare staff for gathering or using VGI, underlining the distinct difference in adoption between these two technologies. As discussed earlier, the spatial analysis skills used in GIS are not directly transferrable to the systems administration and Web development skills used in gathering and using VGI. Considering this, for the introduction of VGI within government, it must be noted that skills and experience using GIS are not directly transferrable.

Second, though VGI is often promoted in terms of resource benefits and cost savings, in our partnership with MRC Acton, we discovered that there are opportunity costs as well. We have postulated that VGI could reduce the political distance between the state and citizens, but it could also simultaneously increase the political distance. We are working with a community-based organization to provide a Geoweb platform to support the identification and management of riverbank erosion. Farmers and land owners would report erosion as a first step to government action mitigating that erosion. However, the reporting of erosion itself was revealed to possibly identify farmer malfeasance in their land management practices, which may lead to increased costs for the farmer who reported the erosion. In this way, erosion monitoring becomes a delicate negotiation of protecting identification of the farmer as well as revealing location-based information. There are numerous informal practices that occur face-to-face that would be impeded in this digital broadcast of problems.

5.3.2 The Challenge for Governments of Accepting Non-expert Data

As a data type contributed by a variety of individuals, VGI fundamentally differs from traditional forms of data collected by experts that is often used in government GIS (Budhathoki et al. 2008). Goodchild (2007) highlights this difference between data that is voluntarily asserted first-hand, by an individual without formal qualifications in that domain, and data that is authoritative, or collected within a formalized framework, often at a distance, by an expert in the subject, as part of their paid work. The latter is the realm of GIS, where it is unquestioned that data is generated by experts in their field; the data can still have errors but at least the source is unquestioned. Governments face a formidable challenge in accepting VGI when they shift from the use of only expert data to a mixed model that can evaluate and incorporate citizen volunteered data. This shift requires that governments engage with several aspects of the VGI creation process, including the individual contributors of VGI, a step towards widespread participation that many in government may not be ready to take.

One of the most significant shifts between traditional forms of expert-collected data and VGI is the individual who is doing the data collection. Rather than a team of trained experts that collect data using specialist tools such as remote sensing, or formalized methodologies, such as a government census, VGI is largely contributed by individuals as a leisure or non-paid work activity (Goodchild 2007; Newman et al. 2010; Tulloch 2008). Citizens instead are considered holders of valuable local knowledge (Elwood and Ghose 2004). Despite evidence that non-experts can indeed contribute information (Budhathoki et al. 2010; Haklay 2010; Parsons et al. 2011), the characterization of VGI as an informal data source, one created by non-experts or 'neogeographers' as a hobby (Hudson-Smith et al. 2009; Turner 2006), may in fact prevent governments from considering VGI as a serious source of data. This terminology and phrasing can often be saddled with negative connotations when compared to authoritative and expert data provided by government agencies (Harvey 2007). To better integrate VGI into government and decision-making, the continued use of these terms may serve to marginalize VGI as a data type, regardless of its fitness relative to authoritative sources.

The perception of VGI as varying significantly in quality compared to authoritative sources is a constraint on government adoption. This aspect of constraint is based on the legal implications of government error—essentially, who will get blamed if data are wrong. Compared to authoritative data, two assessments of Open Street Map, one of the largest VGI collection platforms, have found variable levels of congruence with authoritative data sets, ranging from poor to excellent, with populated areas often displaying improved coverage (Girres and Touya 2010; Haklay 2010). The challenge to the use of VGI is that it could be considered unscientific due to data quality issues surrounding its collection. For example, information about sitings of endangered species could then be ignored in areas proposed for logging. Without a firm view of the quality of VGI, it can be difficult for a government to know how much weight should be given to citizen opinions or comments.

A defining characteristic of VGI is that it is often contributed within multiple informal, casual, and unstructured contexts (Elwood 2009; Flanagin and Metzger 2008). The data may or may not be sufficiently complete, that is, seamless over a spatial extent. Because of this informality, the quality of VGI can vary considerably. For example, unstructured or qualitative data, such as Twitter postings or freeform text in online review sites, can be rich sources of information but simultaneously difficult to incorporate into a decision-making process (Johnson et al. 2012). Concerns over the quality and accuracy of VGI serve as a significant disincentive for governments (Haklay 2010; Seeger 2008). VGI has potential to support governance because the data can help correct errors, refine the data through precision, and fill in gaps where there are no government employees. However, governments at all levels are reminded of their legislated obligations for due diligence in planning. Whether or not VGI is considered an acceptable data source for decision-making, support can depend on the credibility of the source, the presence of a mass volume of like contributions, and favourable comparison to other traditional data types (Flanagin and Metzger 2008; Haklay 2010).

Governments face challenges in utilizing non-expert data. Establishing the credibility of a source may demand that government knows who exactly is providing the data, to ensure it reflects the constituency in question and not the result of outside agendas. Due to the largely anonymous nature of many online activities, governments may never be able to fully verify even the general characteristics of those who contribute VGI (Budhathoki et al. 2010). Can governments be confident that VGI is the product of individuals with first-hand knowledge of a given phenomenon? Or is data contribution driven by a specific agenda? This becomes particularly salient if VGI is collected to assess public perceptions. VGI can be considered a convenience sample of one particular subset of the population, rather than representative of the whole. This contrasts markedly to official data sets gathered with random samples of the population that can therefore support rigorous statistical analysis. These concerns can be partially addressed through creating a VGI collection framework with a strong emphasis on identifying individual contributors (Seeger 2008). Techniques used to verify identity include logins, mail-outs with access codes, and IP logging to ensure participants are from within a certain geographic area. Each of these techniques does come with a risk of alienating or otherwise reducing participation, either because participants want to be anonymous or have difficulty navigating extra layers of technology (Brewer 2006; Vonk et al. 2005).

Tulloch (2008) discusses the use of VGI gathered in Second Life as a support for park design that raises several questions. First, is gathering citizen feedback in this way sufficient to fulfil community involvement requirements? Should input from VGI be balanced with traditional forms of citizen input, such as town hall style meetings, and where should it fit compared to forms of citizen participation such as steering committees and citizen design teams? These legal obligations inherent to governance at all levels are real constraints to the government acceptance of VGI as a data source, as they privilege the status quo data sources that are known entities, or at least have failings that are acknowledged. This mismatch between VGI as a product of often unknown provenance with a variable degree of data quality should be considered as a significant barrier for government adoption of VGI.

Our own research has found that a lack of complete government control over the data collected by and displayed on a VGI platform creates considerable anxiety over deployment. Concerns were present over both the contributed data and the base map data. With our government partners, we used the proprietary Google Maps platform, which provided several advantages, including free satellite imagery, built-in geocoder, and a popular user interface. However, there were also trade-offs, with inaccuracies present in the base map, and coarse resolution imagery, especially in rural areas. There was substantial criticism levied at the accuracy of the Google Maps base map. For example, partners with local knowledge found many mislabelled roads and names of pre-amalgamation hamlets and villages that were no longer commonly used. Officials assumed that this data shown in Google Maps was not authoritative and had questions about provenance and update frequency. Later, when it was discovered that the Canadian federal government base map contained the same errors, government partners had a more positive view of Google base map coverage. This positive view was reinforced when change requests submitted to Google were reflected in the base map within weeks, compared to government base map change requests that could take substantially longer to be reflected, often requiring changes to be published on an annual basis. This example demonstrates how the provenance of data (the 'known' federal data vs. the 'unknown' private company data) can affect willingness to use, regardless of any actual difference between the data.

5.3.3 The Jurisdiction of VGI

It is easy to think of geospatial technologies not simply as place representative but place bound. However, as an online technology, VGI can cross spatial scales ranging from the local to the global. This has significant implications for how governments interact with citizens and whether the directionality of power flows can be rewired (Crampton 2009; Sieber 2004). This phenomenon is termed jumping scale (Cox 1998; Smith 1993) and is relevant to a governance context in use of VGI. Jumping scale refers to an action where individuals operating at one scale (e.g. community scale) circumvent or bypass an intermediary scale of decision-making (e.g. municipal government) to argue their issue at a 'high' level (e.g. provincial or federal level) (Cox 1998; Swyngedouw 2004). For example, VGI may be generated in response to a local issue but then communicated to provincial or federal level decision-makers in a call for intervention. This brings pressure from both the local and national level on other levels of government. This cross-scale aspect means that citizens can use VGI to circumvent the traditional pathways of public participation, though this in turn may not fit will with formal decision-making structures.

The cross-scale nature of VGI presents an obstacle to governments in several ways. First, this type of activity can result in a government losing some control over a particular issue, as VGI can be communicated without regard to political boundaries. As VGI is communicated to other levels of government, more players become implicated in solving or answering an issue. This may be a positive factor, such as in the case of a resource-constrained local government looking to secure greater funding from provincial or federal levels to combat a particular problem. This loss of control also means that a government can be overruled or removed from the decision-making process. There is a danger that after asking citizens to contribute on a particular issue, governments may not be able to properly respond to the citizen feedback, as the required action may be beyond the mandate or geographic region of the government. For example, a municipal government may ask for citizen input on land rezoning, yet this rezoning may require provincial or regional approval. This can create a situation where expectations are raised as to the type of result that will be delivered. This issue is long-standing in planning; in that if citizens are to be asked to contribute, a government must be willing and able to act on that advice (Wittig and Schmitz 1996). Though VGI contributions may be cross-scale, the political decision-making and mandate of governments are not equally flexible.

Our own research in Acton has uncovered the use of VGI to jump scale. We are working with a community-based watershed management organization in Acton to deploy a Geoweb site as a conduit for citizen reporting of erosion and other environmental problems. This process is occurring at a watershed scale, which has only recently become a decision-making boundary in Quebec. The provincial government has created this new jurisdiction for issues like erosion; however, few mechanisms are in place to support integration of citizen or community organization perspectives with the provincial-level policy development process. The community-based watershed management organization is using the Geoweb to collect its own VGI, which then is submitted to provincial ministries to support funding grants at the watershed level. This demonstrates how VGI that is reflective of a local perspective can be leveraged to impact other scales. Local VGI gathered in this manner is used to cross scales, allowing the community-based organization to argue for improved watershed-level support from provincial government.

5.4 Situating Government to Adopt VGI

The generation of VGI and its acceptance by governments is an emerging phenomenon, and as such, the adoption challenges to its use and application are fluid. Through the identification of constraints to the use of VGI by governments, we aim to provide increased clarification as to how the use of VGI can be negotiated. From this analysis we define three ways that governments can situate themselves to more fully participate in gathering and using VGI: increased formalization of VGI collection, encouraging collaboration between governments, and reviving the role that VGI can play in seeing citizens as partners in knowledge generation and improved decision-making.

5.4.1 Increasing Formalization of VGI Collection

For governments to accept and use VGI, one blockage is the value—both real and perceived—of the data. Without confidence that VGI represents citizen input, governments will have a difficult time justifying the use of VGI within their decision-making tasks. Due to the legal framework in which official decision-making must occur, particularly around issues with a geographic context, such as facility siting, land use, and property rights, there must be a defensible process followed to justify taking a course of action that may have negative implications for a certain group of citizens or that is based on input from citizens. In balancing the needs and desires of society as a whole, which can involve reconciling many contradictory opinions and viewpoints, decision-makers must rely on data and information that can be defended as valuable input from citizens and reflective of real citizen concerns. As a new technology, there are still many questions surrounding the value of VGI compared to traditional methods of citizen input. For example, what weight should a decision-maker give to a perspective supported by VGI compared to a perspective supported by citizens who attended a town hall meeting? Is the method of participation (digital vs. in person) indicative of the strength of agreement or opinion? Though each of these questions requires significant follow-up research, issues concerning the value of VGI are reflective of government concern or focus on process, rather than the issue itself.

Due to the structured and formalized way in which government operates, a more formalized VGI collection process, with a focus on data quality and strict controls to contribution, and crowdsourced verification may prove beneficial. For government to adopt VGI, linking it to official government structures and decision-making processes, this may require the institution of specific rules and regulations that can constrict or even eliminate participation. One example of this may be the requirement to officially register on a site using one's real name or other identifying characteristic. The user who feels comfortable contributing VGI anonymously may not feel the same when asked for identifiable information. The identification of individual contributors on some level, not necessarily by name, but as a resident of jurisdiction may impose a constraint on participation in an official context that would not be present in a more informal VGI implementation. Despite these constraints on participation, for VGI to be accepted as a legitimate data source for use in decision-making, a collection framework that enforces some degree of identification and places a frame on types of participation can begin to address some of the government concerns surrounding contributed data.

5.4.2 Encourage Collaboration Across Governments

We identified the cross-scale nature of VGI to be a challenge for government acceptance of VGI. One of the main challenges is that governments are restricted to a certain mandate and constrained geographic area, whereas VGI may be contributed by individuals outside of the area and on topics that are outside of the mandate of any one government agency. With increased government collaboration, VGI can be directed to effect change at the appropriate decision-making level. For example, VGI collected by a municipal government that indicates action required on the part of a provincial government is more likely to be acted upon if the provincial government is involved with or at least aware of its collection. Stronger collaboration between governments will facilitate the ability of citizens to jump scale. Though this type of process has benefits for citizens, it may not have similar benefits for governments and indeed may be actively resisted by governments.

There are practical reasons for increased collaboration between governments, such as realizing cost savings in the collection of VGI. This is relevant at the municipal level and in rural or remote locations, where the IT support staff required to operate a VGI collection framework may not exist. This type of collaboration already exists in many places, as groups of municipal governments will contribute to the shared development of technologies or the joint funding of IT systems, such as enterprise GIS (Budic 1994; Harvey 2003). Sharing expertise on VGI development also can involve collaboration with private enterprises, universities, and non-profits through the use of open source technology (Hall et al. 2010). Similarly, increased collaboration between governments can ease the diffusion of VGI technology. One of the main mechanisms through which technology is transferred in government and organizations in general is through a community of users (Budic 1994; Onsrud and Pinto 1991). In many instances, a driving factor in GIS adoption in municipal planning agencies has been the use in other, similar type agencies. The lessons learned by one agency can make introduction and adoption in another agency easier. This type of diffusion can occur over similar government levels (such as municipal) and also to different scales of government.

5.4.3 Investigating the Participation Potential of VGI

Fundamentally, the process of citizen generation of VGI, government acceptance of this input, and resulting action can represent a variety of forms of participation in governance. Participation can be limited, with citizens treated as passive sensors, feeding data to higher-level decision-makers in a one-way process. Alternately, citizens can be engaged as partners, contributing information as part of a two-way dialogue surrounding an issue and providing an opportunity for direct democracy, enabled by information technology. In situating government to adopt VGI, the use of this approach to facilitating public participation provides one of the most compelling arguments for its adoption and use. In a political climate increasingly defined by microtargeting of communities of interest and much hyperbole about government openness and accountability, the acceptance of VGI as an input to decision-making can position governments as responsive and directly connected to the electorate. Key to realizing this vision is the translation of citizen VGI into actionable policy, a process and transformation that is still very much untested. As such, there exists the potential for VGI generation, with its novelty and experimental status, to be an unintentional (or regrettably intentional) distraction from conventional and possibly more effective forms of citizen participation in decision-making.

The use of VGI within a participatory process can give flexibility to governments, providing a new media with which it can distribute information (e.g. in the form of KMLs) or make more transparent its practice. Used in this way, VGI presents an opportunity for governments to both accept large amounts of data directly from citizens, but also to use that conduit to allow access to government data sets, and even direct discussions with civil servants and decision-makers. This draws on PPGIS research in which the process of place-making via a digital form matters as much if not more than the output—the resulting map or database. Governments can, with examples like the city-based Apps for Democracy contests (http://www.appsfordemocracy.org), spur innovation and entrepreneurship. It also can set new avenues for engagement, reaching out to under-represented groups and reformatting the directionality of power flows. This can bring into contact dissimilar groups, generating conversation, agreement, and eventually action. It is through this type of communication that deep and lasting changes to governance structures and communities are created (Wittig and Schmitz 1996).

5.5 Conclusions

Like the introduction of ICT and GIS into government, there are many possible constraints to the adoption of VGI. These may be technical, organizational, or otherwise based on the local context and VGI implementation process. The negotiation of these constraints requires that governments identify potential bottlenecks and proactively position themselves to address them. This is a significant challenge and one that we have aimed to emphasize. There are many ways in which VGI can add value to government operations. At the most basic level, it represents citizen input, and when incorporated into governance, there is the potential for VGI to represent the kind of direct democracy that defines a vibrant civil society, with citizens engaged as partners in the co-production of decision support information. In a more activist fashion, VGI can be considered an expression of citizen perspective that is often circulated outside of conventional avenues of public participation in governance. Does VGI have a significant role to play as a way to undermine or circumvent governments, replacing defective governance processes with citizen-led initiatives? With the increasing devolution of federal and provincial responsibility to the municipal and community level, or with the wholesale shrinking of governments due to neoliberal policies, can VGI serve multiple roles, as both a response to, and as an outcome of retrenchment? Much as the social economy seeks to fill the gaps of failed neoliberal policies, can VGI and the communities that create it be considered a product or service produced without regard to the private and public economies (Amin et al. 2002; Carpi 1997)?

The path towards greater adoption and use of VGI in governance has many barriers. One significant issue that bears further investigation is the integration of VGI into the government decision-making process, with a focus on identifying the reasons why decision-makers would reject or accept VGI for a specific decision. Implicit in this assessment is to compare the level of trust that decision-makers or planners would have in a VGI data set, compared to an authoritative data set, provided that there is an acceptable level of congruence between the two. With the support of an authoritative data set, would the decision-maker trust the VGI data set? Would this trust extend to a situation where there is no congruent authoritative data set? And similarly, if a VGI data set is in direct conflict with an authoritative data set, is this sufficient for a decision-maker to question the authoritative data set? What would lead the decision-maker to trust the VGI data set over the authoritative data set? Identifying other factors outside of simple congruence may illuminate essential components of the decision-making process that are equal to, or perhaps more important than the simple accuracy or quality of a data set. For example, does the currency of the data set (presuming VGI is more current) matter, particularly in rapidly changing political landscapes? Does the fact that VGI represents citizen (and elector) voices hold sway with decision-makers, particularly elected officials? These questions and more related to them are essential avenues of future work in determining the fit or failure of VGI within the process of governance and decision-making.

Acknowledgements This research has been funded by the Quebec Ministère des services gouvernementaux program "Appui au passage à la société de l'information" and the Canadian GEOIDE Network of Centres of Excellence in Geomatics.

References

Al-Kodmany, K. (2000). Using Web-Based technologies and geographic information systems in community planning. *Journal of Urban Technology, 7*(1), 1–30.
Amin, A., Cameron, A., & Hudson, R. (2002). *Placing the social economy*. London: Routledge.
Brewer, G. A. (2006). Designing and implementing E-Government systems: Critical implications for public administration and democracy. *Administration & Society, 38*(4), 472–499.
Budhathoki, N., Bruce, B., & Nedovic-Budic, Z. (2008). Reconceptualizing the role of the user of spatial data infrastructure. *GeoJournal, 72*(3), 149–160.
Budhathoki, N., Nedovic-Budic, Z., & Bruce, B. (2010). An interdisciplinary frame for understanding volunteered geographic information. *Geomatica, 64*(1), 11–26.
Budic, Z. D. (1994). Effectiveness of geographic information systems in local planning. *Journal of the American Planning Association, 60*(2), 244–263.
Carpi, T. (1997). The prospects for the social economy in a changing world. *Annals of Public and Cooperative Economics, 68*(2), 247–279.
Carver, S., Evans, A., Kingston, R., & Turton, I. (2001). Public participation, GIS, and cyberdemocracy: Evaluating on-line spatial decision support systems. *Environment and Planning B: Planning and Design, 28*(6), 907–921.

Chadwick, A. (2009). Web 2.0: New challenges for the study of E-democracy in era of informational exuberance. *I/S: A Journal of Law and Policy for the Information Society, 5*(1), 9–41.

Cox, K. R. (1998). Spaces of dependence, spaces of engagement and the politics of scale, or: Looking for local politics. *Political Geography, 17*(1), 1–23.

Crampton, J. (2009). Cartography: Maps 2.0. *Progress in Human Geography, 33*(1), 91–100.

De Longueville, B., Annoni, A., Schade, S., Ostlaender, N., & Whitmore, C. (2010). Digital earth's nervous system for crisis events: Real-time sensor web enablement of volunteered geographic information. *International Journal of Digital Earth, 3*(3), 242–259.

Dovey, T., & Eggers, W. (2008). *National issues dialogues Web 2.0: The future of collaborative government*. Washington, DC: Deloitte Research.

Drummond, W., & French, S. (2008). The future of GIS in planning. *Journal of the American Planning Association, 74*(2), 161–174.

Elwood, S. (2008). Volunteered geographic information: Future research directions motivated by critical, participatory, and feminist GIS. *GeoJournal, 72*(3), 173–183.

Elwood, S. (2009). Geographic Information Science: New geovisualization technologies-emerging questions and linkages with GIScience research. *Progress in Human Geography, 53*, 256–263.

Elwood, S., & Ghose, R. (2004). PPGIS in community development planning: Framing the organizational context. *Cartographica, 38*(3/4), 19–33.

Elwood, S., & Leszczynski, A. (2011). Privacy, reconsidered: New representations, data practices, and the geoweb. *Geoforum, 42*(1), 6–15.

Flanagin, A., & Metzger, M. (2008). The credibility of volunteered geographic information. *GeoJournal, 72*(3), 137–148.

Ganapati, S. (2010). *Using geographic information systems to increase citizen engagement* (pp. 1–46). Washington, DC: IBM Center for The Business of Government.

Ganapati, S. (2011). Uses of public participation geographic information systems applications in E-government. *Public Administration Review, 71*(3), 425–434.

Geertman, S. (2006). Potentials for planning support: A planning-conceptual approach. *Environment and Planning B: Planning and Design, 33*, 863–880.

Girres, J. F., & Touya, G. (2010). Quality assessment of the French OpenStreetMap dataset. *Transactions in GIS, 14*(4), 435–459.

Goelman, A. (2005). Technology in context: Mediating factors in the utilization of planning technologies. *Environment and Planning A, 37*, 895–907.

Goodchild, M. (2007). Citizens as sensors: The world of volunteered geography. *GeoJournal, 69*, 211–221.

Goodchild, M., & Glennon, J. (2010). Crowdsourcing geographic information for disaster response: A research frontier. *International Journal of Digital Earth, 3*(3), 231–241.

Haklay, M. (2010). How good is volunteered geographical information? A comparative study of OpenStreetMap and Ordnance Survey datasets. *Environment and Planning B: Planning and Design, 37*(4), 682–703.

Haklay, M., Singleton, A., & Parker, C. (2008). Web mapping 2.0: The Neogeography of the Geoweb. *Geography Compass, 2*(6), 2011–2039.

Hall, G., Chipeniuk, R., Feick, R., Leahy, M., & Deparday, V. (2010). Community-based production of geographic information using open source software and Web 2.0. *International Journal of Geographical Information Science, 24*(5), 761–781.

Harvey, F. (2003). Developing geographic information infrastructures for local government: The role of trust. *Canadian Geographer/Le Géographe canadien, 47*(1), 28–36.

Harvey, F. (2007). Just another private–public partnership? Possible constraints on scientific information in virtual map browsers. *Environment and Planning B: Planning and Design, 34*, 761–764.

Hudson-Smith, A., Crooks, A., Gibin, M., Milton, R., & Batty, M. (2009). NeoGeography and Web 2.0: Concepts, tools and applications. *Journal of Location Based Services, 3*(2), 118–145.

Johnson, P. A., & Sieber, R. E. (2011a). Motivations driving government adoption of the Geoweb. *GeoJournal*, 1–14. doi:10.1007/s10708-011-9416-8

Johnson, P. A., & Sieber, R. E. (2011b). Negotiating constraints to the adoption of agent-based modeling in tourism planning. *Environment and Planning B – Planning and Design, 38*(2), 307–321.

Johnson, P. A., Sieber, R. E., Magnien, N., & Ariwi, J. (2012). Automated web harvesting to collect and analyse user-generated content for tourism. *Current Issues in Tourism*, 15(3), 293–299.

Miller, C. (2007). A beast in the field: The Google maps mashup as GIS/2. *Cartographica, 2*(3), 187–199.

Nedovic-Budic, Z. (1998). The impact of GIS technology. *Environment and Planning B – Planning and Design, 25*(5), 681–692.

Newman, G., Zimmerman, D., Crall, A., Laituri, M., Graham, J., & Stapel, L. (2010). User-friendly web mapping: Lessons from a citizen science website. *International Journal of Geographical Information Science, 24*(12), 1851–1869.

Onsrud, H., & Pinto, J. (1991). Diffusion of geographic information innovations. *International Journal of Geographical Information Science, 5*(4), 447–467.

Parsons, J., Lukyanenko, R., & Weirsma, Y. (2011). Easier citizen science is better. *Nature, 471*(7336), 37.

Pickles, J. (Ed.). (1995). *Ground truth: The social implications of geographic information systems*. New York: Guilford.

Rouse, J. L., Bergeron, S. J., & Harris, T. M. (2007). Participating in the Geospatial Web: Collaborative mapping, social networks and participatory GIS. In A. Scharl & K. Tochterman (Eds.), *The geospatial web: How geobrowsers, social software and the Web 2.0 are shaping the network society* (pp. 153–158). London: Springer.

Saebo, O., Rose, J., & Flak, L. S. (2008). The shape of eParticipation: Characterizing an emerging research area. *Government Information Quarterly, 25*, 400–428.

Schuurman, N. (2000). Trouble in the heartland: GIS and its critics in the 1990s. *Progress in Human Geography, 24*(4), 569–590.

Seeger, C. (2008). The role of facilitated volunteered geographic information in the landscape planning and site design process. *GeoJournal, 72*, 199–213.

Sieber, R. (2004). Rewiring for a GIS/2. *Cartographica, 39*(1), 25–39.

Sieber, R. (2006). Public participation geographic information systems: A literature review and framework. *Annals of the Association of American Geographers, 96*(3), 491–507.

Smith, N. (1993). Homeless/global: Scaling places. In J. Bird, B. Curtis, T. Putnam, G. Robertson, & L. Tickner (Eds.), *Mapping the futures* (pp. 87–119). London: Routledge.

Swyngedouw, E. (2004). Globalisation or 'glocalisation'? Networks, territories and rescaling. *Cambridge Review of International Affairs, 17*(1), 25–48.

te Brömmelstroet, M., & Bertolini, L. (2008). Developing land use and transport PSS: Meaningful information through a dialogue between modelers and planners. *Transport Policy, 15*, 251–259.

te Brömmelstroet, M., & Schrijnen, P. (2010). From planning support systems to mediated planning support: A structured dialogue to overcome the implementation gap. *Environment and Planning B: Planning and Design, 37*(1), 3–20.

Tulloch, D. (2008). Is VGI participation? From vernal pools to video games. *GeoJournal, 72*(3), 161–171.

Turner, A. (2006). *Introduction to neogeography*. Sebastopol: O'Reilly.

Vonk, G., Geertman, S., & Schot, P. (2005). Bottlenecks blocking widespread usage of planning support systems. *Environment and Planning A, 37*, 909–924.

Vonk, G., Geertman, S., & Schot, P. (2007). A SWOT analysis of planning support systems. *Environment and Planning A, 39*, 1699–1714.

Wittig, M. A., & Schmitz, J. (1996). Electronic grassroots organizing. *Journal of Social Issues, 52*(1), 53–69.

Zook, M., & Graham, M. (2007). The creative reconstruction of the Internet: Google and the privatization of cyberspace and DigiPlace. *Geoforum, 38*(6), 1322–1343.

Zook, M., Graham, M., Shelton, T., & Gorman, S. (2010). Volunteered geographic information and crowdsourcing disaster relief: A case study of the Haitian earthquake. *World Medical and Health Policy, 2*(2), 7–33.

Chapter 6
When Web 2.0 Meets Public Participation GIS (PPGIS): VGI and Spaces of Participatory Mapping in China

Wen Lin

Abstract While existing studies provide important insights into power relations and spatial knowledge production impacted by volunteered geographic information (VGI), this chapter argues that more research is needed to investigate how these new geospatial technologies have constituted the actor's subjectivities and the politics of citizen participation. Drawing upon public participation GIS (PPGIS) studies, critical GIS research and critical social theory, this chapter examines the mutual and complex relationships between subject formation and geospatial technology development and their implications for spaces and politics of citizen participation in a variety of contexts. A case study in China is presented with three examples of VGI mapping drawn from ethnographic fieldwork. These VGI practices in China have constituted multiple "DigiPlaces," a notion proposed by Matt Zook and Mark Graham that is characterized by greater visibility with automatic production, increased individualism, and dynamism. Furthermore, these practices are simultaneously impacted by the complex process of subject constitution, informed by Mark Poster's notion of "the mode of information," marked by the proliferation of electronic communications that helps to constitute multiple subjectivities. In particular, coupling with rapid Internet and new communication technology developments, Chinese citizenship witnessed growing awareness of individual rights and more decentered self-identities compared to two decades ago. As such, new spaces of citizen participation are constructed by these VGI practices; however, significant challenges remain regarding the intersection of possibilities and existing economic and sociopolitical inequalities.

W. Lin (✉)
School of Geography, Politics and Sociology, Newcastle University,
Newcastle Upon Tyne, UK
e-mail: wen.lin@ncl.ac.uk

6.1 Introduction

Online mapping is not new. But the recent emergence of a wide array of geovisualization technologies combined with Web 2.0 has in the past 5 years enabled greater user-generated spatial data creation and distribution and has drawn increasing attention from GIS scholars (cf. Elwood 2009; Goodchild 2007; Haklay et al. 2008; Sui 2008; Crampton 2009). Such spatial data provisions, referred to as *volunteered geographic information* (VGI) practices (Goodchild 2007) here[1] and carried out by those who usually do not have formal training in GIS or cartography, often incorporate multimedia representations, including photographs, texts, and sounds that are tagged with locational information (Elwood 2009). Research has started to examine the social and political impacts of VGI practices on citizen science and participatory democracy, as well as how these practices might constitute new forms of surveillance, exclusion, and intrusion into privacy (Elwood 2009, 2010).

In particular, there have been insightful studies investigating different types of VGI and how they overlap with and differ from traditional public participation GIS (PPGIS) practices (Miller 2006; Tulloch 2008; Boulton 2010). In this discussion, researchers have emphasized greater accessibility and user-friendliness of these technologies that might enable public participation through mapping. Meanwhile, existing power relations may also be reinforced and reconfigured in VGI production (Obermeyer 2007; Crutcher and Zook 2009). In addition, various forms of representation and constitution of community through VGI have been acknowledged (Tulloch 2007).

Nonetheless, as compared to many PPGIS practices that tend to revolve around goals of particular organizations or communities (Sieber 2006; Elwood and Ghose 2004), VGI production tends to be much more individualized and dynamic (Zook and Graham 2007). As such, some key questions remain regarding the intersection between VGI and PPGIS. In particular, how should we conceptualize the complex interrelations between the broader socioeconomic, political conditions (which may also be changing) and participatory VGI practices? For example, who is the participating "community" in the face of increasingly individualized VGI practices and possible "remote" participation enabled by the Internet and other information technologies? Are the sociopolitical meanings of "participation" in flux as a result? How might the purposes of participation and interests of various individuals and groups have been transformed by the wider availability of information technologies and locational information? What are the convergences and divergences of participatory VGI practices in different contexts that might shape and give new meanings to local data production and citizen participation?

This study is part of my ongoing project that seeks to contribute to the existing discussions regarding the abovementioned questions. Three streams of research frame my argument here. First, the PPGIS literature provides important insights into understanding how power relations are mediated through GIS usage and spatial knowledge

---

[1] See Elwood (2010) for a summary of other terms such as neogeography, geoweb, map 2.0, etc.

production by citizen and grassroots groups. However, these discussions in PPGIS have been largely framed within the organizational context, which is not adequate for examining the more individualized VGI practices. Second, critical GIS research has addressed the issue of subject constitution complicated by geospatial technologies and pervasive computing. The mutual constitution between the social and the technological at the level of the body may give new meanings to forms of technology-mediated participation, community composition, and power relations in civic engagement. Third, I further draw upon Mark Poster's "the mode of information" from critical social theory in cultural studies and communications studies to situate these cultural shifts.

Through this synthesized framework, I provide an empirical investigation of participatory VGI practices in China, which are informed by the complex and dynamic subjectivity constitution and citizenship transition in increasingly urbanized and globalized China. I suggest that despite the lack of organized citizen participation on various issues, strong state control of spatial knowledge production, and Internet censorship in China, VGI practices open new spaces for civic engagement and citizens' contestation of official discourses in dynamic and subtle ways. At the same time, I argue that within this context, citizen participation, even through arguably "open" approaches such as VGI, is inherently limited to practices of "exercising micro-power and organizing without organization" that carry with them new forms of exclusion. As such, while there are significant differences regarding political context and democratic participation in China when compared to many Western polities, these Chinese VGI practices show some interesting shared traits of VGI elsewhere in the way they constitute self-identities and in their blending of virtual and physical spaces, which may also give new meanings to participation and the politics of spatial knowledge production.

In what follows, the second section discusses the theoretical background of this research. In the third section, I illustrate the dynamics of Chinese citizenship as an introduction to the socioeconomic and political conditions of the VGI practices in China. This is followed by a case study of three VGI examples in China in the fourth section. Finally, the last section provides conclusions and discussions.

6.2 Theoretical Background

6.2.1 VGI and PPGIS: Convergences and Divergences

A growing body of work has sought to examine the purposes of VGI production, values of such data generation, and associated technological transformation as well as social and political implications (Goodchild 2007, 2008; Haklay et al. 2008; Sui 2008; Elwood 2008, 2009, 2010). Overall, it is recognized that VGI authors are simultaneously data users and that VGI production is mobile, ubiquitous (Perkins 2008; Haklay et al. 2008), and often times collaborative in a manner similar to "wikification" (Sui 2008). In particular, the more decentralized mode of VGI data provision comprised by the local data of lay persons has drawn significant attention from GIS scholars examining implications for citizen science and participatory democracy (Goodchild 2007; Tulloch 2007; Boulton 2010; Elwood 2010). A number of studies have addressed the overlap between VGI and PPGIS (Tulloch 2007, 2008; Miller 2006). Due to the scope of this study, I focus on VGI practices in which the data creators knowingly generate and share their data publicly, what I call "participatory VGI" practices.

The rich body of work in PPGIS derives from critiques of GIS in the early 1990s, which argued that GIS technologies embody a positivist epistemology and prioritize instrumental rationality over other forms of knowledge production. In response, significant efforts have emerged to broaden the accessibility of data and technology to citizens and grassroots groups. Researchers have investigated ways of integrating and representing local knowledge of marginalized groups using conventional GIS technologies or re-coding GIS into more user-centered packages (GIS/2) (cf. Sieber 2006; Craig et al. 2002). A key inquiry in the PPGIS literature is to examine in what ways spatial knowledge is produced, by whom, and how this process may shape power relations among different social groups (Sieber 2006; Tulloch 2007; Ghose 2007; Elwood 2009). PPGIS practices have contradictory outcomes, simultaneously empowering and disempowering with shifting boundaries of inclusions and exclusions (Weiner and Harris 2003; Elwood 2004; Ghose 2005). There are also discussions of possible spatial data abuse and invasion of personal privacy (Pickles 2004; Sui 2006). While there are analyses of power relations among individuals in PPGIS practices (e.g., Elwood 2004; Kyem 2004), existing discussions on social relations embedded in PPGIS have been largely focused on the roles of organizations and groups (Sieber 2000; Elwood and Ghose 2004; Tulloch 2007).

In line with efforts to expand technology accessibility in the PPGIS literature, a number of claims have been made to the effect that VGI can provide a new form of participatory mapping, which engages with a broader audience of lay persons through crowdsourcing, greater user-friendliness, and space and time flexibility (Miller 2006; Tulloch 2007, 2008; Kreutz 2010). Miller (2006) investigates the emergence of Google Maps mashups and their impacts on PPGIS. He emphasizes that the ease, speed, and high interactivity of online map usage along with the mashability of Google Maps have enabled a participating public. Such a mapping platform and the example of mashup creation responding to Hurricane Katrina underline the ideals of a user-centered GIS/2 (ibid). Kreutz (2010) discusses the issue of "maptivism," which can be seen as the explicit usage of VGI in activism. He suggests that these volunteered maps are powerful because they can provide a feel of connection, connect topics on complex issues, and trigger engagement.

Meanwhile, researchers have also identified several limitations of VGI practices as they may embed existing social inequality and introduce new forms of technical and social barriers for civic participation (Tulloch 2007; Crutcher and Zook 2009). Tulloch (2007) notes that while these online mapping applications are much less constrained geographically, they are more limited generationally. Crutcher and Zook (2009) examine in what ways existing social and economic inequalities are further intertwined with, and in turn reconfigure, the digital mapping landscape.

They particularly investigate the post-Katrina Google Earth to illustrate how race has shaped the way people use (or do not use) Google Earth. Kreutz (2010) also addresses the challenges in maptivism, including possible intrusion of privacy, embedded propaganda and discrimination, and lack of attention.

Tulloch (2008) provides an insightful examination of overlaps and distinctions between PPGIS and VGI. In particular, with respect to participation, a core issue of the overlap between PPGIS and VGI lies in the investigation by individuals of locations important to them. In the context of PPGIS, it is likely the case that individuals strive to utilize public datasets to participate in decision-making processes that impact the places they care about. In the case of VGI, it might be that individuals create their own datasets rather than using existing public datasets of their beloved places. In addition, VGI has a casual and entertaining side, which cannot fit easily within the existing PPGIS theorization of participation. Tulloch addresses two important distinctions between VGI and PPGIS. One is that technologies used in VGI are usually outside the conventional GIS software. Second, VGI is largely about mapping more than decision making, while PPGIS tends to focus on decisions and explicitly seeks social change through mapping. Associated with this second distinction is that VGI authors may create and share their data unknowingly, which can be a form of "(geo)slavery" (Obermeyer 2007). Yet the line of distinctions may not always be fixed, as VGI authors may acquire a more influential position in a policymaking process through data creation and sharing.

Several studies from critical GIS and critical cartography also examine the divergences between VGI and GIS practices (cf. Elwood 2010). In particular, burgeoning VGI practices indicate an emergence of spatial dataset infrastructure as "patchwork" (Goodchild 2007), which can result in a changing role for citizen and grassroots groups, who might move from being data petitioners to data providers (Elwood 2008; Perkins 2008; Dormann et al. 2006). Second, new private-sector actors such as advertisers embedded in these free online mapping tools may get increasingly involved (Zook and Graham 2007), while there are also significant efforts in providing open-source software (Haklay et al. 2008). Third, VGI may reinforce existing social inequalities, and it may also constitute new forms of exclusion and surveillance, such as the variable technical capacity to code with open application programming interfaces (APIs) and possible data abuse made possible by the availability of massive databases (Elwood 2010; Williams 2007).

6.2.2 Subjectivities and DigiPlaces

Some scholars from critical GIS have investigated new dimensions of subjectivities emerging from pervasive computing and geospatial technologies (Elwood 2010), such as the data-borg (Schuurman 2004), geocoded citizen (Wilson 2009), and digital self (Dodge and Kitchin 2007). For example, employing the notion of cyborg that points to the complex mutual constitution of human and machine (Haraway 1991), Schuurman (2004) suggests a variant of cyborg in the twenty-first century, the data-borg, as the collection and usage of data on our bodies have become much more significant in everyday life. Noting parallel technology advancements that enable massive database constructions in GIS analysis, Schuurman calls for attention to an emerging new cyborg that is rich in data at the individual level (ibid). In particular, through the example of Virtual Coach, which collects athletes' logging of daily data to provide correspondent training schedule, Schuurman illustrates how these daily and long-term data collections can be an extension of the self. Such a self-reconfiguration opens up a variety of opportunities as well as risks. On the one hand, it is empowering for individuals who might adopt a more individually based training plan. On the other hand, it can be used by powerful conglomerations for population control and exploitation.

Kingsbury and Jones (2009) seek to go beyond a fear-hope dialectic often associated with critical studies of geospatial technology. They argue that Google Earth is too often seen as "an Apollonian entity composed of control, order, and calculation" (ibid, p.503). Underscored by this Apollonian view, discussions of this technology tend to be divided into two opposing views of fear and hope regarding its sociopolitical implications. For example, the authors point out that responses to the US Holocaust Memorial Museum initiative, which shows high-resolution images and photos of the crisis in Darfur, tend to fall in two opposing lines: applause for the efforts of educating a broader audience about the geographical context of the crisis or critiques of possible voyeuristic pleasure gained from viewing these images. Both perspectives are limiting. Rather, Google Earth is also a Dionysian entity that is uncertain, alluring, and frenzied. This Dionysian interpretation helps to better understand the multiple ways in which Google Earth is being used in different contexts, frequently reflected in the surfing of Google Earth's alluring and oftentimes bizarre images. This theorization also recognizes the "casual and entertaining side" of VGI (Tulloch 2008, p.165). Moreover, the authors suggest that the Apollonian involves a politics with a subjectivity that is sober, rational, calculated, and sincere, while the Dionysian involves a politics of the artist, anarchist, or hacker that may seem apolitical at first glance. But the Dionysian is also "the place where new ways of political and ethical thinking emerge" (Kingsbury and Jones 2009, p.509).

Gerlach (2010) also suggests that emerging Web 2.0 mapping practices go beyond the subject-object dualism in traditional cartography. He proposes a concept of "vernacular mappings" to describe the everydayness of mapping practices cultivated by, for example, OpenStreetMap. He argues that vernacular mapping consists of "a politics of the aesthetic whereby creative potential is valorized as a series of political interventions, but not necessarily in a subversive or angst-ridden manner" (ibid, p.166). Together, these studies underscore the increasingly ubiquitous nature of these VGI mapping practices, which embody a series of complicated self-expressions and porous boundaries between multiple subjectivities in spatial knowledge production and politics.

Zook and Graham (2007) examine the hybridization of the digital and the material in "daily lived geographies" (p.1323) through the notion of "DigiPlace," which is useful for exploring VGI practices and their implications for spatial knowledge production and representation. The authors argue that the algorithms embedded in and data used by Google Maps and Google Earth can significantly influence how physical places are perceived and used. Recognizing that DigiPlace also shares the power of physical maps in shaping interactions and experiences with place, the authors suggest three important new characteristics of DigiPlace. First, *visibility* in DigiPlace is *automatically* produced and filtered by code dependent on an entity's online presence. Second, DigiPlace is highly *individualized* and defies static representations. Third, DigiPlace significantly increases the *dynamism* of digital cartographic visualization and is constantly evolving. The construction of DigiPlace thus points to the increasing blend of code and place. It is crucial to recognize that there are variable contexts and abilities implicated in creating the spaces of DigiPlace. Each user interacts differently with the different contexts, and these interactions in turn influence their cognition of physical places (ibid).

6.2.3 Mode of Information and Spatial Narratives

Employing a poststructuralist perspective, Poster (1990) seeks to analyze the intersections between the historical emergence of a decentered subject and the massive changes in new communications systems through the notion of "the mode of information." Poster designates three stages in the mode of information: face-to-face, orally mediated exchange, characterized by symbolic correspondences; written exchange, characterized by the representation of signs; and electronically mediated exchange, characterized by informational simulations. Each stage constitutes a particular form of subjectivity. In the oral stage, the self is constituted as a position of enunciation through its embeddedness in a totality of face-to-face relations. In the print stage, the self is constructed as an agent centered in rational/imaginary autonomy. In the electronic stage, the self is decentered, dispersed, and multiplied in continuous instability (ibid). These stages are not sequential but coterminous in the present. He further points out that electronically mediated exchange enables increased distance between addresser and addressee, which "allows a reconfiguration of the relation between emitter and receiver, between the message and its context, between the receiver/subject and representations of him or herself" (Poster 1990, p.14). Such configurations in turn influence and transform social relations among different institutions, communities, and individuals. Without doubt, many modern institutions and practices still dominate social space. Yet Poster maintains that the mode of information is an emergent phenomenon that affects small but important aspects of everyday life (ibid).

In particular, with the advent of the Internet, subject constitution occurs through the mechanism of *interactivity*. These interactive communications lead to the formation of "virtual community" and networks of social relations with characteristics that are new when compared to historical constructions of community. First, this form of electronic communication is associated with a certain fluidity of identity. Second, virtual and real communities *mirror* and *constitute* each other, such that participants code "virtual" reality through categories of "normal" reality (ibid, pp.191–192, emphasis added). Poster suggests that new media provide hopeful possibilities for resistance to modernity through complications of subjecthood that denaturalize the process of subject formation and put into question the interiority of the subject and its coherence. However, these possibilities are not guaranteed by the diffusion of new media communication technologies. Indeed, information technologies can just as well provide totalitarian control rather than the decentralized, multiple "little narrativity" of postmodern culture (ibid, p.198).

As such, in resonance with Kingsbury and Jones's (2009) argument for a Dionysian view of Google Earth, the notion of the mode of information extends beyond the dualism of fear-hope that is so often taken up when one views electronic communications. Moreover, this notion "mode of information" stresses the role of multiple forms of media in shaping subjectivities and social relations, which I argue is an important addition for the discussion of VGI and PPGIS practices.

Media has indeed been considered to play an important role in knowledge production (Flew and Liu 2011). In particular, Habermas's (1989) notion of the "public sphere" characterized by dialogic conversations and facilitated by the mass media has been an important lens for understanding the emergence of liberal-capitalist institutions in Europe (Flew and Liu 2011). Recently, how public spheres and civic engagement have been shaped and constituted by Internet communications has been widely debated. A number of writers have discussed the potential of enabling a virtual public sphere (e.g., Rheingold 1994; Poster 1997). On the other hand, some have argued that the Internet mainly reinforces preexisting social relations between the state and society (e.g., Drezner 2005). Still some others have acknowledged the complex mixture of control and liberation tendencies of the Internet (e.g., Warf 2011). Meanwhile, geographers have examined GIS through the perspective of viewing it as a form of media (e.g., Sui and Goodchild 2003). Nonetheless, how the new media and VGI might have shaped subject formation and in turn impacted the meaning of political community and civic engagement has not been investigated, which this chapter seeks to explore through the case of China. In particular, it is important to take into account a broader context of Chinese citizenship in transition when investigating VGI and public participation in China.

6.3 Dynamics of Chinese Citizenship

Citizenship here refers to "a range of legal, political, social, and economic links between the state and members of society" (Goldman and Perry 2002, p. 3). In particular, citizenship is viewed as an instituted process (Woo 2002), which is fragmented, constantly formed and negotiated, and disrupted at multiple sites (Staeheli 2010). In the Maoist era (1949–1976), the notion of citizenship was filled with the rhetoric of class struggle, collectivism, and altruism, becoming "cultural templates for a Chinese 'socialist' subjectivity" (Keane 2001, p.3). In this discourse, individual rights are viewed as economic, social, and cultural benefits (Keane 2001). The view that framed rights as concessions rather than as entitlements has undergone some modifications in the post-Mao era (Goldman and Perry 2002). Keane (2001) observes that during the 1990s, economic governance (effective at mobilization of the individual) has superseded nationalism (characterized by mass mobilization) as the major mechanism of social organization. Certain groups of intellectuals have begun to think of citizenship as the assertion of political and civil rights. Concrete actions in lawsuits, village elections, and entrepreneurial organizations have also been taken to exercise and expand these rights (Goldman and Perry 2002).

Meanwhile, the past decade also saw the state's increasing interests in constructing e-governance in China (Lin 2008). The stated goals of these state programs include facilitating e-commerce and increasing citizen participation. While many have argued that the latter may largely remain on paper (Lin and Ghose 2010), the e-governance discourse has fueled the state's efforts in building infrastructures for digital communications. Also, the economic reforms have significantly transformed China's media landscape, which has stimulated the need for the assessment of public opinion on issues that are directly related to the reform process. As such, the past decade witnessed a change of the role of public opinion in national policymaking that may go beyond state control (Yang 2009).

Most recently, observers argue that with the rapid growth of Internet users in China, the Internet has played a much bigger role for public opinion discussions (Gao 2009; Tai 2006), as the state remains tight control over the mass media (Yang 2009). No doubt that the Internet has been heavily regulated and censored in China (Yang 2009; MacKinnon 2010). However, as Yang (2009, p.45) notes, "[a]s power seeks domination, it incurs resistance." Many Chinese netizens have learned to use a range of strategies and technologies, such as using alternative proxies or VPN (virtual private network) services to circumvent the Internet censorship (MacKinnon 2010). China had 485 million Internet users in June 2011 (CNNIC 2011), rising from 22.5 million in 2000. The mobile telephone has become an important device for Internet access. However, there remain significant inequalities between urban regions and rural areas (CNNIC 2011; Michael and Zhou 2010). The Internet has transformed the arena of public opinion in Chinese society in at least three ways. First, it creates a new platform for Chinese Internet users to express their opinions online on many issues. Second, it generates "a steady, core cohort of opinion leaders" that constantly guides public opinion in cyberspace. Third, it allows an increasing number of Chinese Internet users to be exposed to other net users' opinions (Tai 2006, p.188). Associated with these transformations is greater awareness of individual rights and strong contestations against hegemonic and authoritarian governance (Yang 2009).

The changing citizenship can be reflected in various forms, including active online activism in China in the past decade (Yang 2009; Tai 2006).[2] Another such form is the phrase "onlookers to change China," which emerged in 2010. This phrase first appeared in an editorial opinion in Southern Weekend (Xiao 2010), known as a

---

[2] Despite China's increasingly sophisticated censorship system, a number of studies contend that Chinese Internet users have engaged in numerous forms of activism (Yang 2009; Tai 2006). In this way, dynamic communities have emerged that can be mobilized by the Internet to act in various issues, such as mobilizing the public for environmental protection by Web-based environmental volunteer groups (Yang 2003) and organizing workers' strikes through the Internet and mobile technologies (Qiu 2009).

liberal newspaper in China. This editorial was widely cited, and the phrase "onlookers to change China" was then picked up quickly by many Chinese netizens.[3] Hu (2011) describes this as the "surrounding gaze," illustrating its historical root in modern Chinese literature and culture and its new meanings in the information age. The surrounding gaze from the "crowd" exerts strong pressure on those being watched. In particular, the recent emergence of social networking websites such as microblog sites (e.g., www.weibo.com) has been acknowledged to attract a much wider audience for hot topic discussions. The onlookers' gaze often takes the digital form of online posts regarding particular social issues that are then forwarded to circulate information widely on social networking sites and blogs. This form of participation may represent one of the bottom levels of citizen participation depicted by Arnstein (1969). But in some respects, the usage of these Internet sites can construct multiple DigiPlaces and, in some cases, form a multitude of public spheres. This evolving hybrid digital citizenship has important imprints on the emerging VGI practices in China, which are discussed below. The empirical investigation is built on document analysis and personal interviews with VGI practitioners by the author.

6.4 VGI Practices in China

On January 15, 2009, Ogle Earth posted an article that asked, "Is China opening up to neogeography?" (http://www.ogleearth.com/2009/01/is_china_openin.html). This post responded to the news that China has started to build up a Chinese version of "Google Earth." It pointed out an important issue regarding state control over spatial resources as well as the Internet censorship constraining the possibilities of VGI practices in China, despite a more open and relaxed attitude by China's leadership when it comes to mapping tools. Nonetheless, this article does not address the non-state practices in online mapping and geographic knowledge production. There have been active grassroots efforts in generating VGI on a variety of issues including environmental protection, crisis mapping, and political contestation and resistance in China (Lin 2010). In the following, I discuss three examples of such participatory VGI practices to show in what ways VGI can reconfigure and create spaces of civic participation in China.

6.4.1 Map of Relief Support and Needs in the Sichuan Earthquake

One example of these practices is a Google Maps mashup map created on May 17, 2008, after the May 12 earthquake in Sichuan (Fig. 6.1). One creator of this map noted that the reason to create this map was that, previously, it was too complicated

---

[3] For example, a search of this editorial opinion from Google site in China (www.google.com.hk) returned 328,000 results on July 31, 2011. The number of results increased to 648,000 on October 29, 2011.

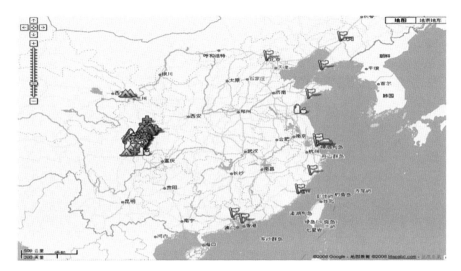

Fig. 6.1 The map of relief support and needs in Sichuan earthquake (retrieved May 21, 2008)

to show such information in tabular form (Ding,[4] April 2010, personal communication). There were around 23 volunteers to help update this map (ibid). A post to announce the creation of this map and call for volunteers to submit any related data regarding the relief efforts was posted on Douban (www.douban.com), a popular social networking site in China. This post provided links to detailed instructions of how to make contributions to the map. In particular, a specific format of map data submission was identified as follows: location, time, verification (yes/no), and content. Information shown on the map eventually was from two major sources: one was the collection by volunteers from radio and TV reports and nongovernmental organizations on site, while the other was from the data directly announced by the government.

Ding has a background in human-computer interactions, and he considered using an IBM open-source package. But eventually, Google Maps was chosen, as "it was convenient and very user-friendly" (Ding, April 2010, personal communication). Ding noted that engineers from Google also collaborated in this project at a later stage. The left part of the map site provides information on the symbols used and a note on the sequence of the listed items. There are eight symbols, of which red exclamation marks are for trapped people, bottles and apples for water and food, green triangles for tents, baskets for a variety of other materials, trucks for vehicles, wrenches for engineers, and flags for volunteers. While the specific process of selecting these symbols is unknown, this use of symbology is more sophisticated than the two other examples.

There were 82,539 hits within a week after this map's creation, and it soon reached a million hits. Moreover, some NGOs on the site of disaster also used this

---

[4] Pseudonyms are used for the interviewees.

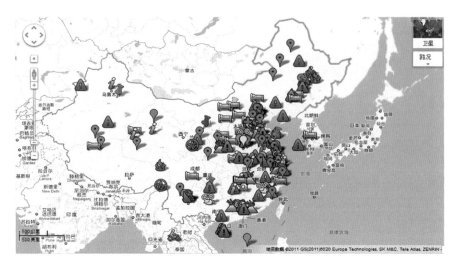

Fig. 6.2 Map of violent evictions (retrieved January 19, 2012)

map to assist their relief efforts (Ding, April 2010, personal communication). It is also notable that this is a project that involved a large number of people in a very short period of time. As noted by Ding, this scale and the level of collaboration and coordination reflected in this map was quite rare at the time: "It is a range of different factors together that contributed to this map. The impact of this earthquake is really astonishing" (ibid).

Indeed, this Sichuan earthquake relief efforts map in part represents the outpouring of volunteer efforts in the aftermath of this earthquake. Meanwhile, this map has paralleled with the Hurricane Katrina map (Miller 2006), the crisis mapping of Haiti's earthquake in 2010, and many other crisis mapping activities around the world. While crisis mapping efforts have continued to evolve and grow worldwide, such crisis mapping efforts in China, based on evidence at hand, have not evolved into the relatively more stable mapping community noted in Meier (2011). Nonetheless, this Sichuan map marks one of the earliest efforts of relatively large scale collaborations of VGI provision in China. A most recent example of such a scale of collaboration is a map on violent evictions in China, which was generated in October 2010 and subsequently drew significant attention from the mainstream media (Fig. 6.2). Part of the goal of the violent eviction map is to dissuade people from purchasing houses connected to such evictions.

6.4.2 Map of China's Mining Accidents

The second example is a map titled "Map of China's Mining Accidents in 2010" (Fig. 6.3), created by Wang. This map is an example of mapping by an individual to address a particular social issue in China. The mapping was derived from Wang's

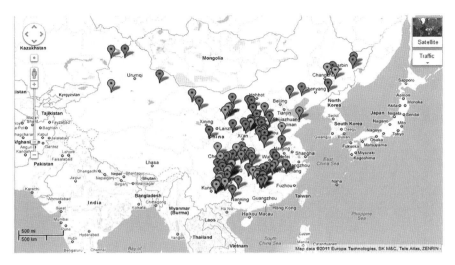

Fig. 6.3 Map of China's mining accidents in 2010 (retrieved January 19, 2012)

strong concern about the mining accidents in China. Wang created a mashup for each mining accident reported in the news. The major information in the description of each mashup includes the following fields: time of accident, location, cause, casualty, and media report source. Each color of the mashup indicates a particular range of casualties, with purple representing more than 20 deaths, red 10–20 deaths, green 5–10 deaths, and blue below 5.[5] The VGI author noted that, technically, it was easy for him to create a map like this. However, it was quite time consuming, especially in the beginning of this process. He read news during the daytime and usually searched related news when he had some spare time in the evening to see if any reported mining accident might have been missed. Another time-consuming issue is that sometimes it was difficult to obtain accurate coordinates using the address reported in the news. He noted that when Google Maps was zoomed in to the county level in China, it was in some cases hard to pin down the accurate location of a village or a county.

Wang first posted this map on his blogs in mid-January and kept updating his map to cover the accidents reported in 2010. He also posted a link on Twitter. The map soon received a significant number of hits after its creation. "Just a few days, there were about thousands of views" (Wang, July 2010, personal communication). Many viewers left comments responding to his blog posts or replied to his Twitter posts. Many were shocked to see the number of mining accidents and associated casualties; some thanked the VGI author for making this map. There were also a few errors that were pointed out by his Twitter followers (he had around 7,000 Twitter

---

[5] Initially, this map used only one color for the symbols, along with description of the number of deaths in the pop-up window of each mashup. During my interview with the VGI author, I noted the possibility of using different colors to indicate different classes of values. This map in turn adopted the current set of legend.

Fig. 6.4 Map of sale/lease ratio (retrieved January 19, 2012)

followers at the time of this mapping practice). "Mostly they were shocked… [They said that the accidents were] as thick as huckleberries [in the map]" (ibid).

When Wang started to make this map, he had already created several similar maps in the recent past. His experience of creating Google Maps mashup maps can be traced back to the first one he made on impacts of environmental pollution in June 2009. In particular, the issue of "cancer villages" due to severe environmental pollution was first reported by *Phoenix Weekly*, a Hong Kong-based news magazine, and a Twitter user posted a tweet calling for a map on these villages. Wang volunteered to make the map. Since then, he made a few more maps, including one map for Greenpeace on water quality issues and another one on lead poisoning in China.

Wang noted that this way of mapping was meant to record ongoing events, especially problems that have emerged in the process of China's economic development. "It is something like my personal documentation of the history… While the sources are from the official reports, they won't compile such information in this way and provide it to the public" (Wang, July 2010, personal communication). As such, this map, along with other Google Maps applications he created, provides an important representation of his concerns about social problems. These online maps are the VGI author's spatial narratives. They were disseminated quickly through some major social networking sites, especially Twitter in this case.

6.4.3 Map of Sale/Rent Ratio

The third example is a map of the ratio of sale to lease price to illustrate the high housing price in urban China, which was created using unofficial data through crowdsourcing (Fig. 6.4). In this Google Maps representation, first created on February 26, 2010, there are five ranges of values for a point symbology. If the ratio is below 20 to 1, it is marked as green, 20–30 to 1 as blue, 30–40 to 1 as purple, 40–50 to 1 as yellow, and above 50 to 1 as red.

Liang, the author of this map, is a photographer whose work has long focused on urban life and urbanization in China. He also used his blog and Twitter to broadcast this project. In one of his tweets on February 26, 2010, he wrote:

> Continue my survey of the ratio of sale to lease price in China on Twitter. I use google spreadsheets (you need to hop over the wall to see them) to collect data. The web address for data submission is […]. All materials collected will be integrated into a google map [map link…] All data are open to the public, @(another Twitter user) helps to tweet on this. (translation by author)

In another tweet soon followed, he noted:

> Continue to use Twitter to collect the information of the ratio of sale to lease price. I don't seek to gain comprehensive information; rather, I'm interested in knowing a handful of relevant information. It is the Twitter friends that make up the statistics bureau. Twitter friends from Hong Kong and Taiwan are welcome to participate. The web address [for the survey] is […] (you need to hop over the wall to see this); the web address for the result summary is […]. (translation by author)

These tweets are quite telling, not only in terms of how Liang collected the data, but in their use of metaphors that capture how Liang constructed the map. First, it is apparent that Liang used a range of available online venues to collect information to address a particular issue, which can be seen as a type of crowdsourcing. When asked later how he could ensure the quality of the submitted data, Liang said that he knew there would be lots of random information submitted and that he would judge the quality by himself. If there was something that was too ridiculous, he would not use that data (Liang, August 2010, personal communication). Second, there were several metaphors employed, such as the term "wall," which referred to a firewall installed by the state for Internet censorship. In particular, the term "Twitter friends" referred generally to a broad range of Twitter users, known and unknown to Liang. This term is also frequently used by other Twitter users in China. Because Twitter is blocked in China, it is recognized that discussions on Twitter by Chinese users are often more political compared to those of other users elsewhere. In this sense, these Chinese Twitter users form certain political communities on Twitter, which are fluid and have constantly changing boundaries. It is not uncommon to see online comments on many hotly debated social issues like "Twitter sightseeing group come here to witness this issue," requesting onlookers to show their concern. Third, the purpose of conducting this survey and making this map, as noted by Liang, is due to the lack of trust of the official data and an attempt to depict the pressure of urban life by the citizens themselves. As such, this map allows spatial narratives that represent resistance to the state, not only speaking against the official data source, but also questioning the burden citizens bear in the process of China's rapid urbanization (seen often to be stimulated by state-led land speculation).

This map was once open to the public for edits. However, there were some lines added such as lines between two random locations with a note suggesting how to get from one location to the other. In addition, the title of the map was changed to "Far East Map." These edits were noticed during the time of my fieldwork from July to August 2010. Liang was surprised to see these abovementioned edits, which were obviously not related to the original theme of this map (Liang, August 2010, personal communication). Upon the most recent visit of this map's site, this map receives more than 53,000 hits and is not available for public editing, with the latest update marked on September 14, 2010. As such, this is at least an 8-month project with 64 records collected. It is not the first VGI map made by Liang. His first map was on locations of his exhibits, after which he made a few maps including one on a tour of photographing Chinese county governments' main office buildings, which involved 23 counties from 11 provinces. Explaining why he made these maps, Liang noted, "First, it's fun. Second, it's useful." Liang further described that mapping in this way was like "writing diaries on the map" (Liang, August 2010, personal communication).

In sum, there are several characteristics of these VGI constructions in the process of data collection and delivery, representation, and analysis. First, these maps are usually initiated by individuals, which may evolve into a collaborative project, remain an individual commitment, or continue as a dynamic mix of these two. Moreover, these VGI authors are quite savvy in using a variety of online venues to circulate their maps. In particular, social networking sites play an important role. Second, because of the topics covered in these VGI constructions, accuracy of locations does not seem to be a dominant concern in the mapping processes. The cartographic design may also be rudimentary. Yet the mapping is not a static process, and the authors may learn from their past experiences, reflected in the second example regarding the change of its legend. Third, the purposes behind these mapping constructions and associated analyses vary, ranging from personal interest in documenting daily practices to addressing broader social issues. Therefore, such VGI constructions engage participants from different backgrounds, each construction forming a loose and constantly changing community engaged in the use and development of these maps. However, the VGI authors being interviewed in this project (with a total number of 12 at the time of this writing) are all men, mostly in their 30s, who have used the Internet intensively.

Furthermore, the acknowledgment of "personal documentation of the history" and "writing diaries on the map" indicates that these mapping practices provide important spatial narratives for these VGI authors. That the attention to social and political issues through VGI mapping represents a gaze from the participants and map viewers underscored changing citizenship in China with greater awareness of political rights that are exercised and negotiated through contestations of and resistances to dominant state power. A VGI author may have more than one map created. As such, a VGI practitioner may navigate from a map of interested places to a map of environmental pollution sites in China, traversing the boundaries between private and public, personal and political. These practices act as a "mode of information" (Poster 1990) mediated by Web 2.0 and geospatial technologies, and they construct multiple DigiPlaces (Zook and Graham 2007) that are hybrid, interactive, and mobile. They can be seen as a form of micro-politics, which can be described as "organizing without organization" (Hu 2011).

Meanwhile, with these forms of dynamic inclusion addressed above, there are new forms of exclusion. First, while there has been a significant increase of the number of netizens in China in the past decade, the issue of a digital divide remains a significant challenge, especially among the urban users and rural ones (e.g., Michael and Zhou 2010). Second, the examples cited here suggest that active VGI authors are overwhelmingly young and male, with a great deal of experience using the Internet. More research is needed on the demographic underpinnings regarding VGI practices.

6.5 Conclusion

While this case study by no means depicts the whole picture of VGI practices in China, the above analysis shows that a synthesized theoretical framework from PPGIS, critical GIS, and critical social theory helps to understand complex mapping practices and their sociopolitical implications situated in different contexts. VGI practices in China discussed here simultaneously share some traits with, and differ from, existing VGI practices in the West regarding spatial knowledge production and meanings of community, participation, and civic engagement. In particular, these VGI practices share the major characteristics of DigiPlaces, marked by increasing individualism, dynamism, and digitally generated visibility (Zook and Graham 2007). As shown in the above VGI examples, these online, interactive mapping practices are usually initiated by individuals. Subsequent mapping may be conducted by a singular individual or by a larger group of volunteers. These visualizations are usually prompted by these VGI authors' concerns for particular issues, rather than proposed from a particular organizational agenda. These mapping constructions evolve over time. As such, these VGI visualizations are highly dynamic.

These VGI practices create new participatory spaces but also new exclusions. Through these mapping practices and constructions of DigiPlaces, new participatory spaces are produced, transcending the virtual and the physical in a complex and nuanced way. The specific goals of setting up these maps may vary greatly depending on their particular content. Yet these practices all strive to utilize the power of visualization to broadcast a particular concern to a broader audience and to send a political message with these visualizations. These mapping platforms therefore provide new spaces for the VGI authors to express such concerns. These concerns are shared by a larger community, evident in relatively high total counts of hits for these VGI products. These practices do not stay digital only. Rather, they are actively constructed and reconstructed with actions in non-digital spaces, such as the relief efforts in the first example, or the violent eviction map attempting to persuade people from purchasing houses with history of evictions. It is difficult to measure the actual extent of impacts of such mapping in mobilizing a broader community. Nonetheless, such seemingly mundane actions of documenting, watching, and monitoring through visualization, which can be disseminated quite widely, are indicative of an important form of participation and engagement from the bottom and grassroots in China. On the other hand, multiple levels of exclusion emerge in these processes. Such mapping practices require little formal training in cartography or GIS. However, they still require a certain level of familiarity with the tool and with the online mapping interface, as well as a certain level of time commitment on top of the existing digital divide regarding Internet access.

Moreover, the boundaries and composition of the "community" involved in these VGI practices have been influenced by the dynamic interplay between Web 2.0 technologies and political and social landscapes in China. On the one hand, these VGI practices change the traditional power relations of spatial data provision and representation, resulting in patchworks of spatial data provision (Goodchild 2008). Yet such spatial data productions in China might embody an additional, if not different, layer of political meanings when compared to other VGI practices such as OpenStreetMap. These multiple patchworks of spatial data productions in China constitute micro-power and a way of organizing without organization. In an environment of strong state control characterized by hegemony, such mapping practices by these "onlookers" are informed by and intersected with a changing citizenship identity that is also influenced by a mode of information (Poster 1990). To borrow James Scott's term, this mode of participation can be seen as a "weapon of the weak" (Scott 1987). In addition, these participatory VGI practices are also intersected with the entertaining side of mapping and playfulness of spatial narratives, constituting a mixture of the Apollonian and the Dionysian dimensions of politics. On the other hand, these participatory efforts are intersected by lines of class, age, and gender in the shifting social fabric in China in the context of urbanization and globalization.

More broadly, the dynamics of sociopolitical contexts in China that have shaped China's VGI practices might differ greatly from other societal contexts. But the complexities of DigiPlaces construction, intersected with Web 2.0 technologies, share some important traits with instances in the West, as noted above; these include increasing engagements at the individual level, exemplified in these VGI practices. If Habermas is right about the impact of mass communication media, which made the reception of media products a form of privatized appropriation, and in turn shaped the construction of public spheres and citizen participation (Thompson 1995), then what does it mean that through the new media of the Internet, and Web 2.0 in particular, spatial narratives constantly merge private traits into publicly available platforms and individual bodies into networks? How can these various modes of information come into play in constituting spatial knowledge production and civic engagement? It is through this perspective that perhaps PPGIS and VGI are deeply intertwined with respect to implications for spatial politics and, consequently, possible forms of technology-, data-, and media-mediated public participation in the so-called information age.

References

Arnstein, S. (1969). A ladder of citizen participation. *Journal of the American Institute of Planners, 35*(4), 216–224.

Boulton, A. (2010). Just maps: Google's democratic map-making community? *Cartographica, 45*(1), 1–4.

CNNIC (China Internet Network Information Center). (2011). The 28th statistical report on internet development in China. http://www.cnnic.net.cn/dtygg/dtgg/201107/W020110719521725234632.pdf. Accessed November 15, 2011.

Craig, W. J., Harris, T. M., & Weiner, D. (Eds.). (2002). *Community participation and geographical information systems*. London: Taylor and Francis.

Crampton, J. (2009). Cartography: Performative, participatory, political. *Progress in Human Geography, 33*(6), 840–848.

Crutcher, M., & Zook, M. (2009). Placemarks and waterlines: Racialized cyberscapes in post-Katrina Google Earth. *Geoforum, 40*, 523–534.

Dodge, M., & Kitchin, R. (2007). Outlines of a world coming into existence: Pervasive computing and the ethics of forgetting. *Environment and Planning B: Planning and Design, 34*, 431–445.

Dormann, C., Caquard, S., Woods, B., & Biddle, R. (2006). Role-playing games in cybercartography: Multiple perspectives and critical thinking. *Cartographica, 41*, 47–58.

Drezner, D. (2005). Weighing the scales: The internet's effect on state-society relations. Paper presented March 2005 at conference: "Global Flow of Information," Yale Information Society Project, Yale Law School. http://islandia.law.yale.edu/isp/GlobalFlow/paper/Drezner.pdf. Accessed November 15, 2011.

Elwood, S. (2004). Partnerships and participation: Reconfiguring urban governance in different state contexts. *Urban Geography, 25*(8), 755–770.

Elwood, S. (2008). Volunteered geographic information: Future research directions motivated by critical, participatory, and feminist GIS. *GeoJournal, 72*, 173–183.

Elwood, S. (2009). Geographic information science: New geovisualization technologies – emerging questions and linkages with GIScience research. *Progress in Human Geography, 33*(2), 256–263.

Elwood, S. (2010). Geographic information science: Emerging research on the societal implications of the geospatial web. *Progress in Human Geography, 34*(3), 349–357.

Elwood, S., & Ghose, R. (2004). PPGIS in community development planning: Framing the organizational context. *Cartographica, 38*(3–4), 19–33.

Flew, T., & Liu, R. (2011). Who's a global citizen? Julian Assange, WikiLeaks and the Australian media reaction. Paper presented at Australian and New Zealand Communication Association (ANZCA) 2011 annual conference, Communication on the Edge: Shifting Boundaries and Identities, University of Waikato, Hamilton, New Zealand, July 6–8, 2011.

Gao, B. (2009). Observations of China's civil society in 2009: The formation of the social field. *Bolan Qunxhu, 3*, 2010.

Gerlach, J. (2010). Vernacular mapping, and the ethics of what comes next, guest editorial. *Cartographica, 45*(3), 165–168.

Ghose, R. (2005). The complexities of citizen participation through collaborative governance. *Space and Polity, 9*(1), 61–75.

Ghose, R. (2007). Politics of scale and networks of association in public participation GIS. *Environment and Planning A, 39*, 1961–1980.

Goldman, M., & Perry, E. (Eds.). (2002). *Changing meanings of citizenship in modern China*. Cambridge, MA: Harvard University Press.

Goodchild, M. (2007). Citizens as sensors: The world of volunteered geography. *GeoJournal, 69*, 211–221.

Goodchild, M. (2008). Commentary: Whither VGI? *GeoJournal, 72*, 239–244.

Habermas, J. (1989). *The structural transformation of the public sphere*. Cambridge, MA: MIT Press.

Haklay, M., Singleton, A., & Parker, C. (2008). Web mapping 2.0: The neogeography of the GeoWeb. *Geography Compass, 2*(6), 2011–2039.

Haraway, D. (1991). *Simian, cyborgs and women: The reinvention of nature*. New York: Routledge.

Hu, Y. (2011). The surrounding gaze. http://huyongpku.blog.163.com/blog/static/1243594962011 012113731103/. Accessed November 15, 2011.

Keane, M. (2001). Redefining Chinese citizenship. *Economy and Society, 30*(1), 1–17.

Kingsbury, P., & Jones, J. P. (2009). Walter Benjamin's Dionysian adventures on Google Earth. *Geoforum, 40*, 502–513.

Kreutz, C. (2010). Maptivism – Maps for activism transparency and engagement, speech given at Re:publica 2010, Berlin. http://www.youtube.com/watch?v=47zn9sz1DcQ&feature=player_embedded. Accessed September 15, 2011.

Kyem, P. (2004). Of intractable conflicts and participatory GIS applications: The search for consensus amidst competing claims and institutional demands. *Annals of the Association of American Geographers, 94*(1), 37–57.

Lin, W. (2008). GIS development in China's urban governance: A case study of Shenzhen. *Transactions in GIS, 12*(4), 493–514.

Lin, W. (2010). Emerging neogeographic practices in China: (New) spaces of participation and resistance? Paper presented at the Annual Meeting of the Association of American Geographers, Washington, DC, April 14–18, 2010.

Lin, W., & Ghose, R. (2010). Social constructions of GIS in China's changing urban governance: The case of Shenzhen. *Cartographica, 45*(2), 89–102.

MacKinnon, R. (2010). Networked authoritarianism in China and beyond: Implications for global internet freedom. Paper presented at Liberation Technology in Authoritarian Regimes, Stanford University, October 11–12, 2010. http://iis-db.stanford.edu/evnts/6349/MacKinnon_Libtech.pdf. Accessed November 15, 2011.

Meier, P. (2011). What is crisis mapping? An update on the field and looking ahead. http://irevolution.net/2011/01/20/what-is-crisis-mapping/. Accessed November 15, 2011.

Michael, D., & Zhou, Y. (2010). *China's digital generation 2.0: Digital media and commerce go mainstream*. Boston Consulting Group Report.

Miller, C. (2006). A beast in the field: The Google maps mashup as GIS/2. *Cartographica, 41*(3), 187–199.

Obermeyer, N. (2007). Thoughts on volunteered (geo)slavery. http://www.ncgia.ucsb.edu/projects/vgi/participants.html. Accessed June 24, 2011.

Perkins, C. (2008). Cultures of map use. *The Cartographic Journal, 45*(2), 150–158.

Pickles, J. (2004). *A history of spaces: Cartographic reason, mapping and the geo-coded world*. New York: Routledge.

Poster, M. (1990). *The mode of information: Poststructuralism and social context*. Chicago: University of Chicago Press.

Poster, M. (1997). Cyberdemocracy: Internet and the public sphere. In D. Porter (Ed.), *Internet culture* (pp. 202–214). London: Routledge.

Qiu, L. (2009). *Working-class network society: Communication technology and the information have-less in urban China*. Cambridge, MA: The MIT press.

Rheingold, H. (1994). *The virtual community*. New York: Harper.

Schuurman, N. (2004). Databases and bodies – A cyborg update. *Environment and Planning A, 36*, 1337–1340.

Scott, J. (1987). *Weapons of the weak: Everyday forms of peasant resistance*. New Haven: Yale University Press.

Sieber, R. (2000). GIS implementation in the grassroots. *Journal of Urban and Regional Information Systems Association, 12*(1), 15–29.

Sieber, R. (2006). Public participation geographic information systems: A literature review and framework. *Annals of the Association of American Geographers, 96*(3), 491–507.

Staeheli, L. (2010). Political geography: Where's citizenship? *Progress in Human Geography, 35*(3), 393–400.

Sui, D. (2006). The Streisand lawsuit and your stolen geography. *GeoWorld*, (December Issue), 26–29.

Sui, D. (2008). The wikification of GIS and its consequences: Or Angelina Jolie's new tattoo and the future of GIS. *Computers, Environment and Urban Systems, 32*, 1–5.

Sui, D., & Goodchild, M. (2003). A tetradic analysis of GIS and society using McLuhan's law of the media. *The Canadian Geographer, 47*(1), 5–17.

Tai, Z. (2006). *The internet in China: Cyberspace and civil society*. New York: Routledge.

Thompson, J. (1995). *The media and modernity: A social theory of the media.* Cambridge: Polity Press.
Tulloch, D. (2007). Many, many maps: Empowerment and online participatory mapping. *First Monday, 12*(2). http://firstmonday.org/htbin/cgiwrap/bin/ojs/index.php/fm/article/view/1620/1535. Accessed November 15, 2011.
Tulloch, D. (2008). Is VGI participation? From vernal pools to video games. *GeoJournal, 72*(3–4), 161–171.
Warf, B. (2011). Geographies of global Internet censorship. *GeoJournal, 76*(1), 1–23.
Weiner, D., & Harris, T. (2003). Community-integrated GIS for land reform in South Africa. *URISA Journal, 15*, 61–73.
Williams, S. (2007). Application for GIS specialist meeting. http://www.ncgia.ucsb.edu/projects/vgi/participants.html. Accessed January 28, 2011.
Wilson, M. (2009). Coding community. Unpublished PhD dissertation, Department of Geography, University of Washington.
Woo, M. (2002). Law and gendered citizen. In M. Goldman & E. Perry (Eds.), *Changing meanings of citizenship in modern China* (pp. 308–329). Cambridge, MA: Harvard University Press.
Xiao, S. (2010). Paying attention is power, Onlooking to change China. News editorial of *Southern Weekend*. January 13, 2010. http://www.infzm.com/content/40097. Accessed July 29, 2011.
Yang, G. (2003). Weaving a green web: The internet and environmental activism in China. China Environment Series, Issue 6. http://bc.barnard.columbia.edu/~gyang/Yang_GreenWeb.pdf. Accessed November 15, 2011.
Yang, G. (2009). *The power of the internet in China: Citizen activism online.* New York: Columbia University Press.
Zook, M., & Graham, M. (2007). The creative reconstruction of the Internet: Google and the privatization of cyberspace and DigiPlace. *Geoforum, 38*, 1322–1343.

Chapter 7
Citizen Science and Volunteered Geographic Information: Overview and Typology of Participation

Muki Haklay

Abstract Within volunteered geographic information (VGI), citizen science stands out as a class of activities that require special attention and analysis. Citizen science is likely to be the longest running of VGI activities, with some projects showing continuous effort over a century. In addition, many projects are characterised by a genuine element of volunteering and contribution of information for the benefit of human knowledge and science. They are also tasks where data quality and uncertainty come to the fore when evaluating the validity of the results. This chapter provides an overview of citizen science in the context of VGI – hence the focus on geographic citizen science. This chapter highlights the historical context of citizen science and its more recent incarnation. It also covers some of the cultural and conceptual challenges that citizen science faces and the resulting limitation on the level of engagement. By drawing parallels with the Participatory Geographic Information Systems (PGIS) literature, the chapter offers a framework for participation in citizen science and concludes with the suggestion that a more participatory mode of citizen science is possible.

7.1 Introduction

Volunteered geographic information (VGI), coined by Goodchild (2007), encompasses a wide range of activities and practices, ranging from the 'fun' activities of locating summer holiday photographs (Turner 2006) to focused surveying in the aftermath of an earthquake (Zook et al. 2010). Within these practices, there is a subset that falls into the category of citizen science – the involvement of non-professional

---

M. Haklay (✉)
Department of Civil, Environmental and Geomatic Engineering,
University College London, London, UK
e-mail: m.haklay@ucl.ac.uk

scientists in data collection and, to some extent, its analysis. While it is possible to try to formulate a definition that delineates the boundaries of what should or should not be considered citizen science, a much more fruitful approach is to understand the general properties of citizen science and its overlap with VGI. As we shall see, not all citizen science is geographic, that is, involving a project in which a location on Earth plays an important role. Once the differentiation between the geographic and non-geographic projects is clarified, we can focus on the former, as they include an element of VGI, by necessity.

Before turning to citizen science itself, it is worth noticing, in the context of this book, how the different contextualisation of VGI illuminates aspects that might go unnoticed or unexplored otherwise. Thus, VGI can be seen as a way of producing geographical information and as a tool for updating national geographical databases (Goodchild 2007; Antoniou et al. 2010) in which case the appropriate context is spatial data quality and the production of geographical information. When it is viewed within the context of critical, participatory or feminist Geographic Information Systems (GIS) (Elwood 2008), questions about the nature of the participation, power relations and other societal aspects of VGI are opened – thus, the process of creating VGI is becoming as important as analysing the product. When Budhathoki et al. (2010) look at the reasons for participation in leisure, volunteering and contribution to open source projects in the context of VGI, they highlight that the behaviour of participants and the reasons that lead to their involvement are important elements of VGI. Other interpretations and contextualisation of VGI provide their own lens onto the activities of participants and the resulting products. Because VGI is an area that requires a socio-technical analysis, these prisms are valuable in helping to understand the phenomena and to consider the relevant applications of its products. The use of citizen science provides another such prism, highlighting how VGI operates within scientific knowledge production.

In the current chapter, we start with an overview of the field and its changes in the past decade following the growth of the Internet in general, and the World Wide Web (Web) in particular, as a global communication platform. Next, the enabling factors and trends are briefly outlined and explained. Following this review, the characteristics of geographic citizen science are reviewed based on current evidence. We then discuss the intriguing cultural challenge to existing scientific practices that citizen science puts forward. Finally, by drawing on established practices in participatory GIS, a framework for participation in citizen science is offered and its implications are analysed.

7.2 Citizen Science

As noted before, while the aim here is not to provide a precise definition of citizen science with rigid boundaries, a definition and clarification of the core characteristics of citizen science is needed. We define citizen science here as scientific activities in which non-professional scientists voluntarily participate in data collection, analysis and dissemination of a scientific project (Cohn 2008; Silvertown 2009). People who participate in a scientific study without playing some part in the study itself – for example, volunteering in a medical trial or participating in a social science survey – are not included in this definition. At the same time, the core issue of 'who is a scientist' is left deliberately blurred. This is because it is easier to identify professional scientists as those that are employed to carry out scientific work or investigation. With unpaid scientists, the situation is more complex – many will not define or identify themselves as scientists even if they are carrying out significant work within the scientific frameworks of data collection and interpretation. Others will use the qualification of amateur scientist to describe themselves, or a similar definition such as bird watcher. However, for our purposes, scientists are all the active participants in a scientific project.

It is important to notice that there are boundaries, albeit fuzzy, of what should be considered a citizen science project. While it is easy to identify a citizen science project when the aim of the project is the collection of scientific information, as in the recording of the distribution of plant species, there are cases where the definition is less clear-cut. For example, the process of data collection in OpenStreetMap or Google Map Maker is mostly focused on recording verifiable facts about the world that can be observed on the ground. The tools used by OpenStreetMap mappers – such as remotely sensed images, GPS receivers and map editing software – are scientific instruments. With their attempt to locate observed objects and record them on a map accurately, they follow in the footsteps of surveyors who followed scientific principles in their work such as Robert Hooke, who carried out an extensive survey of London following the fire of 1666 using scientific methods (though, unlike OpenStreetMap volunteers, Hooke was paid for his effort). Finally, cases where facts are collected in a participatory mapping activity, such as the one that Ghose (2001) describes, should probably be considered citizen science only if the participants decided to frame it as such. The framing of the activity is important because in citizen science the expectations are that the data collection will follow a certain protocol and that data analysis and visualisation will be carried out according to established practices. Under a citizen science framing, the activity will focus on recording observations rather than highlighting community views or opinions.

Notice also that, by definition, citizen science can only exist in a world in which science is socially constructed as the preserve of professional scientists in academic institutions and industry because, otherwise, any person who is involved in a scientific project would simply be considered a contributor and, potentially, a scientist. As Silvertown (2009) noted, until the late nineteenth century, science was mainly developed by people who had additional sources of employment that allowed them to spend time on data collection and analysis. Famously, Charles Darwin joined the *Beagle* voyage, not as a professional naturalist but as a companion to Captain FitzRoy. Thus, in that era, almost all science was citizen science albeit mostly by affluent gentlemen and gentlewomen scientists. While the first professional scientist is likely to be Robert Hooke, who was paid to work on scientific studies in the seventeenth century, the major growth in the professionalisation of scientists was mostly in the latter part of the nineteenth and throughout the twentieth century.

Even with the rise of the professional scientist, the role of volunteers has not disappeared, especially in areas such as archaeology, where it is common for enthusiasts

Fig. 7.1 British Trust of Ornithology listing of potential engagement projects and level of engagement

to join excavations, or in natural science and ecology, where non-professionals collect and send samples and observations to national repositories. These activities include the Christmas Bird Watch that has been ongoing since 1900 and the British Trust for Ornithology Survey, which has collected over 31 million records since its establishment in 1932 (Silvertown 2009) – see Fig. 7.1. Astronomy is another area in which amateurs and volunteers have been on a par with professionals when observation of the night sky and the identification of galaxies, comets and asteroids are considered (BBC 2006). Finally, meteorological observations have also relied on volunteers since the early start of systematic measurements of temperature, precipitation or extreme weather events (WMO 2001).

This type of citizen science provides the first class of 'classic' citizen science – the 'persistence' parts of science where the resources, geographical spread and the nature of the problem mean that volunteers sometimes predate the professionalisation and mechanisation of science. There activities usually require a large but sparse network of observers who work as part of a hobby or leisure activity. This type of citizen science has flourished in specific enclaves of scientific practice, and the progressive development of modern communication tools has made the process of collating the results from the participants easier and cheaper while keeping many of the characteristics of data collection processes close to their origins.

A second type of citizen science activity is environmental management within the context of environmental justice campaigns. Modern environmental management includes strong technocratic- and scientific-oriented management practices (Bryant and Wilson 1998; Scott and Barnett 2009) and environmental decision making is heavily based on scientific environmental information. As a result, when an environmental conflict emerges – such as community protest over a noisy local factory or planned expansion of an airport – the valid evidence needs to be based on scientific data collection. This aspect of environmental justice struggle is encouraging communities to carry out 'community science' in which scientific measurements and analysis are carried out by members of local communities so they can develop an evidence base and set action plans to deal with problems in their area. A successful example of such an approach is the 'Global Community Monitor' method for allowing communities to deal with air pollution issues (Scott and Barnett 2009). Air pollution issues are diagnosed with a simple method of sampling air using plastic buckets followed by analysis in an air pollution laboratory before, finally, the community is provided with instructions on how to understand the results. This activity is termed 'Bucket Brigade' and was used across the world in environmental justice campaigns. Similarly, in London, community science was used to collect noise readings in two communities that are impacted by airport and industrial activities. The outputs were effective in bringing environmental problems to the attention of decision makers and regulatory authorities (Haklay et al. 2008a). As in 'classic' citizen science, the growth in electronic communication enabled communities to identify potential methods – for example, through the 'Global Community Monitor' website – as well as find the details of international standards, regulations and scientific papers that can be used together with the local evidence.

However, the emergence of the Internet and the Web as a global infrastructure has enabled a new incarnation of citizen science, which has been termed 'citizen cyberscience' by Francois Grey (2009). As Silvertown (2009) and Cohn (2008) noted, the realisation by scientists that the public can provide free labour, skills, computing power and even funding and the growing demands from research funders for public engagement have motivated the development and launch of new and innovative projects. These projects utilise personal computers, GPS receivers and mobile phones as scientific instruments. Within citizen cyberscience, it is possible to identify three subcategories: volunteered computing, volunteered thinking and participatory sensing.

Volunteered computing was first developed in 1999, with the foundation of SETI@home (Anderson et al. 2002), which was designed to distribute the analysis of data that was collected from a radio telescope in the search for extraterrestrial intelligence. The project utilises the unused processing capacity that exists in personal computers and uses the Internet to send and receive 'work packages' that are analysed automatically and sent back to the main server. Over 3.83 million downloads were registered on the project's website by July 2002. The system on which SETI@home is based, the Berkeley Open Infrastructure for Network Computing (BOINC), is now used for over 100 projects, covering physics, processing data from the Large Hadron Collider through LHC@home; climate science with the running of climate models in climateprediction.net; and biology in which the shape of proteins is calculated in Rosetta@home.

While volunteered computing requires very little from the participants, apart from installing software on their computers, in volunteered thinking the volunteers are engaged at a more active and cognitive level (Grey 2009). In these projects, the participants are asked to use a website in which information or an image is presented to them. When they register onto the system, they are trained in the task of classifying the information. After the training, they are exposed to information that has not been analysed and are asked to carry out classification work. The Stardust@home project (Westphal et al. 2006), in which volunteers were asked to use a virtual microscope to try to identify traces of interstellar dust, was one of the first projects in this area, together with the NASA ClickWorkers that focused on the classification of craters on Mars. Galaxy Zoo (Lintott et al. 2008), a project in which volunteers classify galaxies, is now one of the most developed ones, with over 100,000 participants and with a range of applications that are included in the wider Zooniverse set of projects (see http://www.zooniverse.org/).

Participatory sensing is the final and most recent type of citizen science activity. Here, the capabilities of mobile phones are used to sense the environment. Some mobile phones have up to nine sensors integrated into them, including different transceivers (mobile network, WiFi, Bluetooth), FM and GPS receivers, camera, accelerometer, digital compass and microphone. In addition, they can link to external sensors. These capabilities are increasingly used in citizen science projects, such as Mappiness in which participants are asked to provide behavioural information (feeling of happiness) while the phone records their location to allow the linkage of different locations to well-being (MacKerron 2011). Other activities include the sensing of air quality (Cuff et al. 2007) or noise levels in the application NoiseTube (Maisonneuve et al. 2010) by using the mobile phone's location and the readings from the microphone.

Before turning to the context and drivers of the currently resurgent citizen science, it is worth noting that other typologies of citizen science are offered by Cooper et al. (2007), Wilderman (2007), Bonney et al. (2009) and Wiggins and Crowston (2011). These classifications highlight aspects such as the level of informal science education, the involvement of participants in various aspects of research activity or the purpose of the project. This is expected in an emerging field, where similarly to VGI, researchers are scanning the field and suggesting ways to understand the landscape.

7.3 Context and Drivers

The general trends that ushered in the Web Mapping 2.0 era (Haklay et al. 2008b; Goodchild 2007) are also part of the drivers that allowed the recent growth in citizen science in general and citizen cyberscience in particular. These factors include the increased availability of higher-capacity domestic Internet connections and the reducing costs of computers and sophisticated mobile devices; the reduced costs of computer storage and the availability of extra storage capacity on personal computers that make it possible for participants to store and process large amounts of data without major implications to their own activities; the continued development of Internet technologies and standards such as eXtensible Markup Language (XML) that facilitate the transfer of information between computers; the increased accuracy of the Global Positioning System (GPS) since 2000 and the subsequent reduction in cost of receivers; and the growth of sophisticated Web applications that enable rich interaction by their users, allowing for applications such as the virtual microscope that was used in Stardust@home.

However, these general and mostly technological factors are only part of the trends that led to the blossoming of citizen science. As important are the social trends that enabled it. The main factor that needs to be considered is the growth in the population of well-educated individuals, with many millions of people who have advanced degrees or some level of higher education in science or engineering but do not use their scientific knowledge in their daily life. Moreover, with the increase in the demands of secondary education, many who have dropped out of school even are equipped with basic scientific knowledge that is sufficient to make them effective participants in citizen science projects. For many of these people, education provided a starting point for an interest in science, which is not fulfilled in their daily activities. Thus, citizen science provides an opportunity to explore this dormant interest – while the educational attainment means that the scientists who design the project can assume a basic understanding of principles by the participants.

To this, we should add the increase in leisure activities and the reduction in working hours that has occurred in many advanced economies over the past four decades. Before the growth in electronic communication, hobbies were limited to private activities or occasional gatherings of a small group of enthusiasts. The ability of the Web to accommodate narrow interests and to allow a highly distributed network of individuals to share and discuss their interests is especially important for such activities. It allows for the creation of websites, mailing lists and other ways in which these enthusiasts can come together and share experience or join forces in working on scientific data collection or analysis.

Notice that together, the drivers point to an inherent bias in the socio-economic make-up of citizen science. Participants are highly likely to be living in an advanced economy and to be a member of the middle class, thus to have the education, technical skills, access to resources and infrastructure that facilitates participation in these activities.

7.4 Geographical Citizen Science

Against the backdrop of general citizen science, we can identify a specific subtype of activities that can be termed 'Geographical Citizen Science' and therefore fall within the definition of VGI. Geographical citizen science includes projects where the collection of location information is an integral part of the activity. It should become clear from the overview above that a significant part of the information that was and is collected through 'classic' citizen science, community science and citizen cyberscience projects is geographical – as in the location of observations in the Christmas Bird Watch or the recording of noise along a given route in NoiseTube (Maisonneuve et al. 2010). Yet, in the past, the location was usually approximated and sometimes given in grid co-ordinates of only 100-m, or even 1-km, accuracy, which meant that linking observations to a location could be tricky and highly uncertain. Even though location technology is increasingly available through Personal Navigation Devices (PND), GPS receivers and mobile phones, it is important to remember that because of their affordance (in the sense of familiarity and ease of use, as in Norman 1990), paper maps remain a very effective medium for data collection. This has been shown in citizen science studies in London as well as in the integration of paper mapping activities in OpenStreetMap through Walking Papers (Migurski 2009). As long as the data collection media supports accurate geographic location, the project can result in high-quality geographical citizen science.

The first important characteristic of geographical citizen science is to understand the role of the volunteer. The role can be active or passive. An active contribution happens when the participant is expected to consciously contribute to the observation or the analysis, as in the case of taking an image of an observed species, tagging it and sending it electronically to the project's hub. A passive data collection can happen when the contributor is acting more as an observation platform and the data are gathered without active engagement, for example, when a person volunteers to be tagged by a GPS receiver to monitor daily walking activity or replaces a memory card and batteries in an automatic digital camera that is installed on a deer track (Cohn 2008). A further differentiation can be made between geographically explicit and implicit citizen science projects (Antoniou et al. 2010). In geographically explicit projects, the activity is *aimed* at collecting geographical information – for example, in the British Trust of Ornithology project a participant is required to record a specific location of where the observation is taking place. In geographically implicit projects, the aim might be to collect images of different species; some of the images may arrive with location information (i.e. geotagged), but the aim of the project is *not* to collect geographical information.

These different schemes will have an impact on the need to motivate and engage the participants and the level of training and knowledge that is required from them. In addition, different schemes will influence the ability to secure quantitative and qualitative information. While all forms can support the collection of some quantitative information, only the active and geographically explicit projects are likely to provide meaningful qualitative information such as descriptions of personal perceptions or textual descriptions of the place where the observation was recorded.

Research into the motivation and the spatial characteristics of VGI (Budhathoki et al. 2010; Coleman et al. 2009; Haklay 2010), as well as studies of Wikipedia and citizen science projects (Nov et al. 2011), is providing some guidance on what it is possible to achieve with geographical citizen science. What we know so far is that citizen science is a 'serious leisure' activity and that the most likely participants will join with some existing interest in the subject and will be keen to learn more. They will be predominantly male, well educated and from higher income brackets, which gives them both the time to participate in the activity and the financial resources for specialist equipment and/or participation in field work. There will be 'participation inequality' with some participants contributing a lot, while many others contribute a little.

The areas that can be covered well by geographical citizen science will be areas with a high population concentration or high level of outdoor activity such as popular national parks. Coverage of other areas requires special planning and creation of suitable motivational schemes, including monetary ones, or reliance on a smaller pool of volunteers. There will also be temporal aspects to data collection, with summer months, weekends and daytime being the more popular times in which participants engage in data collection activities.

All these means that the data are essentially heterogeneous, and it is important to remember that the data quality will vary according to the number of volunteers who work on the data and the particular knowledge of each volunteer. Thus, the data should not be treated as homogeneous and complete where a statement of quality can refer to the whole assemblage; rather, localised measures must be used. While all these aspects of geographical citizen science have an impact on the type of research questions and scientific challenges that can be answered through utilisation of suitable schemes, there is a potentially more significant challenge hindering this use of citizen science. At issue is more the contemporary culture of science than the putative value of citizen science itself.

7.5 Culture of Science Problems

The story of modern science is often told by highlighting the increased precision and accuracy with which information is obtained and analysed (e.g. Bryson 2004). In experimental science, the instruments and devices were designed over the years to provide better accuracy, and complex experimental protocols were created to ensure that the level of uncertainty associated with measurements was reduced.

While it is well recognised that different academic disciplines have their own culture and specific practices (Latour 1993), such as the practice of double-blind studies in medicine but not in other areas that research human subjects, the issue of dealing with uncertainty has been central to many areas of science, including geographic information science (see Couclelis 2003).

Interestingly, the attempt to eliminate uncertainty is especially prevalent in areas in which science is used in practical applications such as engineering or environmental management. Due to organisational reasons and policy (King and Kraemer 1993), protocols are enshrined in regulations and become 'the correct way of doing science' with rigid protocols, some parts of which are arbitrary.

The strive to eradicate uncertainty (or at least reduce it) and the development of complex protocols are at the heart of the cultural issue that is leading to suspicion, derision and dismissal of citizen science as a valuable method of scientific data collection. The mistrust of citizen science is, as noted in the opening, based on the view that science is best left to scientists and that it requires rigour, knowledge and skills that only professional scientists develop over time. As Silvertown (2009) noted, 'The apparent underrepresentation of citizen science in the formal literature probably has two causes. First, the term itself is relatively recent, and in fact hundreds of scientific papers have resulted from the data collected in Christmas Bird Counts and other long-running volunteer monitoring programmes. Second, **projects that fit uneasily into the standard model of hypothesis-testing research are written about only in the grey literature, or even not at all**' (p. 471, emphasis added). This is also highlighted in Holling (1998) who emphasised that there are two cultures of science and that citizen science by necessity belongs to the type of science that incorporates uncertainty and highlights integrative approaches. Interestingly, suspicion of VGI follows along the lines of what Holling identified in the area of ecology over a decade ago. Despite the evidence that VGI can be as accurate as professional data (Haklay 2010; Girres and Touya 2010), mistrust of VGI as a useful source of geographical information is common among professional users (e.g. Flanagin and Metzger 2008).

What is forgotten by those who oppose citizen science is the development of instrumentation and its impact on the balance of knowledge and skills that are required by the operator, as well as the level of motivation, dedication and attention to detail of volunteers. Scientific instrumentation has evolved tremendously over the past 350 years if we take the invention of instruments such as Hooke's microscope in 1663 as an indication of the development of modern scientific instruments. Since then, the instruments have improved enormously in terms of their observational power, accuracy and precision. However, until fairly late in the twentieth century, they demanded a significant amount of theoretical and practical knowledge to operate effectively. A great deal of professional judgement was required to balance the accuracy of a calculation with the practical aspects of conducting research. Consider, for example, that many calculations for NASA's moon missions were carried out with slide rules, where experience and judgement is necessary to decide if the calculations are satisfactory.

The computerisation and miniaturisation of scientific instruments, especially in the past 20 years, have changed this equation. Now, the humble GPS receiver encapsulates knowledge and procedures that are highly complex and is capable of calculating them precisely without any intervention from the user. The GPS satellites themselves also encapsulate significant amounts of scientific knowledge and understanding. For example, William Roy spent about 6 weeks and engaged a team of 12 men to measure the 5-mile baseline of the English triangulation system with an accuracy of 2.5 inches. Today, a single person, equipped with a good quality GPS receiver and a mobile phone, can achieve a similar fit in less than a day. The sophistication of the equipment and, more importantly, the science that is integrated into the computational parts of it enable this.

The ability of equipment to provide accurate and precise measurement is central to the ability of volunteers to provide reliable scientific information, especially when these instruments are used in tandem with their personal knowledge and commitment. For example, research in the USA has shown that citizen scientists identified crab types correctly 95% of the time (Cohn 2008). Importantly, the basic understanding of scientific principles and methods, which are now routinely taught at school level, means that the participants in the research have an understanding of what is required of them and what is needed to take a reliable scientific measurement. What is more, because of this basic knowledge, they can carry out the observation without supervision and with very little training. Citizen scientists show significant commitment to the topic and are as capable as the best researchers in many cases. Thus, the information that they produce should be trusted.

7.6 Geographical Citizen Science as Participatory Science

Against the technical, social and cultural background of citizen science, we offer a framework that classifies the level of participation and engagement of participants in citizen science activity. While there is some similarity between Arnstein's (1969) 'ladder of participation' and this framework, there is also a significant difference. The main thrust in creating a spectrum of participation is to highlight the power relationships that exist within social processes such as urban planning or in participatory GIS use in decision making (Sieber 2006). In citizen science, the relationship exists in the form of the gap between professional scientists and the wider public. This is especially true in environmental decision making where there are major gaps between the public's and the scientists' perceptions of each other (Irwin 1995).

In the case of citizen science, the relationships are more complex, as many of the participants respect and appreciate the knowledge of the professional scientists who are leading the project and can explain how a specific piece of work fits within the wider scientific body of work. At the same time, as volunteers build their own knowledge through engagement in the project, using the resources that are available on the Web and through the specific project to improve their own understanding, they are more likely to suggest questions and move up the ladder of participation.

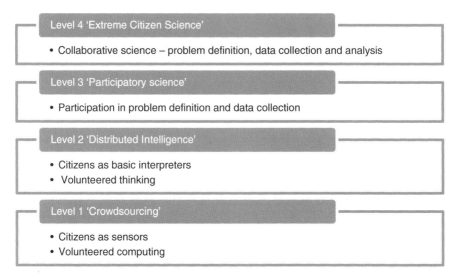

Fig. 7.2 Levels of participation and engagement in citizen science projects

In some cases, the participants would want to volunteer in a passive way, as is the case with volunteered computing, without full understanding of the project as a way to engage and contribute to a scientific study. An example of this is the many thousands of people who volunteered for the Climateprediction.net project, where their computers were used to run global climate models. Many would like to feel that they are engaged in one of the major scientific issues of the day, but would not necessarily want to fully understand the science behind it.

Therefore, unlike Arnstein's ladder, there should not be a strong value judgement on the position that a specific project takes. At the same time, there are likely benefits in terms of participants' engagement and involvement in the project to try to move to the highest level that is suitable for the specific project. Thus, we should see this framework as a typology that focuses on the level of participation (Fig. 7.2).

At the most basic level, participation is limited to the provision of resources, and the cognitive engagement is minimal. Volunteered computing relies on many participants that are engaged at this level, and following Howe (2006), this can be termed 'crowdsourcing'. In participatory sensing, the implementation of a similar level of engagement will have participants asked to carry sensors around and bring them back to the experiment organiser. The advantage of this approach, from the perspective of scientific framing, is that as long as the characteristics of the instrumentation are known (e.g. the accuracy of a GPS receiver), the experiment is controlled to some extent, and some assumptions about the quality of the information can be used. At the same time, running projects at the crowdsourcing level means that despite the willingness of the participants to engage with a scientific project, their most valuable input – their cognitive ability – is wasted.

The second level is 'distributed intelligence' in which the cognitive ability of the participants is the resource that is being used. Galaxy Zoo and many of the 'classic' citizen science projects are working at this level. The participants are asked to take some basic training and then collect data or carry out a simple interpretation activity. Usually, the training activity includes a test that provides the scientists with an indication of the quality of the work that the participant can carry out. With this type of engagement, there is a need to be aware of questions that volunteers will raise while working on the project and how to support their learning beyond the initial training.

The next level, which is especially relevant in 'community science' is a level of participation in which the problem definition is set by the participants, and in consultation with scientists and experts, a data collection method is devised. The participants are then engaged in data collection, but require the assistance of the experts in analysing and interpreting the results. This method is common in environmental justice cases and goes towards Irwin's (1995) call to have science that matches the needs of citizens. However, participatory science can occur in other types of projects and activities – especially when considering the volunteers who become experts in the data collection and analysis through their engagement. In such cases, the participants can suggest new research questions that can be explored with the data they have collected. The participants are not involved in detailed analysis of the results of their effort – perhaps because of the level of knowledge that is required to infer scientific conclusions from the data.

Finally, collaborative science is a completely integrated activity, as it is in parts of astronomy where professional and non-professional scientists are involved in deciding on which scientific problems to work on and the nature of the data collection so it is valid and answers the needs of scientific protocols while matching the motivations and interests of the participants. The participants can choose their level of engagement and can be potentially involved in the analysis and publication or utilisation of results. This form of citizen science can be termed 'extreme citizen science' and requires that scientists act as facilitators, in addition to their role as experts. This mode of science also opens the possibility of citizen science without professional scientists, in which the whole process is carried out by the participants to achieve a specific goal.

This typology of participation can be used across the range of citizen science activities, and one project should not be classified only in one category. For example, in volunteer computing projects, most of the participants will be at the bottom level, while participants that become committed to the project might move to the second level and assist other volunteers when they encounter technical problems. Highly committed participants might move to a higher level and communicate with the scientist who coordinates the project to discuss the results of the analysis and suggest new research directions.

This typology exposes how citizen science integrates and challenges the way in which science discovers and produces knowledge. Questions about the way in which knowledge is produced and truths are discovered are part of the epistemology of science. As noted above, throughout the twentieth century, as science became more specialised, it also became professionalised. While certain people were employed as scientists in government, industry and research institutes, the rest of the population – even if they graduated from a top university with top marks in a scientific discipline – were not regarded as scientists or as participants in the scientific endeavour unless they were employed professionally to do so. In rare cases and following the tradition of 'gentlemen/women scientists', wealthy individuals could participate in this work by becoming an 'honorary fellow', or affiliated with a research institute, which, inherently, brought them into the fold. This separation of 'scientists' and the 'public' was justified by the need to access specialist equipment, knowledge and other privileges such as a well-stocked library. It might be the case that the need to maintain this separation is a third reason that practising scientists shy away from explicitly mentioning the contribution of citizen scientists to their work in addition to those identified by Silvertown (2009).

However, like other knowledge professionals who operate in the public sphere, such as medical experts or journalists, scientists need to adjust to a new environment that is fostered by the Web. Recent changes in communication technologies, combined with the increased availability of open access information and the factors that were noted above, mean that processes of knowledge production and dissemination are opening up in many areas of social and cultural activities (Shirky 2008). Therefore, some of the elitist aspects of scientific practice are being challenged by citizen science, such as the notion that only dedicated, full-time researchers can produce scientific knowledge. For example, surely, it should be professional scientists who can solve complex scientific problems such as long-standing protein-structure prediction of viruses. Yet, this exact problem was recently solved through a collaboration of scientists working with amateurs who were playing the computer game Foldit (Khatib et al. 2011). Another aspect of the elitist view of science can be witnessed in interaction between scientists and the public, where the assumption is of unidirectional 'transfer of knowledge' from the expert to lay people. Of course, as in the other areas mentioned above, it is a grave mistake to argue that experts are unnecessary and can be replaced by amateurs, as Keen (2007) eloquently argued. Nor is it suggested that because of citizen science, the need for professionalised science will diminish; in many citizen science projects, it seems that the participants accept the difference in knowledge and expertise of the scientists who are involved in these projects (Bonney et al. 2009). At the same time, scientists need to develop respect towards those who help them beyond the realisation that they provide free labour, which was noted above.

Given this tension, the participation hierarchy can be seen to be moving from a 'business as usual' scientific epistemology at the bottom to a more egalitarian approach to scientific knowledge production at the top. The bottom level, where the participants are contributing resources without cognitive engagement, maintains the hierarchical division of scientists and the public. The public is volunteering its time or resources to help scientists, while the scientists explain the work that is to be done but without expectation that any participant will contribute intellectually to the project. Arguably, even at this level, the scientists will be challenged by questions and suggestions from the participants, and if they do not respond to them in a sensitive manner, they will risk alienating participants. Intermediaries such as the IBM World

Community Grid, where a dedicated team is in touch with scientists who want to run projects as well as a community of volunteered computing providers, are cases of 'outsourcing' the community management and thus allowing, to an extent, the maintenance of the separation of scientists and the public.

As we move up the ladder to a higher level of participation, the need for direct engagement between the scientist and the public increases. At the highest level, the participants are assumed to be on equal footing with the scientists in terms of scientific knowledge production. This requires a different epistemological understanding of the process, in which it is accepted that the production of scientific insights is open to any participant while maintaining scientific standards and practices such as systematic observations or rigorous statistical analysis to verify that the results are significant. The belief that, given suitable tools, many lay people are capable of such endeavours is challenging to some scientists who view their skills as unique. As the case of the computer game that helped in the discovery of new protein formations (Khatib et al. 2011) demonstrated, such collaboration can be fruitful even in cutting-edge areas of science. However, it can be expected that the more mundane and applied areas of science will lend themselves more easily to the fuller sense of collaborative science in which participants and scientists identify problems and develop solutions together. This is because the level of knowledge required in cutting-edge areas of science is so demanding.

Another aspect in which the 'extreme' level challenges scientific culture is that it requires scientists to become citizen scientists in the sense that Irwin (1995), Wilsdon et al. (2005) and Stilgoe (2009) advocated. In this interpretation of the phrase, the emphasis is not on the citizen as a scientist, but on the scientist as a citizen. It requires scientists to engage with the social and ethical aspects of their work at a very deep level. Stilgoe (2009, p. 7) suggested that, in some cases, it will not be possible to draw the line between the professional scientific activities, the responsibilities towards society and a fuller consideration of how a scientific project integrates with wider ethical and societal concerns. However, as all these authors noted, this way of conceptualising and practising science is not widely accepted in the current culture of science.

Therefore, we can conclude that this form of participatory and collaborative science will be challenging in many areas of science. This will not be because of technical or intellectual difficulties but mostly because of the cultural aspects that were mentioned throughout this chapter. This might end up being the most important outcome of citizen science as a whole, as it might eventually catalyse the education of scientists to engage more fully with society.

7.7 Conclusions

Geographical citizen science has clearly grown in recent years and is showing significant potential in areas such as biodiversity, air pollution or recording the changing shapes of cities. There are, however, two issues that are critical when considering the research directions that link VGI, participatory GIS and citizen science.

First and foremost, there is a need to consider which scientific questions can be answered by citizen science according to the patterns of data collection, the ability to recruit and train volunteers, the suitable participation level and other aspects of VGI. Second, there is a need to overcome the cultural issues and to develop an understanding and acceptance of citizen science within the scientific community. This will require challenging some of the deeply held views in science, such as viewing uncertainty not as something that can be eliminated through tighter protocols but as an integral part of any data collection and therefore developing appropriate methods to deal with it during analysis. Moreover, the view of science as separate from societal and ethical concerns is also a challenge – especially at higher levels of engagement between scientists and participants.

One intriguing possibility is that citizen science will work as an integral part of participatory science in which the whole scientific process is performed in collaboration with the wider public. Some examples are already emerging in geography (Pain 2004) and might provide direction for the future development of citizen science projects.

Acknowledgement This chapter was written with the support of the 'Extreme' Citizen Science – ExCiteS grant, funded by the EPSRC (Engineering and Physical Sciences Research Council), EPSRC reference EP/I025278/1 and the EveryAware project, funded by the EC FP7. Special thanks go to the participants of the VGI workshops in GIScience 2010, AAG 2011, Vespucci Summer Institute 2011 and members of the ExCiteS group for their comments and questions.

References

Anderson, D. P., Cobb, J., Korpela, E., Lebofsky, M., & Werthimer, D. (2002). SETI@home: An experiment in public-resource computing. *Communications of the ACM, 45*(11), 56–61.

Antoniou, B., Haklay, M., & Morley, J. (2010). Web 2.0 geotagged photos: Assessing the spatial dimension of the phenomenon. *Geomatica, The Journal of Geospatial Information, Technology and Practice, 64*(1), 99–110.

Arnstein, S. R. (1969). A ladder of citizen participation. *Journal of the America Institute of Planners, 35*(4), 216–224.

BBC (2006). Citizen science, Radio 4 series. http://www.bbc.co.uk/radio4/science/citizenscience.shtml. Accessed June 2011.

Bonney, R., Ballard, H., Jordan, H., McCallie, E., Phillips, T., Shirk, J., & Wilderman, C. (2009). Public participation in scientific research: Defining the field and assessing its potential for informal science education. In *A CAISE Inquiry Group Report*, Center for Advancement of Informal Science Education (CAISE), Washington, DC (Technical Report).

Bryant, R. L., & Wilson, G. A. (1998). Rethinking environmental management. *Progress in Human Geography, 22*(3), 321–343.

Bryson, B. (2004). *A short history of almost everything*. London: Black Swan.

Budhathoki, N. R., Nedovic-Budic, Z., & Bruce, B. (2010). An interdisciplinary frame for understanding volunteered geographic information. *Geomatica, The Journal of Geospatial Information, Technology and Practice, 64*(1), 11–26.

Cohn, J. P. (2008). Citizen science: Can volunteers do real research? *Bioscience, 58*(3), 192–197.

Coleman, D. J., Georgiadou, Y., & Labonte, J. (2009). Volunteered geographic information: The nature and motivation of producers. *International Journal of Spatial Data Infrastructures Research, 4*, 332–358.

Cooper, C. B., Dickinson, J., Phillips, T., & Bonney, R. (2007). Citizen science as a tool for conservation in residential ecosystems. *Ecology and Society 12*(2), 11. http://www.ecologyandsociety.org/vol12/iss2/art11/. Accessed January 5, 2012.

Couclelis, H. (2003). The certainty of uncertainty: GIS and the limits of geographic knowledge. *Transactions in GIS, 7*(2), 165–175.

Cuff, D., Hansen, M., & Kang, J. (2007). Urban sensing: Out of the woods. *Communications of the Association for Computing Machinery, 51*(3), 24–33.

Elwood, S. (2008). Volunteered Geographic Information: Future research directions motivated by critical, participatory, and feminist GIS. *GeoJournal, 72*(3–4), 173–183.

Flanagin, A. J., & Metzger, M. J. (2008). The credibility of volunteered geographic information. *GeoJournal, 72*(3–4), 137–148.

Ghose, R. (2001). Use of information technology for community empowerment: Transforming geographic information systems into community information systems. *Transactions in GIS, 5*(2), 141–163.

Girres, J.-F., & Touya, G. (2010). Quality assessment of the French OpenStreetMap dataset. *Transactions in GIS, 14*(4), 435–459.

Goodchild, M. (2007). Citizens as sensors: The world of volunteered geography. *GeoJournal, 69*, 211–221.

Grey, F. (2009). The age of citizen cyberscience, CERN Courier, April 29, 2009. http://cerncourier.com/cws/article/cern/38718. Accessed July 2011.

Haklay, M. (2010). How good is OpenStreetMap information? A comparative study of OpenStreetMap and Ordnance Survey datasets for London and the rest of England. *Environment and Planning B, 37*, 682–703.

Haklay, M., Francis, L., & Whitaker, C. (2008a). Citizens tackle noise pollution. *GIS Professional*, August issue.

Haklay, M., Singleton, A., & Parker, C. (2008b). Web mapping 2.0: The Neogeography of the Geoweb. *Geography Compass, 3*, 2011–2039.

Holling, C. S. (1998). Two cultures of ecology. *Conservation Ecology 2*(2), 4. http://www.consecol.org/vol2/iss2/art4/. Accessed January 5, 2012.

Howe, J. (2006). The rise of crowdsourcing. *Wired Magazine,* June 2006.

Irwin, A. (1995). *Citizen science*. London: Routledge.

Keen, A. (2007). *The cult of the amateur: How blogs, MySpace, YouTube, and the rest of today's user-generated media are destroying our economy, our culture, and our values* (1st ed.). New York: Doubleday.

Khatib, F., DiMaio, F., Foldit Contenders Group, Foldit Void Crushers Group, Cooper, S., Kazmierczyk, M., Gilski, M., Krzywda, S., Zabranska, H., Pichova, I., Thompson, J., Popovic, Z., Jaskolski, M., & Baker, D. (2011). Crystal structure of a monomeric retroviral protease solved by protein folding game players. *Nature Structural & Molecular Biology*. Published online September 18, 2011. doi:10.1038/nsmb.2119.

King, J. L., & Kraemer, L. K. (1993). Models, facts, and the policy process: The political ecology of estimated truth. In M. F. Goodchild, B. O. Parks, & L. T. Steyaert (Eds.), *Environmental modeling with GIS* (pp. 353–360). New York: Oxford University Press.

Latour, B. (1993). *We have never been modern*. Cambridge: Harvard University Press.

Lintott, C. J., Schawinski, K., Slosar, A., Land, K., Bamford, S., Thomas, D., Raddick, M. J., Nichol, R. C., Szalay, A., Andreescu, D., Murray, P., & van den Berg, J. (2008). Galaxy zoo: Morphologies derived from visual inspection of galaxies from the Sloan Digital Sky Survey. *Monthly Notices of the Royal Astronomical Society, 389*(3), 1179–1189.

MacKerron, G. (2011). mappiness.org.uk. In *LSE Research Day 2011: The Early Career Researcher*, May 26, 2011, London School of Economics and Political Science, London, UK (Unpublished).

Maisonneuve, N., Stevens, M., & Ochab, B. (2010). Participatory noise pollution monitoring using mobile phones. *Information Polity, 15*(1–2), 51–71.

Migurski, M. (2009). Walking papers. walking-papers.org. Accessed 5 January 2012.

Norman, D. A. (1990). *The design of everyday things*. New York: Doubleday.

Nov, O., Arazy, O., & Anderson, D. (2011). Technology-mediated citizen science participation: A motivational model. In *Proceedings of the AAAI International Conference on Weblogs and Social Media (ICWSM 2011)*, Barcelona, Spain, July 2011.

Pain, R. (2004). Social geography: Participatory research. *Progress in Human Geography, 28*(5), 652–663.

Scott, D., & Barnett, C. (2009). Something in the air: Civic science and contentious environmental politics in post-apartheid South Africa. *Geoforum, 40*(3), 373–382.

Shirky, C. (2008). *Here comes everybody: The power of organizing without organizations.* New York: Penguin Press.

Sieber, R. (2006). Public participation and geographic information systems: A literature review and framework. *Annals of the American Association of Geographers, 96*(3), 491–507.

Silvertown, J. (2009). A new dawn for citizen science. *Trends in Ecology & Evolution, 24*(9), 467–471.

Stilgoe, J. (2009). *Citizen scientists – Reconnecting science with civil society.* London: Demos.

Turner, A. J. (2006). *Introduction to neogeography.* Sebastopol: O'Reilly Media, Inc.

Westphal, A. J., von Korff, J., Anderson, D. P., Alexander, A., Betts, B., Brownlee, D. E., Butterworth, A. L., Craig, N., Gainsforth, Z., Mendez, B., See, T., Snead, C. J., Srama, R., Tsitrin, S., Warren, J., & Zolensky, M. (2006). Stardust@home: Virtual microscope validation and first results. In *37th Annual Lunar and Planetary Science Conference*, League City, Texas, March 2006, Abstract no. 2225.

Wiggins, A., & Crowston, K. (2011). From conservation to Crowdsourcing: A typology of citizen science. In *Proceedings of the Forty-fourth Hawaii International Conference on System Science (HICSS-44)*, Koloa, HI, January 2011.

Wilderman, C. C. (2007). Models of community science: Design lessons from the field. In C. McEver, R. Bonney, J. Dickinson, S. Kelling, K. Rosenberg, & J. L. Shirk (Eds.), *Citizen science toolkit conference*. Ithaca: Cornell Laboratory of Ornithology.

Wilsdon, J., Wynne, B., & Stilgoe, J. (2005). *The public value of science or how to ensure that science really matters.* London: Demos.

World Meteorological Organisation. (2001). *Volunteers for weather, climate and water.* Geneva: World Meteorological Organisation. WMO No. 919.

Zook, M., Graham, M., Shelton, T., & Gorman, S. (2010). Volunteered geographic information and crowdsourcing disaster relief: A case study of the Haitian earthquake. *World Medical & Health Policy, 2*(2). doi:10.2202/1948–4682.1069.

Part II
Geographic Knowledge Production and Place Inference

Chapter 8
Volunteered Geographic Information and Computational Geography: New Perspectives

Bin Jiang

Abstract Volunteered geographic information (VGI), one of the most important types of user-generated web content, has been emerging as a new phenomenon. VGI is contributed by numerous volunteers and supported by web 2.0 technologies. This chapter discusses how VGI provides new perspectives for computational geography, a transformed geography based on the use of data-intensive computing and simulations to uncover the underlying mechanisms behind geographic forms and processes. We provide several examples of computational geography using OpenStreetMap data and GPS traces to investigate the scaling of geographic space and its implications for human mobility patterns. We illustrate that the field of geography is experiencing a dramatic change and that geoinformatics and computational geography deserve to be clearly distinguished, with the former being a study of engineering and the latter being a science.

8.1 Introduction

The field of geographic information science (GIScience) is currently benefiting from the increasing availability of massive amounts of volunteered geographic information (VGI) (Goodchild 2007; Sui 2008) contributed by individuals in the form of user-generated content supported by Web 2.0 technologies. The emergence of VGI represents something of a paradigm shift in terms of geographic data acquisition from the conventional top-down approach, mainly dominated by national mapping agencies, to the bottom-up approach, in which data are contributed by individual volunteers through crowdsourcing—a massive collective of amateurs

B. Jiang (✉)
Division of Geomatics, University of Gävle, Gävle, Sweden
e-mail: bin.jiang@hig.se

performing functions that were previously performed by trained professionals (Howe 2009). Massive amounts of VGI of various types and the computations performed with these data constitute a significant part of eScience or data-intensive computing, which is being characterized as the fourth paradigm in scientific discovery (Hey et al. 2009). Among many others, OpenStreetMap (OSM) is one of the most successful examples of VGI.

OSM is a wiki-like collaboration, or a grassroots movement, that provides an editable map of the world using data from portable GPS devices, aerial photography, and other free sources (Bennett 2010). Currently, there are more than 400,000 registered OSM contributors or users, and this number has been growing exponentially in the past few years. OSM is not owned by anyone, which is both amazing and unprecedented. For the first time in human history, researchers can obtain street data of the entire world for analysis and computation. This analysis and computation can provide deep analytical insights into cities and our environments for sustainable development. This opportunity is significant and is very different from what is possible with Google Maps. Google Maps allows mashups, but its licensed and copyrighted data prevent us from obtaining analytical insights. We cannot learn how cities or regions have been sprawling outward by exploring only Google Maps. Instead, we need to perform analysis and computation to quantify the level of urban sprawl. In this regard, OSM, in addition to Google Maps, provides a rich data source for researchers (free of charge) to use to better understand our cities and environments through advanced spatial analysis and computing. This understanding can further be used for spatial planning, for example, redeveloping parts of a city or restricting further development of some parts of a country. In other words, OSM data can be analyzed to obtain knowledge in various forms of patterns, structures, relationships, and rules for spatial decision making. For instance, how is urban sprawl related to economic activities, population density, and public health issues such as obesity?

This chapter will discuss how geospatial analysis and computation of OSM data will lead to some hidden and surprising findings about the structures and patterns of geographic space. Our discussion is based on the assumption that OSM data, or VGI in general, are good enough to be used for computing and analysis. Although there are quality issues with VGI, this does not prevent us from uncovering hidden or surprising patterns about cities, environments, and human activities. The accumulation of evidence, as well as Linus' law—*"given enough eyeballs, all bugs are shallow"* (Raymond 2001, p. 30)—indicates that the quality of OSM data matches that of the data provided by mapping agencies, mirroring the results of studies (Giles 2005) on other user-generated content such as Wikipedia.

This chapter will briefly review computational geography and its evolution along with other related notions emerging in the field of geographic information systems (GIS). We provide a new definition of computational geography, and we differentiate it from geoinformatics. We then present some exemplars of computational geography that rely on OSM data to uncover the underlying structures and patterns of earth surface processes. This chapter concludes with a few remarks on future research.

8.2 What Is Computational Geography?

The notion of computational geography first appeared in 1994, when the Centre for Computational Geography was established as an interdisciplinary initiative at the University of Leeds. Two years later, in 1996, an international conference series on computational geography (geocomputation) was established. This conference has since been held more than ten times. As reflected in the literature, geocomputation has apparently become a favored term, even though geocomputation and computational geography refer to the same scientific undertakings and have been used interchangeably.

What is computational geography? This question has been intensively studied and hotly debated in the GIS/geocomputation community (e.g., Longley et al. 1998; Gahegan 1999; Ehlen et al. 2002). The same question has been answered and examined by various scholars on numerous occasions. To summarize, there are three basic views that were put forward by early pioneers on what computational geography is. The first view, mainly held by Stan Openshaw (2000) and Mark Gahegan (1999), recognizes the impact of increasing computing power and complex computational methods on geography or on geosciences in general. This view stresses dealing with unsolvable geographic problems using, for example, high-performance computing, artificial intelligence, data mining, and visualization. The second view is more concerned with the science of geography in a computationally intensive environment and expects geocomputation to offer a means of explaining geographic phenomena. This view is mainly held by Helen Couclelis (1998) and Bill Macmillan (1998). Paul Longley and his associates, such as Mike Goodchild (Longley et al. 2001), hold the third view that geocomputation is synonymous with GIScience, the science behind GIS technologies, which deals with fundamental questions raised by the use of geographic information and technologies.

The emergence of computational geography occurred at a time when GIS/geoinformatics as a tool underwent rapid development after a few decades of evolution and applications. Many GIS pioneers had started to think of certain fundamental issues surrounding the development of GIS. Along with computational geography, the widely recognized terms of GIScience (Goodchild 1992) and spatial information theory (Frank et al. 1992) appeared at almost the same time in the 1990s. It is no wonder that many GIS researchers see an overlap between GIS, geoinformatics, geomatics, GIScience, and spatial information theory. The coemergence of these terms is a clear indicator that this field has been under rapid development and evolution. Every term tried to capture some essential part of the development. In what follows, we offer a slightly new definition of computational geography which captures the impact of data-intensive computing or eScience, and we state how it is different from, for example, geoinformatics.

Computational geography is a transformed data-driven geography that aims to understand the underlying mechanisms of geographic forms and processes via simulations of complex geographic phenomena and based on data-intensive computing. Along with the emerging field of computational social science (Lazer et al. 2009), computational geography is both data and computationally intensive. Computational geography is a science of geography that focuses on geographic forms and processes and that offers explanations through simulations. In other words, what computational geography seeks to address is not only how the world looks but also how the world *works*. This is in contrast to the focus of GIScience on how the world *looks* rather than how the world works (Goodchild 2004). In contrast, geoinformatics usually takes an engineering or geoscience approach, with the goal of developing tools and models for geospatial data acquisition, management, analysis, and visualization to deal with real-world problems. Despite the difference, both computational geography and geoinformatics are closely related in terms of geospatial information and developed tools. This view on the difference between geoinformatics and computational geography reflects a similar view about "bioinformatics" being a field of engineering and "computational biology" being a science.

Much of the appeal of computational geography in the twenty-first century lies in the increasing availability of massive amounts of data about our environments and human activities in both physical and virtual spaces. With the increasing volume of data being generated from all types of scientific instruments, often acquired on a 24/7 basis, computational geography should adopt data-intensive geospatial computing to practice the science of geography. In this regard, the deployment of high-performance computing, grid/cloud computing, and geographically distributed sensors provides a powerful means of computing. At the same time, the emerging VGI contributed by volunteers and gathered via social media constitutes a valuable and unprecedented data source for researchers in computational geography. In the next section, we will draw upon some of our recent studies to illustrate what computational geography is and how VGI can support computational geography research.

8.3 Examples of Computational Geography

Here, we describe a few recent computational geography studies that use VGI or OSM in particular. Central to these studies are two basic concepts: topology and scaling. Topology refers to the topological relationships of numerous geographic units, while scaling is often characterized by a power-law distribution or a heavy-tailed distribution in general. We illustrate in this section that both topology and scaling help uncover the underlying structures and patterns of geographic space, but first we must further clarify these two concepts.

8.3.1 Concepts of Topology and Scaling

Topology, initially a branch of mathematics, can be defined as the study of qualitative properties that are invariant under distortion of geometric space. For this reason, topology is also called "rubber geometry." In the GIS literature, the concept of topology has appeared on at least two occasions. The most familiar is probably topologically

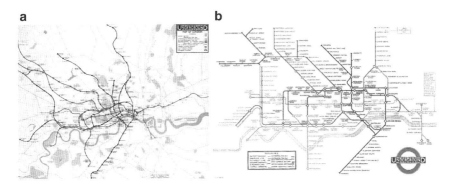

Fig. 8.1 Two versions of the London underground map: (**a**) a geometric map and (**b**) a topological map

integrated geographic encoding and referencing (TIGER). The TIGER data structure or database was created by the US Census Bureau in the 1970s. The concept of topology also appeared in the GIS literature with Max Egenhofer and Robert Franzosa's (1991) formulation of the topological relationship. Although the essence of topology is the same (which is about relationships), we adopt the notion of topology to refer to topologically based geospatial analysis. Topology, in contrast to geometric aspects such as locations, orientations, sizes, and shapes, is concerned with the relationship of geographic objects or units. To further elaborate the difference between topology and geometry, let us examine the London underground map as an illustrative example.

Figure 8.1 illustrates two versions of the London underground map: the left is geometrically corrected (a geometric map), while the right is topologically retained but geometrically distorted (a topological map). As one can see, all the locations and links of the stations are completely distorted in the topological map, with the exception of the relative orientations between the stations. This topological map is much more informative than the geometric map in terms of navigation along the tube lines. However, if we want to obtain in-depth structures or patterns, this topological map provides little information. For example, how many lines must be passed to get from one station to another? One can simply figure out the answer with the 12 tube lines. What if there are hundreds of lines? Put more generally, how many intermediate streets must be crossed in one city to go from one street to another? This is a basic question that, for example, is relevant to a taxi driver who is seeking optimal routes.

Figure 8.2 presents two versions of a topological map. Figure 8.2b illustrates a topological map showing the intersection or topology of tube lines, from which we can see certain in-depth structures. Among the 12 lines, 10 form an interconnected core in which nearly everyone is connected to everyone else, forming a sort of complete graph. Two lines, the East London line and the Waterloo & City line, are outside the core, with few connections to the others. This map indicates that if someone travels from the East London line to the Waterloo & City line, he or she must pass through another intermediate line; the two lines are not directly connected. In comparison

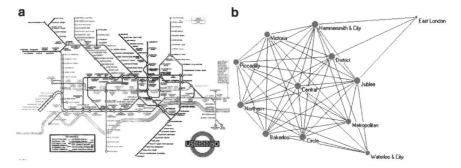

Fig. 8.2 Two versions of topological maps: (**a**) topology of stations and (**b**) topology of lines (Note: the node sizes in the right image show how many other lines intersect, i.e., the degree of connectivity)

with the geometric map, the topological map of the stations still retains certain geometric aspects, such as the relative positions and/or orientations of the lines and stations. In this regard, the topological map of the lines is purely topological: there are neither geometric locations for the nodes nor geometric distance for the links.

Scaling, or more specifically the scaling of geographic space in the context of this chapter, refers to the fact that there are far more small things than large ones in a geographic space. For example, there are far more small cities than large ones, far more short streets than long ones, and far more low buildings than high buildings. This phenomenon of "far more small events than large ones" is widespread, so it is said to be "more normal than normal." Scaling is the regularity behind many geographic phenomena. That there are far more small things than large ones also underscores a kind of spatial heterogeneity, i.e., there is no average thing in a geographic space. Because of the lack of an average thing, geographic space can also be said to be scale free. Note that scale in "scale free" means size, an average size or an arithmetic mean. "Scale free" implies that a notion of average size or mean makes little sense in characterizing a variable that exhibits a power-law distribution. The variation of things in a geographic space is highly heterogeneous or diverse. A major difference between the scaling of geographic space and spatial heterogeneity is that the former is characterized by a power-law distribution, while the latter by a normal distribution. In general, scaling must be characterized by heavy-tailed distributions such as a power-law or lognormal and exponential functions.

8.3.2 The First Example: Street Pattern of Sweden

The interconnected streets of a country constitute its basic infrastructure or backbone. Streets form a connected whole stretching across the county. Unfortunately, although the graph representation has found many applications in the computation of distance, routing, and tracing, the underlying structure and pattern of streets cannot be simply illustrated with the conventional street networks using junctions and street

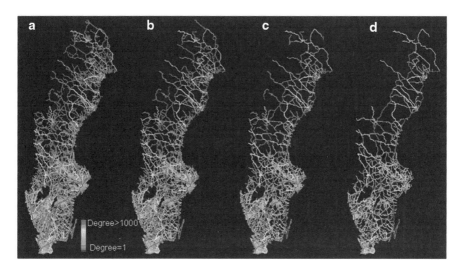

Fig. 8.3 Four levels of detail showing the hierarchical structure of the 160,000 streets of Sweden: (**a**) source map, (**b**) first level, (**c**) second level, and (**d**) third level

segments, respectively, as the nodes and links of a graph. We call the conventional street network a geometric network in the sense that (1) every junction has a unique geographic location and (2) every street segment is assigned a geometric distance. The geometric network embeds the connectivity of street junctions or that of street segments. Structurally, the conventional street network illustrates a monotonous pattern because every junction is connected by almost the same number of other junctions, or equivalently, every segment is connected by almost the same number of other segments. However, the topology of streets exhibits a very interesting pattern, which can be said to be universal for all types of street networks all over the world.

We retrieved the entire Swedish street network database from the OSM databases and generated individual streets to assess how they are connected to each other. Note that the streets can be put into two categories: streets identified by unique names (Jiang and Claramunt 2004) and natural streets generated by joining principles (Jiang et al. 2008). In this study, we first merged adjacent street segments with the same names to create street units and then adopted some principle to join the street units into natural streets. This procedure was performed because of the missing names for many street segments in the OSM databases. The resulting natural streets are very close to named streets. Eventually, we obtained over 160,000 streets from over 600,000 arcs. Figure 8.3 illustrates the hierarchical levels of the street network and indicates that there are far more short streets (blue) than large ones (red). The least connected streets have a degree of 1, while the most connected streets have a degree of 1,040. This very high ratio of the most connected degree to the least connected degree is a clear indicator of a heavy-tailed distribution.

This finding of the scaling pattern of the street network has profound implications for understanding other phenomena such as traffic flow. For example, the majority of traffic flow occurs on only in a few of the most connected streets, while a vast majority of less connected streets accommodate only a small amount of the traffic flow (Jiang 2009). Eventually, traffic flow and human mobility patterns also demonstrate this scaling pattern. We can further claim that it is the scaling of geographic space that shapes human movement patterns. This is the type of mechanism that we seek to discover through computational geography.

In addressing why human activities show the scaling pattern, Barabási (2010) tried to seek an answer from the perspective of people rather than that of space. He explains that we conduct our affairs in bursts because we set a priority for them. In terms of human movement, we spend most of our time (e.g., 90% of our time) in one place near our home, city, or nearby, and only occasionally (e.g., 10% of our time or less) do we travel somewhere far from where we usually are. This is a traditional way of thinking—society is complicated because every individual person is complicated; in fact, we can think of individuals as molecules or atoms (Buchanan 2007). In a recent study, Jiang and Jia (2011a) created two types of moving agents (random and purposive) and simulated their movement patterns in a street network. It was found that moving behaviors have little effect on the overall traffic patterns.

Given the scaling pattern or property, map generalization or mapping in general can be conducted in a simple manner. The head/tail division rule that we formulated can be applied in this case. The head/tail division rule states that anything with the scaling pattern can be divided into two imbalanced parts: a low percentage of larger items in the head and a high percentage of smaller items in the tail (Jiang and Liu 2012). In fact, Fig. 8.3 illustrates an application of the head/tail division rule by simply placing larger streets in the head recursively to create different levels of detail (Jiang 2012). We further conjecture that the scaling of geographic space is some fundamental underlying mechanism of map generalization (Jiang et al. 2011), which is the underlying property that makes generalization and mapping possible.

8.3.3 The Second Example: Street Block Pattern of France

The second exemplar concerns the scaling pattern that emerges from the numerous street blocks of a country. A street block refers to a minimum ring or cycle formed by adjacent street segments, also called a "city block" in an urban environment. By street blocks, we mean both city blocks in cities and field blocks in the countryside. We developed a recursive algorithm to automatically derive a massive number of street blocks from street networks of the three largest European countries (Jiang and Liu 2012). In this chapter, we use the French case as an example to illustrate how the interconnected blocks help uncover underlying patterns. First, we found that the sizes of the blocks show a lognormal distribution, one of several heavy-tailed distributions. This observed distribution implies that there are far more small blocks than large ones. Interestingly, using the head/tail division rule, we can partition all the blocks into categories: those smaller than the mean and those greater than the mean. In fact, the smaller blocks can be clustered into cities, while the larger ones constitute the countryside.

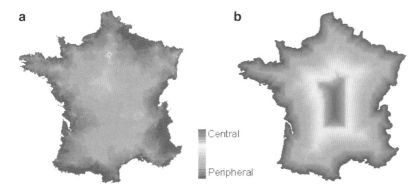

Fig. 8.4 Centers of France: (**a**) topological center and (**b**) geometric center

Second, we defined the notion of border numbers, indicating how far individual blocks are from the outermost border. Inspired by the notion of Bacon numbers, which show how far an actor or actress is from Kevin Bacon in the Hollywood universe ("six degrees of Kevin Bacon"), the border numbers are defined as follows: the blocks on the outermost border have border number 1, those blocks directly connected to blocks with border number 1 have border number 2, and so on. The border number is defined from a topological perspective, which is clearly different from a geometric perspective. Figure 8.4 illustrates the difference. Both geometric and topological distances are colored using a spectral color legend: the farther a block is, the more central it is. There are two centers: a topological center and a geometric center. Clearly, the topological center is the location of Paris. The geometric center is, in fact, a direct application of the medial axis (Blum 1967). The geometric center is not what human beings perceive to be the center of the country but the topological center is.

We can extend this reasoning to define a center in biological organisms. For example, what is the center of the human body? Relying on Blum's medial axis, we would derive the skeleton, but common sense tells us that both the heart and the mind are the two centers of the human body. We conjecture that if we were to take the topological perspective, we would be able to derive these two centers. This is based on the assumption that the sizes of the cells or any subunits, similar to the blocks, are heavy-tail distributed. We have not found any scientific literature to support the above reasoning, but we need new geographic imaginations in the computer age (Sui 2004); data-intensive computing, involving a massive amount of geographic information, facilitates creative imagination in some unique ways.

8.3.4 The Third Example: Verifying Zipf's Law via Natural Cities

The scaling property has several variants, one of which is Zipf's law, formulated by the linguist George Kingsley Zipf (1949). Zipf's law states that the size of any city is inversely proportional to its rank, for example, the second largest city is 1/2 of the

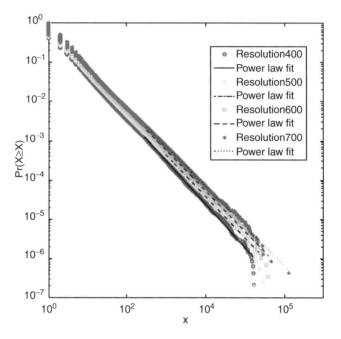

Fig. 8.5 Power-law distributions of natural cities derived with different resolutions

largest one; the third largest city is 1/3 of the largest, and so on. Usually, city sizes are measured by their population or physical extent. Conventionally, cities are defined legally or administratively, for example, as census-designated places, urban areas, or metropolitan areas. The subjective and even arbitrary nature of these definitions poses problems for the verification of Zipf's law. In this regard, there have already been studies seeking more objective definitions of cities or city boundaries (e.g., Holmes and Lee 2009; Rozenfeld et al. 2011). However, such studies still used aggregated data rather than individual-based data for defining cities.

We propose a new approach to defining cities by adopting street nodes as a proxy of population (Jiang and Jia 2011b). We retrieved over 120 GB of OSM data for the USA, and we extracted 25 million street nodes. By applying a clustering algorithm, we grouped the nearest street nodes into individual urban settlements. We define the derived urban settlements as "natural cities" because the clustering was performed recursively and automatically. Eventually, we obtained approximately two to four million natural cities, depending on the chosen resolutions for the clustering process. The resolutions we chose were 400, 500, 600, and 700 m because they were the same magnitude as the city block sizes. Interestingly, the derived natural cities strikingly exhibit a power-law distribution, but the Zipf exponent may deviate from 1.0.

We conducted a comparison study by examining Zipf's law between natural cities and urban areas and found that the law is remarkably stable for all natural cities, ranging from the largest to the smallest (with only one road node) (Jiang and Jia 2011b). Surprisingly, the Zipf exponent remains unchanged for the entire range; one can refer to Fig. 8.5 for the log-log plot of the distribution. This result contrasts sharply with the results for urban areas, which exhibit Zipf's law for some of the largest cities; the Zipf exponent varies from one part to another of the entire range. This behavior may indicate that Zipf's law, or the power law in general, underlies certain self-organized processes.

8.4 Discussions

We have introduced two key concepts around which three studies regarding computational geography have been presented to uncover the underlying scaling property and in particular to illustrate the underlying mechanism of human activity patterns in geographic space. Let us further elaborate the implications of these studies. Current geospatial analysis is very much dominated by two stubborn mindsets: one is geometric thinking in terms of sizes, shapes, orientations, and positions and the other is a Gaussian way of thinking that uses geostatistics to characterize spatial properties such as spatial dependence and spatial heterogeneity. Spatial dependence also goes by the name spatial autocorrelation, which is expressed succinctly by Tobler's (1970) first law of geography—*"everything is related to everything else, but near things are more related than distant things"* (p. 236). Spatial heterogeneity emphasizes that the Earth's surface is heterogeneous, but this heterogeneity is still very much characterized by a normal distribution. Statistical theories based on scaling and heavy-tailed distributions are rarely adopted for the study of geographic phenomena. We, therefore, want to promote two alternative ways of thinking, topological thinking and scaling thinking, to get insights into geographic forms and processes.

Topological thinking is rooted in one of the two fundamental views about space—Leibniz's relative space, which focuses on relationships between individual objects. This is to be understood in contrast to geometric or topographic thinking, which is dominated by Newtonian ideas about absolute space. The topological focus on relationships is not particularly new, since it was well treated in geographic literature a long time ago (e.g., Haggett and Chorley 1969). However, what is considered to be unique or interesting in this chapter is how topological thinking helps us appreciate the scaling pattern. Thinking topologically is part of thinking spatially, and it is becoming increasingly important at a time when most of human activities have shifted to virtual space and when emerging social media is coming to dominate everyday human activities (Allen 2011).

It should be noted that there are at least two factors that prevent us from gaining insight into the scaling of geographic space: (1) how we look at geographic space (perspective) and (2) the size of the study areas we choose (scope). For example, the geometric representation of street networks is unable to demonstrate the scaling property. More importantly, the geometric representation is not what human beings perceive about street networks. The second factor is closely related to the availability of massive amounts of crowdsourced geographic information. In this regard, VGI provides an unprecedented data source enabling us to conduct this type of geospatial analysis and modeling.

One may argue that geographic information maintained by governments or private companies would allow us to achieve the same goal. This would indeed be true if the data were accessible, but this is easily said and only rarely done. First of all, there are many restrictions on access to data. Secondly, the data studied are rarely shared among all interested parties. What if some other researchers want to verify the results with the same data? Often there is no way to do so. This is a big constraint for scientific research. With OSM, for the first time, we can seamlessly integrate publicly available geographic information and continuously keep it updated through volunteered efforts.

There are many other formats of VGI, emerging from a variety of social media such as Facebook, Twitter, and Flickr, which are not addressed in this chapter. These data may go beyond what Goodchild (2007) refer to about VGI, but their potential for studying spaces and places could be enormous. The convergence of GIS and social media is providing new ways of studying interactions among people, space, and place, which are fundamental to geography (Sui and Goodchild 2011). We believe there are needs for much further studies to be conducted along these directions in the future.

8.5 Concluding Remarks

In this chapter, we briefly reviewed the emergence of computational geography, its definitions, and its evolution in the past two decades. We provided an alternative definition of computational geography that deals with simulations of geographic phenomena to uncover the underlying mechanisms of geographic forms and processes. Computational geography is distinct from geoinformatics, which is more concerned with the engineering side of geographic information in terms of data acquisition, management, analysis, and visualization. Governments have put enormous effort into data collection (e.g., population censuses, housing, and economic activity), but these data are seldom made available for research. In this regard, VGI, and OSM in particular, provides a valuable data source for computational geography.

We provided several examples of research in computational geography that relies on VGI and that rests on conceptualizations of relative and relational space. Despite of being an important way to conceptualize space, geographic studies still tend to examine (absolute) space in the geometric sense rather than the topological sense we refer to in this chapter. It is time to rethink the relative view of space in geographic studies, particularly in computational geography. Gatrell's (1984) ideas remain valid and deserve greater attention in geographic research, especially alternative ways of conceptualizing and defining space. We have seen in this chapter that topology and scaling indeed matter in geospatial analysis. We need to shift our mindsets from geometric to topological thinking and from the Gaussian mindset to something that is more "normal" than normal—the scaling property.

Acknowledgement The author would like to thank the book editors for their constructive comments that led to many improvements, in particular Daniel Sui, for bringing my attention to recent references on topological thinking in human geography.

References

Allen, J. (2011). Topological twists: Power's shifting geographies. *Dialogues in Human Geography, 1*(3), 283–298.
Barabási, A. (2010). *Bursts: The hidden pattern behind everything we do*. Boston, MA: Dutton Adult.
Bennett, J. (2010). *OpenStreetMap: Be your own cartographer*. Birmingham: PCKT Publishing.
Blum, H. (1967). A transformation for extracting new descriptors of form. In W. Whaten-Dunn (Ed.), *Models for the perception of speech and visual form* (pp. 362–380). Cambridge, MA: MIT Press.
Buchanan, M. (2007). *The social atom: Why the rich get richer, cheaters get caught, and your neighbor usually looks like you*. New York: Bloomsbury.
Couclelis, H. (1998). Geocomputation in context. In P. A. Longley, S. M. Brooks, R. McDonnell, & B. Macmillan (Eds.), *Geocomputation: A primer*. Chichester: Wiley.
Egenhofer, M., & Franzosa, R. (1991). Point-set topological spatial relations. *International Journal of Geographical Information Systems, 5*(2), 161–174.
Ehlen, J., Caldwell, D. R., & Harding, S. (2002). GeoComputation: What is it? *Computers, Environment and Urban Systems, 26*, 257–265.
Frank, A. U., Campari, I., & Formentini, U. (1992). *Theories and methods of spatio-temporal reasoning in geographic space* (Lecture Notes in Computer Science, Vol. 639). Berlin: Springer.
Gahegan, M. (1999). What is geocomputation? *Transactions in GIS, 3*(3), 203–206.
Gatrell, A. C. (1984). *Distance and space: A geographical perspective*. Oxford: Oxford University Press.
Giles, J. (2005). Internet encyclopedias to head to head. *Nature, 438*, 900–901.
Goodchild, M. F. (1992). Geographical information science. *International Journal of Geographical Information Systems, 6*(1), 31–45.
Goodchild, M. F. (2004). The validity and usefulness of laws in geographic information science and geography. *Annals of the Association of American Geographers, 94*(2), 300–303.
Goodchild, M. F. (2007). Citizens as sensors: The world of volunteered geography. *GeoJournal, 69*(4), 211–221.
Haggett, P., & Chorley, R. J. (1969). *Network analysis in geography*. London: Edward Arnold.
Hey, T., Tansley, S., & Tolle, K. (2009). *The fourth paradigm: Data intensive scientific discovery*. Redmond: Microsoft Research.
Holmes, T. J., & Lee, S. (2009). Cities as six-by-six-mile squares: Zipf's law? In E. L. Glaeser (Ed.), *The economics of agglomerations*. Chicago: University of Chicago Press.
Howe, J. (2009). *Crowdsourcing: Why the power of the crowd is driving the future of business*. New York: Three Rivers Press.
Jiang, B. (2009). Street hierarchies: A minority of streets account for a majority of traffic flow. *International Journal of Geographical Information Science, 23*(8), 1033–1048.
Jiang, B. (2012). Head/tail breaks: A new classification scheme for data with a heavy-tailed distribution. *The Professional Geographer* xx, xx–xx.
Jiang, B., & Claramunt, C. (2004). Topological analysis of urban street networks. *Environment and Planning B: Planning and Design, 31*, 151–162.
Jiang, B., & Jia, T. (2011a). Agent-based simulation of human movement shaped by the underlying street structure. *International Journal of Geographical Information Science, 25*(1), 51–64.
Jiang, B., & Jia, T. (2011b). Zipf's law for all the natural cities in the United States: A geospatial perspective. *International Journal of Geographical Information Science, 25*(8), 1269–1281.
Jiang, B., & Liu, X. (2012). Scaling of geographic space from the perspective of city and field blocks and using volunteered geographic information. *International Journal of Geographical Information Science, 26*(2), 215–229. Preprint, arxiv.org/abs/1009.3635.
Jiang, B., Zhao, S., & Yin, J. (2008). Self-organized natural roads for predicting traffic flow: A sensitivity study. *Journal of Statistical Mechanics: Theory and Experiment*. July, P07008, Preprint, arxiv.org/abs/0804.1630.

Jiang, B., Liu, X., & Jia, T. (2011). Scaling of geographic space as a universal rule for map generalization. Preprint: http://arxiv.org/abs/1102.1561.

Lazer, D., Pentland, A., Adamic, L., Aral, S., Barabási, A.-L., Brewer, D., Christakis, N., Contractor, N., Fowler, J., Gutmann, M., Jebara, T., King, G., Macy, M., Roy, D., & Van Alstyne, M. (2009). Computation social science. *Science, 323*, 721–724.

Longley, P. A., Brooks, S. M., McDonnell, R., & Macmillan, B. (1998). *Geocomputation: A primer*. Chichester: Wiley.

Longley, P. A., Goodchild, M. F., Maguire, D. J., & Rhind, D. W. (2001). *Geographic information systems and science*. Chichester: Wiley.

Macmillan, B. (1998). Epilogue. In P. A. Longley, S. M. Brooks, R. McDonnell, & B. Macmillan (Eds.), *Geocomputation: A primer*. Chichester: Wiley.

Openshaw, S., & Abrahart, R. J. (2000). *GeoComputation*. London: CRC Press.

Raymond, E. S. (2001). *The Cathedral & The Bazaar: Musings on Linux and open source by an accidental revolutionary*. Sebastopol: O'Reilly Media.

Rozenfeld, H. D., Rybski, D., Gabaix, X., & Makse, H. A. (2011). The area and population of cities: New insights from a different perspective on cities. *American Economic Review, 101*(5), 2205–2225.

Sui, D. Z. (2004). GIS, cartography, and the "third culture": Geographic imaginations in the computer age. *The Professional Geographer, 56*(1), 62–72.

Sui, D. Z. (2008). The wikification of GIS and its consequences: Or Angelina Jolie's new tattoo and the future of GIS. *Computers, Environment and Urban Systems, 32*(1), 1–5.

Sui, D. Z., & Goodchild, M. (2011). The convergence of GIS and social media: Challenges for GIScience. *International Journal of Geographical Information Science, 25*(11), 1737–1748.

Tobler, W. (1970). A computer movie simulating urban growth in the Detroit region. *Economic Geography, 46*(2), 234–240.

Zipf, G. K. (1949). *Human behavior and the principles of least effort*. Cambridge, MA: Addison Wesley.

Chapter 9
The Evolution of Geo-Crowdsourcing: Bringing Volunteered Geographic Information to the Third Dimension

Marcus Goetz and Alexander Zipf

Abstract Volunteered geographic information (VGI) describes the collaborative and voluntary collection of any kind of spatial data, and has evolved to become an important source for geo-information. Users participate in VGI communities and share their data with other community members at no charge. The data is based on personal measurements or personal knowledge, as well as on available aerial imagery provided by Bing Maps etc. In the early beginnings, VGI comprised only two-dimensional (2D) data, but now more and more users also contribute 3D-compliant data such as height information. By utilizing such 3D information or 3D-VGI, it is possible to create virtual but increasingly realistic 3D map features and models that can be compared to products such as Google Earth. In this chapter, the evolution of VGI from 2D to 3D is discussed. In particular, the creation of a 3D virtual globe including visualization of 3D building models as well as traffic infrastructure, landuse areas, and points of interest (POIs) is reviewed. Additional data sources and the semantic enrichment of virtual models are also discussed. Crowdsourced geodata can serve as a real alternative data source and VGI can be utilized for generating rich 3D city models.

9.1 Introduction

The terms geo-crowdsourcing, user-generated geographic content, and volunteered geographic information (VGI) describe a quite new phenomenon in geoinformatics, whereby an ever-expanding group of users collaboratively and voluntarily collects different types of spatial data (Goodchild 2007a). That is, both laypeople and

---

M. Goetz (✉) • A. Zipf
GIScience research group, University of Heidelberg, Heidelberg, Germany
e-mail: m.goetz@uni-heidelberg.de; zipf@uni-heidelberg.de

professionals create geographic data based on (personal) measurements (e.g., via GPS devices or personal knowledge) and provide this data in a Web 2.0 community platform to other users of the community. In doing so, the VGI communities create a comprehensive data source of many different types, with the members of these communities acting as remote sensors (Goodchild 2007b). Especially in urban regions, VGI data is often available at a very detailed scale, which is the reason for its increasing use in urban data management (Song and Sun 2010). One of the most popular examples of such a VGI community is the OpenStreetMap (OSM) project, which will be described in more detail later.

In the early beginnings of VGI, the available data mainly comprised two-dimensional (2D) data, but since 2008, people more and more often started to collect 3D data such as height information, roof geometry information, etc., transforming VGI from 2D maps and imagery to a 3D data source. Adding 3D information to VGI projects is an important step, not only due to the fact that we are living in a 3D world, but also because 3D information allows the development and provision of many different applications. For example, by providing 3D models for a city district, it is possible to demonstrate future city development plans to the broad public in a public-participation process, consistent with Sarjakoski's suggestion that, "three-dimensional modeling and photorealistic visualization and animation should be included in public participation GIS for the sake of space-to-feel level experience in urban plans" (Sarjakoski 1998). Not only public-participation initiatives benefit from 3D models, but also (scientific) analysis often achieves more exact results when using 3D data. For example, in the area of visibility analysis in urban areas, it has been demonstrated that "3D visibility indices are more effective than 2D indices" (Yang et al. 2007). Furthermore, 3D information about urban areas supports decisions in emergencies (Lee 2007; Kolbe et al. 2008; Lee and Zlatanova 2008; Schilling and Goetz 2010) and can also be utilized for visualizing topological relations (Lee 2001). To the authors of this chapter, it is therefore evident that 3D information about both urban and rural areas is useful for diverse applications and it is crucial to have access to different sources of 3D data. By demonstrating the richness and diversity of VGI data (and especially OSM data), it will be proven that VGI is a real alternative data source for 3D information. One step towards this, and also for encouraging the members of VGI communities, is the project OSM-3D[1]: a virtual globe for visualizing OSM data as a 3D model. The fundamental ideas and basics behind this project have already been described and demonstrated (Over et al. 2010). However, the project has been recently extended to the whole European region, and additionally, refinements and improvements (especially for building construction) have been made.

The remainder of this chapter is organized as follows: First, there is an introduction to the OSM project. Afterwards, there is an extensive overview about the OSM-3D project with a special focus on a quantitative analysis of 3D-compliant OSM

---

[1] www.osm-3d.org

attributes, as well as the generation of building models. Additionally, this section contains an outlook for the enrichment of OSM data by using additional data sources. Thereafter, there is a discussion about possible semantic enrichment of 3D city models by using VGI from OSM. The last section summarizes the presented work and discusses future research.

9.2 OpenStreetMap: One of the Most Popular Examples for VGI

In the last few years, diverse initiatives (with different user groups, aims, etc.) for the collection of VGI, such as geotagged Flickr images,[2] Wikimapia,[3] Foursquare,[4] Gowalla,[5] etc., have emerged. One of the most popular examples is indisputably the OpenStreetMap project. OSM was initiated in 2004 and rapidly developed into a fast growing Web community with currently more than 400,000 registered users, thus more than 400,000 potential contributors.

Users of OSM are able to contribute data to the community by adding several georeferenced points (i.e., nodes) to the database. These points are created by using personal measurements with a GPS device (e.g., a GPS-enabled cell phone) or by applying personal knowledge about the surrounding areas. Additionally, different providers of fine-resolution aerial images such as Bing Maps have granted permission to use their images for mapping activities (e.g., mapping street segments or building shapes). This decision has also increased the amount of data inside OSM because now it is possible to map data around the world without the need of being there physically. In addition to georeferenced nodes, users within OSM can also combine them into so-called *ways* (i.e., a connection of nodes), allowing the creation of linestring geometries (e.g., for street segments). These ways do not necessarily have to be closed (i.e., the starting point equals the ending point), but if so, they can be further utilized for mapping polygons (e.g., areas) with arbitrary shape. For mapping complex polygons with outer shells and inner holes, it is also possible to create so-called *relations* inside OSM. Additionally, these relations can be used for describing complex relationships between different OSM nodes or ways. The latest OSM dataset (November 2011) contains more than 1.25 billion georeferenced nodes, 114 million ways, and 1,100,000 relations.

But OpenStreetMap does not only contain pure geometric information. Additionally, OSM adapts a concept of open and unlimited key-value pairs for adding different (semantic) information and attributes on top of the geometry. That is, users can map their geometries by using nodes, ways, or relations and enrich those with distinct information by attaching key-value pairs. The key describes a distinct information

---

[2] flickr.com
[3] wikimapia.org
[4] http://foursquare.com
[5] http://gowalla.com

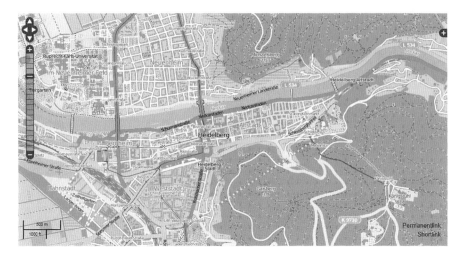

Fig. 9.1 The OSM map perspective on the city of Heidelberg with different types of map features (streets, buildings, etc.) (OSM 2011a)

domain or condition, and the corresponding value describes the information itself. So, for example, a way with the key *highway* describes a street for vehicles and humans. Additionally, the value *motorway* refines this street as a motorway not utilizable by pedestrians. Since the number of key-value pairs is not limited (i.e., a user can add an arbitrary number of key-value pairs), it is possible to further refine the information about the motorway, for example, by adding the key *maxspeed* with the value *130*. Also, users are enabled to map not only streets but also natural areas such as forests or seas, different points of interests (POIs) such as ATMs or letter boxes, building ground shapes, etc. The OSM key-value methodology itself is very open; thus, it is very easy for a user to add any kind of additional information. There are of course some best practices and recommendations for mapping different map features such as the keys *amenity*, *building*, *natural*, *place*, *waterway*, etc., as described on the OSM wiki page (OSM 2011b). A complete list of all keys is also available at Tagwatch (2011).

The data of OSM itself can be either downloaded from different OSM data provider platforms (e.g., Geofabrik[6]) or by using the Web API which is part of the visual Web interface of OSM.[7] By using this Web interface, users (also those who are not OSM members) can access the 2D online map and browse through the world. Figure 9.1 depicts an exemplary screenshot of the OSM map. Furthermore, this Web API also allows the download of OSM data.

---

[6] http://download.geofabrik.de/

[7] http://www.openstreetmap.org

9.3 Three-Dimensional OpenStreetMap

Most applications and research investigations that use OpenStreetMap as a data source are 2D and only visualize the data in 2D from a bird's eye view. That is, according to Over et al. (2010), there is little research on the 3D visualization and usage of OSM data available, and there are basically just two different applications which provide a real 3D perspective with additional digital terrain model (DTM) and 3D buildings: the so-called KOSMOS Worldflier (Brejc 2011) and the OSM-3D project (OSM-3D 2011). A couple of other applications for providing perspective scenes (with flat terrain) and some extruded buildings do also exist, for example, Ziegler (2011), but these are very limited regarding scene display size, application functionality, and 3D visualizations. In contrast, the OSM-3D project is based on public Open Geospatial Consortium (OGC) standards and drafts, such as a Web Map Service (WMS) or Web 3D Service (W3DS) and dedicated client software, and provides a detailed virtual globe with atmospheric visualization effects. A comprehensive number of data types have been selected and used for displaying different 3D map features such as a DTM, POIs, buildings, streets, labels, natural areas, etc.

To the authors' knowledge, OSM-3D is the only project that uses such an extensive amount of OSM data for the generation of 3D models. Furthermore, it is the first approach to generate realistic 3D models (especially building models) by purely using VGI from OSM, thus demonstrating the richness and power of VGI. Figure 9.2a depicts an overview of the 3D globe in XNavigator. Additionally, Fig. 9.2b shows a more detailed perspective of the Riva Del Garda in Italy, where different natural areas as well as streets, waterways, and 3D buildings are visualized. The next sections will describe OSM-3D in more detail, especially focusing on the generation of 3D building models and 3D map features. Before that, we include a discussion about relevant and 3D-compliant OSM keys and values, as well as a brief introduction to the system architecture of OSM-3D.

9.3.1 Analyses of 3D-Compliant OpenStreetMap Attributes

Basically, all geographic data in OSM is 2D. The users within the OSM community measure 2D GPS points or draw geometries based on 2D aerial imagery. Thus, the visible geometry inside OSM, which is in most cases the visible map that can be consumed via the Web interface of OSM, is 2D. Therefore, at first glance, OpenStreetMap does not seem to provide any kind of 3D data. However, when taking a closer and more detailed look at the data structure as well as the best practices and popular key-value pairs within the project, it becomes evident that there is nevertheless plenty of 3D information inside OSM.

The OSM-key *height*, as one could imagine due to the semantic meaning of the key, obviously contains 3D data. It describes the (vertical) height of a map feature, wherein the default length measure (if not explicitly provided by the user) is meters. Table 9.1 contains quantitative and qualitative information about the application of

Fig. 9.2 A virtual globe with atmospheric effects (**a**) and a detailed view of the Riva Del Garda with 3D buildings and map features (**b**). Both are visualized in the client XNavigator of the project OSM-3D.org

the key *height* to the three different OSM data types. In contrast, Table 9.2 contains information about the combinatorial usage of the key *height* with other OSM keys (only those keys which are supposed to describe 3D objects were utilized as combinatorial keys). Some other keys with very little usage are also not considered in the table. The key *building* is used for mapping a building, *man_made* describes everything that is created by people, *tower:type* describes some kind of tower, *landuse* is

Table 9.1 Quantitative and qualitative information about the OSM-key *height* (November 2011)

Classification figure	Value
Absolute number of map features with the key *height*	676,339
Absolute number of nodes with the key *height*	38,945
Absolute number of ways with the key *height*	629,611
Absolute number of relations with the key *height*	7,783
Relative number of map features with the key *height*	0.0492%
Relative number of nodes with the key *height*	0.0031%
Relative number of ways with the key *height*	0.5513%
Relative number of relations with the key *height*	0.6649%

Table 9.2 Investigations of the combinatorial usage of the OSM-key *height* with other keys (November 2011)

Combinatorial key	Absolute count	Relative count compared to *height*	Relative count compared to combinatorial key
building	624,122	92.2795%	1.3931%
man_made	41,122	6.0795%	5.5403%
tower:type	21,840	3.2291%	59.4539%
landuse	18,497	2.7348%	0.3491%
technology	18,150	2.6836%	99.9233%
amenity	8,994	1.3298%	0.2247%
natural	5,871	0.8681%	0.0923%
shelter	4,899	0.7243%	3.2327%
barrier	3,116	0.4607%	0.2734%
building:part	1,292	0.1910%	27.7479%
bridge	164	0.0242%	0.0142%

utilized for describing any kind of usage type (normally areas, but it is also often used for the usage type of buildings), *technology* is used for describing the technology of towers, *amenity* describes any kind of facility, *natural* is used for describing different natural areas, the key *shelter* is used for describing a shelter, the key *barrier* can be utilized for mapping obstacles such as walls or fences, the key *building:part* describes a part of a building, and the key *bridge* can be used for mapping bridges (for streets, railways, etc.). All statistics are based on analyses of the latest OSM dataset.[8] The list is ordered in descending order.

As described in Table 9.2, most map features for which height information is available are buildings. Also, it is interesting that nearly every element with the key *technology* is also enriched with height information. Since it is likely that some of the keys mentioned in Table 9.2 are also combinatorial keys of the key *building*, Table 9.3 shows adjusted values, that is, the analysis refers to map features which are not buildings.

---

[8] Data from 12 November 2011.

Table 9.3 Investigations of map features which are not building but have height information (November 2011)

Combinatorial key	Absolute count	Relative count compared to *height*
man_made	23,984	3.5462%
tower:type	21,695	3.2077%
landuse	127	0.0188%
technology	18,149	2.6834%
amenity	7,600	1.1237%
natural	5,866	0.8673%
shelter	4,895	0.7237%
barrier	3,109	0.4597%
bridge	157	0.0232%

Table 9.4 Comparing the key *height* with *building:height* (November 2011)

Classification figure	Absolute count	Relative count
Buildings with the key *height* (nonexclusive)	624,122	1.3931%
Buildings with the key *building:height* (nonexclusive)	28,323	0.0632%
Buildings with the key *height* (exclusive)	623,857	1.3925%
Buildings with the key *building:height* (exclusive)	28,058	0.0626%
Buildings with both keys	265	0.0006%
Buildings with both keys and equal values	249	0.0005%

It becomes obvious that nearly all map features with height information are buildings or at least closely related to buildings (e.g., building parts or roofs). In some ways, this is not very surprising because (besides a DTM) buildings are indeed 3D, whereas, for example, streets on the ground could also be considered (nearly) 2D. In contrast, it is questionable how height information makes sense for a 2D natural area, for example, although there are more than 5,000 natural areas with height information.

Due to the open key-value methodology in OSM, a different key for height information of buildings, namely, *building:height*, was used in the early mapping activities. Although this key has been declared as obsolete and replaced by the general key *height*, there are still map features with this key available and even some users who still utilize *building:height* instead of *height* for their new building mappings. Table 9.4 contains information about the usage of those keys, as well as a comparison between the provided values. As one can see, there are also a couple of buildings with both keys available but with different values. In most cases, the values only differ by 1–2 m, but there are alsov few cases with differences of 5 or more meters. The biggest difference is 21 m.

9 The Evolution of Geo-Crowdsourcing: Bringing Volunteered Geographic... 147

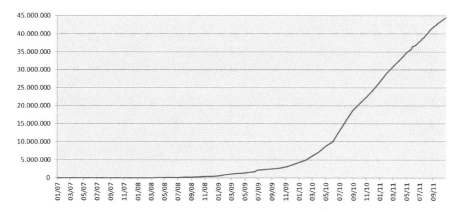

Fig. 9.3 Development of the amount of buildings in OpenStreetMap

Besides the keys *height* and *building:height*, there are also some other keys with height information such as *seamark:light:height* (total usage: 23,009) for describing the height of a sea light, *light:height* (total usage: 1,348) for describing the height of a light (e.g., a street light), or *min_height* (total usage: 1,321) for describing the elevation of a map feature (i.e., the space between the ground and the feature, mostly used in combination with *building*). However, currently they are rarely used and will thus not be investigated in detail within this chapter. Nevertheless, they still might contain 3D-related data and thus might be relevant when creating 3D models from OpenStreetMap.

As discussed above, the key *height* is in most cases applied to a building feature. In the early days of OSM, users concentrated on streets and land-use areas and mapped hardly any buildings. But over the course of time and with an increasing interest in OSM, users started to map more and more buildings. The release of aerial imagery from Bing Maps for OSM has further increased this trend. Figure 9.3 depicts the development of the total number of buildings over the course of time between early 2007 (zero buildings) and end of 2011 (about 45 million buildings). It is particularly impressive that there are nearly as many buildings inside OSM as the streets (currently about 45.6 million) that used to be the biggest fraction of OSM map features.

9.3.2 System Architecture

The main component of the system architecture is the Web 3D Service (W3DS), a Web service for deriving 3D scene graphs in a common 3D file format such as VRML or X3D. It is currently considered as a draft specification (OGC 2005), but it is likely that W3DS will become an Open Geospatial Consortium standard in the near future. Within OSM-3D, a prototypical implementation of W3DS has been developed, allowing the provision of 3D VRML models based on OSM. The models are available as VRML because of strong and wide support in common browsers, as well as good compression rates for the data. To visualize the data, a client software the XNavigator has been developed. This client software allows very intuitive and user-friendly consumption of the data derived from the W3DS because the client automatically sends the location-dependent user requests to the W3DS, so the user does not have to do this on his or her own. Additionally, the client allows the selection of individual layers with different types of data, such as buildings, streets, and natural areas. Furthermore, additional functionalities such as routing, address search, POI search, or GPS track visualization add value to the application. Since all components are based on OGC standards, the application platform is very flexible and scalable. Also, all layers can be styled according to user requirements by using 3D-styled layer descriptors (3D-SLD) (Over et al. 2010). This very short introduction to OSM-3D and its architecture ought to be enough in the context of this chapter. For more detailed information, please refer to common publications such as Zipf et al. (2007), Schilling et al. (2009), and Over et al. (2010). Furthermore, some details about the performance as well as the data processing are described in the following sections, especially Sect. 9.3.5.

9.3.3 Generation of Building Features

Besides the DTM (which is not discussed in detail within this chapter), the main 3D parts of OSM-3D are the 3D building models. As stated above, the whole OSM dataset currently contains nearly 45 million buildings, that is, 45 million footprints which are possibly enriched with geometric or semantic information (see below). Simple footprint geometries, that is, geometries with no holes inside, are mapped as a single closed way within OSM, whereby the way must contain at least 4 nodes with the first node being equal to the last node. For mapping complex polygons (e.g., buildings with holes inside the footprint), users need to utilize OSM relations. These consist of one or more outer members (describing the outer shape of the polygon) and an arbitrary number of inner members (describing holes in the shell geometry).

Obviously, for the generation of 3D building models, the 2D footprint is not enough, thus additional information is required. As stated above, due to the open OSM key-value pair methodology, it is no problem to add further (3D) information to building footprints, and every OSM user can do so. That is, 3D information is not explicitly mapped as a geometry by the community but implicitly provided as key-value pairs on top of the corresponding map feature. Table 9.5 contains the building geometry and building appearance-related keys which are currently available in OSM and are more or less often used within the community. Table 9.5 also demonstrates a disadvantage of the open key-value pair methodology: for some building attributes (e.g., the roof shape), there are several potential OSM keys available; thus when investigating them, all potential keys need to be considered and compared with each other (Fig. 9.4).

How can this information be used to generate 3D building models? One of the most important attributes of a building is its height because it allows the generation of a 3D volumetric body by simply extruding the building footprint (which is available in OSM) with the added height information. Figure 9.5a depicts a quite simple

9 The Evolution of Geo-Crowdsourcing: Bringing Volunteered Geographic...

Table 9.5 OSM keys with geometry-related information

OSM key	Absolute count	Relative count compared to all buildings	Exemplary values
building:architecture	877	0.00196%	Renaissance, gothic
building:cladding	9,123	0.02036%	Brick, panel
building:cladding:colour	3	0.00001%	Black
building:colour/color	915	0.00204%	White, brown
building:façade:colour/color	125	0.00028%	White, brown
building:facade:material	1,083	0.00242%	Glass, wood, brick
building:floors	30	0.00007%	7
building:levels	435,879	0.97294%	12, 56
building:min_height	5	0.00001%	5, 18
building:min_level	3,305	0.00738%	1, 5
building:roof	61,211	0.13663%	Pitched, hipped
building:roof:angle	1,979	0.00442%	30, 20
building:roof:colour/color	2,067	0.00461%	Red, #05ff78
building:roof:extent	19	0.00004%	1 (meter)
building:roof:height	1,167	0.00260%	2, 1.5
building:roof:material	175	0.00039%	Shingles, metal
building:roof:orientation	4,954	0.01106%	Along, across
building:roof:shape	26,870	0.05998%	Pitched, hipped
building:roof:type	358	0.00080%	Pitched, hipped

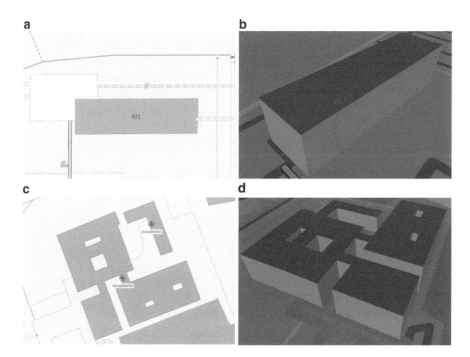

Fig. 9.4 A simple building footprint (OSM 2011a) (**a**) with corresponding 3D building model (created by extrusion) (OSM-3D 2011) (**b**), as well as a complex building footprint with holes in the geometry (OSM 2011a) (**c**) with corresponding 3D building model (OSM-3D 2011) (**d**)

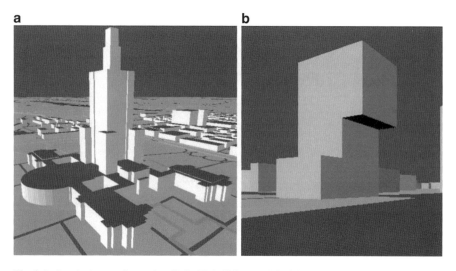

Fig. 9.5 A quite impressive and realistic 3D building model of the *Palace of Culture and Science* in Warsaw, Poland (OSM-3D 2011) (**a**) and a building with overhangs, modeled by using *building:min_level* (OSM-3D 2011) (**b**)

building footprint of a rectangular building, and Fig. 9.5b shows the corresponding 3D building model which has been generated by extruding the footprint with the building height. Also, more complex building footprints, such as those containing holes in the outer shell (Fig. 9.5c) can be transformed into 3D building models by footprint extrusion (Fig. 9.5d).

By utilizing several closed ways for describing different building parts with individual heights, it is even possible to map complex building structures which look quite impressive. Figure 9.5a depicts such a building, which is a 3D model of the *Palace of Culture and Science* in Warsaw, Poland. Again, this model is generated by extruding the individual building parts with their corresponding heights, as well as considering the values of *building:min_level* or *building:min_height* (which describe whether a building is raised in the air or located directly on the ground). Figure 9.5b depicts a building which has been mapped with *building:min_level* so that overhangs can be expressed and visualized in 3D.

For generating even more realistic building models, roof geometries can be added on top of the extruded buildings. That is, the OSM keys which contain roof information (currently *building:roof*, *building:roof:shape*, and *building:roof:type*) are evaluated. Since *building:roof:shape* is defined as a best practice (OSM 2011c) and also most commonly used (Table 9.5), this key is evaluated first and only if the other keys considered are not supplied. Currently, the building generation process of OSM-3D allows the creation of flat roofs, pitched roofs, cross-pitched roofs, hipped roofs, pyramidal roofs, and gambrel roofs (OSM 2011c). Additional roof types such as monopitched roofs will also be implemented in the project in the near future.

The roof-generation process and the required algorithms depend heavily on the geometry of the building footprints. For very simple footprints which only consist of one closed way with five nodes (where the starting node equals the end node),

9 The Evolution of Geo-Crowdsourcing: Bringing Volunteered Geographic...

```
Algorithm createSimplePitchedRoof(G, A)

Input: G = 2D Geometry (Polygon) from OSM
Input: A = Attributes as OSM key/value pairs
 1: RG ← empty
 2: fpp[] ← extractEdgePoints(G)              // fpp[0]-fpp[1] and fpp[2]-fpp[3] are the long sides of the footprint geometry
 3: if exists A[height] then
 4:    height ← A[height]
 5: else if exists A[building:height] then
 6:    height ← A[building:height]
 7: else
 8:    height ← RandomNumer
 9: repeat
10:    increaseHeight(fpp, A[height])
11: for each fpp
12:    if exists A[building:roof:shape] then
13:       rt ← A[building:roof:shape]
14:    else if exists A[building:roof:style] then
15:       rt ← A[building:roof:style]
16:    else if exists A[building:roof:type] then
17:       rt ← A[building:roof:type]
18:    if rt == 'pitched' then
19:       if building:roof:orientation == 'across' then
20:          rp1 ← computeCenter(fpp[0], fpp[1])
21:          rp2 ← computeCenter(fpp[2], fpp[3])
22:          rpl1 ← computePlane(fpp[0], rp1, rp2, fpp[3], fpp[0])   // first plane of the roof
23:          rpl2 ← computePlane(rp1, fpp[1], fpp[2], rp2, rp1)      // second plane of the roof
24:          rt1 ← computeTriangle(fpp[0], rp1, fpp[1], fpp[0])      // first side-triangle
25:          rt2 ← computeTriangle(fpp[2], rp2, fpp[3], fpp[2])      // second side-triangle
26:       else
27:          rp1 ← computeCenter(fpp[1], fpp[2])
28:          rp2 ← computeCenter(fpp[3], fpp[0])
29:          rpl1 ← computePlane(fpp[0], rp1, rp2, fpp[1], fpp[0])
30:          rpl2 ← computePlane(rp1, fpp[2], fpp[3], rp2, rp1)
31:          rt1 ← computeTriangle(fpp[0], rp2, fpp[3], fpp[0])
32:          rt2 ← computeTriangle(fpp[1], rp1, fpp[2], fpp[1])
33:       RG ← computeGeometry(rpl1, rpl2, rt1, rt2)
34:       if exists A[building:roof:extent] then
35:          extendRoofGeometry(RG, A[building:roof:extent])
Output: RG
```

Fig. 9.6 Pseudo code of the algorithm *createSimplePitchedRoof* which creates a pitched roof geometry

the roof generation is straightforward: each node represents a node of the roof geometry, thus these are the basis of the roof. Depending on the roof type, other points can be calculated by utilizing linear algebra. These points are then connected with each other (again depending on the roof type), so finally the roof geometry can be computed. For instance, a pyramidal roof can be computed by calculating the centroid of the building footprint and creating a triangle geometry for every pair of adjacent roof basis nodes and the centroid (which is raised in the air). The resulting four triangle geometries are then the final roof geometry. How far the roof is raised in the air is either explicitly added in OSM with the key *building:roof:height* or implicitly with the key *building:roof:angle* (though, in this case, the real height needs to be computed with trigonometric equations). For a pitched roof, there are some more computations required, but these are also quite straightforward.

The procedure is depicted in Fig. 9.6 as pseudo code. The input parameters of the algorithm are the 2D footprint geometry of the building and all available OSM key-value

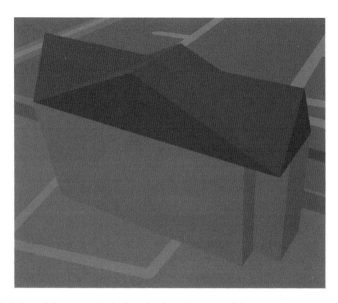

Fig. 9.7 Building with a cross-pitched roof, where the OBB of the footprint was the basis for the roof computation

pairs. If the building has a pitched roof (all potential OSM keys are analyzed), the algorithm computes a proper roof geometry (also considering the roof orientation according to *building:roof:orientation*). Also, if the key *building:roof:extent* is supplied within OSM, the roof geometry is extended accordingly. Other roof types can be computed with similar algorithms.

The above algorithms and procedures are valid for simple geometries consisting of five points, whereas the geometry does not necessarily have to be rectangular. Also, for shifted and rotated geometries, this approach returns valid results.

Some building geometries in OSM are slightly concave, that is, by definition they are concave, but when the geometry is compared with the oriented bounding box (OBB) of the geometry, all points are very close to the OBB. This may be a result of imprecise mapping but may occur also in the case of very small notches in the geometry. For the creation of roof geometries for such building footprints, the approach in OSM-3D first applies a slight simplification to the building footprint so that imprecise mappings are neglected. Additionally, it is assumed that geometries for which every point of the geometry is closer to the OBB than a distinct threshold do also have a simple roof geometry. Currently, this threshold has been defined as 1 m, so if a geometry has a notch of 0.8 m, a simple roof geometry is computed for this building. In this case, the basis for the roof computation algorithms is not the geometry itself but the OBB of the geometry (which can be computed with linear algebra), as depicted in Fig. 9.7. A building footprint with a small notch is the basis of this building model, where the roof geometry (a cross-pitched roof) is computed for the OBB of the building footprint. These roofs are not really equal to the real-world roofs, but they can be considered as a quite good approximation of reality, and they ought to be enough for the sake of a basic VGI-based 3D model. Currently, there is plenty of work aimed at the improvement of roof geometries. Some first prototypical results were obtained by using straight-skeleton algorithms with procedural extrusion (Laycock and Day 2003; Kelly and Wonka 2011), but there is still plenty of work to do for a broad application of these algorithms to OSM data.

Finally, for making the building models even more realistic and appealing, the roof can be colored according to the tags *building:roof:colour* or *building:roof:color*, and the building body can be colored according to *building:facade:colour*, *building:facade:color*, *building:colour*, or *building:color*.

9.3.4 Adding Additional Data Sources

As one can see from the previous sections, there are theoretically many OSM keys which are relevant for building-model creation, but practically, most of them are rarely available in OSM. Therefore, it is also interesting to see how OSM data can be extended by using other data sources.

One such source is that of the aerial images of Bing Maps, which are publicly available, and (even more important) can be legally utilized for mapping activities within OSM. One building property which is rarely available within OSM, but can be easily derived from Bing Maps aerial images, is information about the roof color (currently 1,546 OSM features have *building:roof:colour* and 521 OSM features have *building:roof:color*). That is, by utilizing these images, it is possible to gather additional information which is very useful when creating 3D building models. For an automatic derivation of building roof colors, a tiny program has been developed. It takes the centroid of each building shape (only those buildings without a given roof color) and requests an aerial image from Bing Maps for these distinct coordinates. Within a raster of 100 pixels (10*10 pixels edge length) around the building centroid (under the precondition that the centroid is inside the shape geometry), an RGB color code is derived for every pixel. By computing the average of all RGB color codes, the real roof color can be approximated. Since the Bing Maps API license only allows 50,000 requests per day, the developed application is limited in this respect. Until now, the gathered roof color is only stored in an internal database but not automatically added to the official OSM database. This is due to the fact that an analysis and investigation of the derived results is still missing, so it is not yet known how reliable the computed roof colors are. Nevertheless, this approach is likely to return good results and to add a more realistic variety to the created city models, as seen in Fig. 9.8. Besides aerial images from Bing Maps, also other data sources such as LiDAR data or terrestrial images can be utilized. However, these will not be discussed in more detail within this chapter.

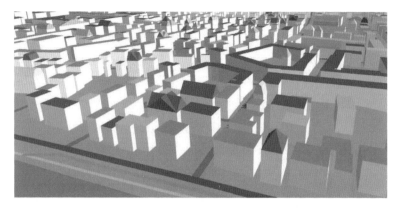

Fig. 9.8 Several building models with varying roof colors

9.3.5 Performance and Statistics

The above mentioned project OSM-3D has already been in operation for a couple of years. During this time, the required data quantity as well as the processing time has increased substantially. The system currently covers 3D map features for the European region; for the rest of the world, all features are 2D (e.g., streets or land-use areas), and buildings are not computed for non-European areas. The European restriction has been chosen on the one hand for data reduction and on the other hand because of the fact that most of the OSM-mapping activities happen in the European area anyway.

Focusing on the buildings, the system currently requires a 12-GB database for 2D building footprints (generated beforehand from raw OSM data) and a 68-GB database for the georeferenced 3D-VRML building models. The two databases (footprints and models) are distributed on two individual database servers. The processing itself is performed weekly on a dedicated workstation with 2.5-GHz CPU and 2-GB RAM and takes about 75 h.

9.4 Adding Semantic Information

Obviously, the above described creation processes for a 3D visualization only focus on geometry; the 3D model is created (as VRML or X3D) and visualized. But what about the semantics of the model? More and more applications from different fields such as urban planning or emergency response require not only pure geometric models but also semantic information about the different model features. Providing comprehensive semantics allows more complex and sophisticated applications and analysis, but a source of semantic information is required.

9 The Evolution of Geo-Crowdsourcing: Bringing Volunteered Geographic...

Table 9.6 Examples for semantic building information within OSM

OSM key	Absolute count	Relative count compared to all buildings	Exemplary values
building:architect	2	0.00001%	Saarinen
building:architecture	285	0.00073%	Renaissance, gothic
building:buildyear	16	0.00004%	1999, 2001
building:condition	60	0.00015%	Preserved, renovated
building:fireproof	119	0.00031%	Yes, no
building:floors	11	0.00003%	7
building:levels	348,013	0.89489%	12, 56
building:roof	31,288	0.08046%	Flat, pitched, hipped
building:roof:material	144	0.00037%	Shingles, metal
building:roof:shape	19,321	0.04968%	Pitched, hipped
building:roof:type	354	0.00091%	Hipped, pitched
building:type	58,559	0.15058%	House, mobile_home
building:use	240,354	0.61805%	Residential, commercial
name	575,435	1.47969%	BST48
addr:country	1,251,101	3.21712%	Germany
add:city	1,602,691	4.12122%	Heidelberg
addr:street	2,483,069	6.38505%	Berliner Straße
addr:housenumber	2,649,836	6.81388%	48
building:architect	2	0.00001%	Saarinen
building:architecture	285	0.00073%	Renaissance, gothic

Due to the open key-value pair methodology of OpenStreetMap, there are many different potential keys which contain not only geometric but also semantic information. It can even be stated that the majority of OSM keys are of a semantic type rather than a geometric type. The semantic information is from various domains, such as the key *access* for describing the legal accessibility of a distinct map feature, the key *landuse* for describing the primary usage of areas of land, the key *tracktype* for providing a classification of tracks, the key *name* for describing the name of a map feature, the key *oneway* for describing one-way characteristics of a street segment, and the key *smoothness* for providing a classification schema for the physical usability of a way for wheeled vehicles. There are many different (semantic) keys and far too many to be listed in this chapter. Nevertheless, the examples given ought to be enough for illustration and demonstration of the great diversity of semantic information within OpenStreetMap, ranging from street properties to building properties to accessibility constraints.

Since this chapter discusses 3D objects and has a special focus on buildings, Table 9.6 provides an overview of the semantic building information that can be found in OpenStreetMap. It depicts various keys which contain semantic information about buildings and additionally provides count information as well as some examples. Generally, in OSM, one must also distinguish between different spellings, especially American English (AE) and British English (BE), for the keys. Users are asked to use BE spelling, but AE spelling is also widely used. The values in Table 9.6 are accumulated regarding their spelling, that is, different spellings describing the same information are combined in one row (e.g., *building:facade:color* and *building:facade: colour*). Again, there is a great diversity, but there are also some keys which are rarely used within OSM (e.g., *building:architect*). Additionally, some of the keys mentioned in Table 9.5 (e.g., *building:roof:shape*) could also be considered as semantic information and not only geometric information.

9.5 Conclusions and Future Work

In this chapter, the potential application of VGI and especially OpenStreetMap for the generation of 3D models has been discussed. A special focus has been laid on the generation of 3D city models and building models. First, it has been demonstrated why it is important to utilize 3D data and 3D models in different applications and analyses. After a brief introduction to OpenStreetMap, a comprehensive analysis of 3D-compliant OSM attributes has been provided, as well as diverse quantitative and qualitative investigations of the current OSM dataset. Afterwards, a very brief introduction to the architecture of OSM-3D has been given, followed by a detailed discussion of the creation of 3D building models. Trying to focus not only on geometry, an overview and discussion on semantic attributes, that is, OSM keys that contain semantic information, have been provided. By conducting this research, it can be demonstrated that VGI (especially from OSM) is a rich and powerful data source for 3D information, which can be utilized for the generation of 3D city models. This chapter demonstrates what kind of information is available. Furthermore, various examples show the computation of 3D building models.

As described above (and also demonstrated in OSM-3D), it is generally possible to create 3D models based on VGI from OSM. However, the main issue about this is the missing data in many places. While many cities in Germany (and France) are mapped to a high degree also with buildings, the situation is different in many other countries. At the moment, the OSM community of users mainly focuses on geometric aspects of streets and natural areas (which they can map with their nodes), but buildings and semantic information about them are not yet typically added. Also, 3D-compliant information is not often added at the moment, which might be the case because of missing measurement methods for height values, but it is likely that future cell phones will also include sensors for such measurements. Figure 9.9 depicts a screenshot from the city of Madrid, Spain – it demonstrates the missing data in some urban areas because Madrid contains few building shapes or even information about buildings. However, we are confident that users will be motivated to add (semantic) information about buildings because projects such as OSM-3D and other future applications demonstrate the power and potential that lies in 3D-VGI. Another example can be seen in Fig. 9.10. It shows a part of Frankfurt (Main) in Germany where, similar to many cities in Europe, nearly all building outlines have been mapped.

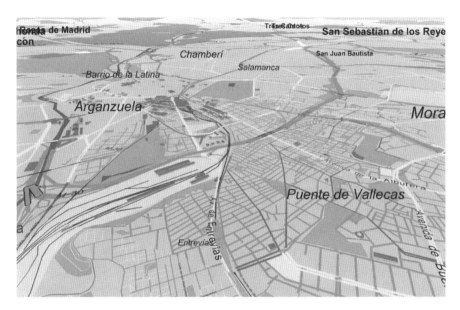

Fig. 9.9 The city of Madrid, Spain in OSM-3D (November 2011)

Fig. 9.10 The city of Frankfurt (Main), Germany in OSM-3D (November 2011)

Focusing on the generation of semantically enriched city models, it is also interesting to see if it is possible to generate City Geography Markup Language (CityGML, an OGC standard for describing and exchanging semantic building models in SDIs (Gröger et al. 2008)) building models from OpenStreetMap. Some early results were already achieved, demonstrating that it is possible to generate low-level CityGML geometries (similar to those in OSM-3D) with semantic properties (Goetz and Zipf 2012).

Acknowledgements The authors would like to thank all members of the chair of GIScience for their proofreading and helpful hints and all contributors to the OSM-3D.org project. Additionally, we would like to thank our intern Daniel Söder for creating the screenshots of OSM-3D scenes for this chapter. This research has been partially funded by the Klaus-Tschira Foundation (KTS) Heidelberg.

References

Brejc, I. (2011). Kosmos WorldFlier. http://igorbrejc.net/category/3d. Accessed November 7, 2011.
Goetz, M., & Zipf, A. (2012). Towards defining a framework for the automatic derivation of 3D CityGML models from volunteered geographic information. *International Journal of 3-D Information Modeling (IJ3DIM), 1*(2), 1–16.
Goodchild, M. F. (2007a). Citizens as sensors: The world of volunteered geography. *GeoJournal, 69*(4), 211–221.
Goodchild, M. F. (2007b). Citizens as voluntary sensors: Spatial data infrastructure in the world of Web 2.0. *International Journal of Spatial Data Infrastructures Research, 2*, 24–32.
Gröger, G., Kolbe, T. H., Czerwinski, A., & Nagel, C. (2008). OpenGIS city geography markup language (CityGML) encoding standard – version 1.0.0. OGC Doc. No. 08–007r1.
Kelly, T., & Wonka, P. (2011). Interactive architectural modeling with procedural extrusions. *ACM Transactions on Graphics, 30*(2), 14–28.
Kolbe, T. H., Gröger, G., & Plümer, L. (2008). CityGML – 3D city models and their potential for emergency response. In S. Zlatanova & J. Li (Eds.), *Geospatial information technology for emergency response* (pp. 257–274). London: Taylor & Francis.
Laycock, R. G., & Day, A. M. (2003). *Automatically generating roof models from building footprints*. Paper presented at the 11th International Conference in Central Europe on Computer Graphics, Visualization and Computer Vision (WSCG' 03), Plzen – Bory, Czech Republic.
Lee, J. (2001). *3D data model for representing topological relations of urban features*. Paper presented at the 21st Annual ESRI International User Conference, San Diego, CA, United States.
Lee, J. (2007). A three-dimensional navigable data model to support emergency response in microspatial built-environments. *Annals of the Association of American Geographers, 97*(3), 512–529.
Lee, J., & Zlatanova, S. (2008). A 3D data model and topological analyses for emergency response in urban areas. In S. Zlatanova & J. Li (Eds.), *Geospatial information technology for emergency response* (pp. 143–168). London: Taylor & Francis.
OGC. (2005). Web 3D service. Discussion paper. Ref No. OGC 05–019.
OSM. (2011a). OpenStreetMap. http://www.openstreetmap.org/. Accessed November 7, 2011.
OSM. (2011b). OpenStreetMapWiki. http://wiki.openstreetmap.org/. Accessed November 7, 2011.
OSM. (2011c). Proposed features/Building attributes. http://wiki.openstreetmap.org/wiki/Proposed_features/Building_attributes. Accessed November 7, 2011.
OSM-3D. (2011). OSM-3D in XNavigator. http://www.osm-3d.org. Accessed November 7, 2011.

Over, M., Schilling, A., Neubauer, S., & Zipf, A. (2010). Generating web-based 3D city models from OpenStreetMap: The current situation in Germany. *Computers, Environment and Urban Systems, 34*(6), 496–507.

Sarjakoski, T. (1998). Networked GIS for public participation – Emphasis on utilizing image data. *Computers, Environment and Urban Systems, 22*(4), 381–392.

Schilling, A., & Goetz, M. (2010). *Decision support systems using 3D OGC services and indoor routing – Example scenario from the OWS-6 testbed*. Paper presented at the 5th 3D GeoInfo conference, Berlin, Germany.

Schilling, A., Over, M., Neubauer, S., Neis, P., Walenciak, G., & Zipf, A. (2009). *Interoperable location based services for 3D cities on the web using user generated content from OpenStreetMap*. Paper presented at the 27th urban data management symposium, Ljubljana, Slovenia.

Song, W., & Sun, G. (2010). *The role of mobile volunteered geographic information in urban management*. Paper presented at the 18th international conference on geoinformatics, Beijing, China.

Tagwatch. (2011). Tagwatch planet-latest. http://tagwatch.stoecker.eu/Planet-latest/En/tags.html. Accessed November 7, 2011.

Yang, P. P.-J., Putra, S. Y., & Li, W. (2007). Viewsphere: A GIS-based 3D visibility analysis for urban design evaluation. *Environment and Planning B: Planning and Design, 34*(6), 971–992.

Ziegler, S. (2011). osm3d Viewer. http://www.osm3d.org. Accessed November 7, 2011.

Zipf, A., Basanow, J., Neis, P., Neubauer, S., & Schilling, A. (2007). *Towards 3D spatial data infrastructures (3D-SDI) based on open standards – Experiences, results and future issues*. Paper presented at the 3D GeoInfo07, ISPRS WG IV/8 International Workshop on 3D Geo-information: Requirements, acquisition, modelling, analysis, visualisation, Delft, Netherlands.

Chapter 10
From Volunteered Geographic Information to Volunteered Geographic Services*

Jim Thatcher

Abstract Volunteered geographic information refers to a range of geo-collaboration projects in which individuals voluntarily collect, maintain, and visualize information. This chapter introduces the related, but distinct, concept of volunteered geographic services (VGS). VGS is based on the organization and exchange of discrete actions through mobile spatially aware devices like smartphones. VGS attempts to extend the limits of VGI by directly linking users together through time and space. VGS' potential uses in crisis response are considered in light of VGI's successful role in recent crises. A specific implementation of VGS, created by the PSUmobile. org team, is presented in detail.

10.1 Introduction

As this volume attests, volunteered geographic information has become the subject of significant study since its coining in 2007 (Goodchild 2007). As a special, geographic case of the more general Web 2.0 phenomenon of user-generated content, VGI as a field of study has remained centered around large-group, collaborative mapping and visualization projects like OpenStreetMap and Ushahidi. While many studies have focused on improving the use, understanding, or accuracy of volunteered geographic *information*, utilizations have remained focused on information as distinct and separated from action. Services like Ushahidi are used in crisis mapping to aggregate, correlate, and present information in a manner useful to rescue services (Morrow et al. 2011). Studies such as those of Zook et al. (2010) and Goodchild and Glennon (2010)

---

*The term volunteered geographic services was coined by Krzysztof Janowicz. The PSUmobile. org project was also headed by Krzysztof (http://psumobile.org/).

J. Thatcher (✉)
Department of Geography, Clark University, PA, USA
e-mail: jethatcher@gmail.com

have examined the accuracy and efficacy of volunteered geographic information (VGI) in crisis situations. There remains an oft-unstated separation of information and action. VGI applications gather and (re)present geographic information from the ground up, but decision making and action coordination are left outside of the system (Savelyev et al. 2011; Thatcher et al. 2011).

This chapter presents a framework to move VGI beyond information and into an integrated system of coordinated, ground-up *actions*. Volunteered geographic services (VGS) collapse the conceptual distance between information representation and action. Present implementations of VGI, such as Ushahidi, are platforms for the provision or analysis of data. The appropriate action may be decided due to this information, but it is not decided *through* the VGI system. VGS, on the other hand, crowd-sources micro-transactions allowing users on the ground to coordinate and initiate requests for and offers of aid. This chapter will present an open-source implementation of VGS by the PSUmobile.org team of designers and developers. The concept of VGS, of directly linking providers and receivers of services through space, is of wide utility. For example, although the PSUmobile.org implementation of VGS focuses on the non-monetary exchange of aid during crises, a VGS system could be used as a geospatially referenced market for the exchange of any broadly defined "good," legal, or otherwise. Throughout this chapter, VGS refers to the concept of volunteered geographic services as distinct from volunteered geographic information, while the specific implementation is referred to as the PSUmobile.org implementation for clarity.

In making the case for volunteered geographic services, this chapter first reviews recent literature on volunteered geographic information. VGI's recent use in crisis response, where reliable, powerful tools for emergency response have been developed and deployed, is situated as the groundwork for a system of *actions*. Two potential uses of a VGS system are presented and then the PSUMobile.org implementation of VGS is described in detail. Finally, VGS is briefly considered within the context of existing VGI applications. Potential problems and dangers of VGS are discussed as issues for future research, addressing the relationship between information and action within an increasingly geospatially aware society. As the ability to locate oneself and communicate with others becomes increasingly mediated through mobile technology, the relationship between what is known and what can be done is of increasing importance. This chapter concludes by reiterating that VGS is meant to replace neither VGI nor traditional crisis-response infrastructure but as a complement to the two.

10.2 VGI and Web 2.0: A Shift in Creator, Presenter, and Interpreter

The increased availability, use, and study of volunteered geographic information can be seen as part and parcel of the coming together of two disparate trends. On the one hand, new geospatial technologies allow individual users to place their physical location within an abstract system of location. The knowledge acquisition, application, and provision allowed through the emerging geospatial infrastructure have broad applications in the academic (Yang et al. 2010), public (Kingston 2007; Ruiz and Remmert 2004), and private (Francica 2012) worlds. While the long process and supporting technologies that led to the creation of modern geospatial infrastructures and technologies, like GPS, cannot be addressed here, it is important to note that their creation allows for placing of individual location within a globalized imperfect system (Schuurman 2009). On the other hand, VGI never achieves importance without the rise of the user-created and user-shared information that lies at the heart of "Web 2.0."

Although there is disagreement on its exact definition, in this chapter, "Web 2.0" refers to an ethos of sharing that has been coupled with O'Reilly's (2005) "Web 2.0" environment in which "network technologies are used to leverage the potential contributions of a wide variety of users" (Flanagin and Metzger 2008). The rapid increase in online users creating and sharing information, up to 35% of all users according to Lenhart (2006) and likely much higher today, is coupled intrinsically with tools created to harness this desire and simplify the process. "Web 2.0" projects like Wikipedia cannot succeed without user creation and sharing of information, but, at the same time, they are also entirely reliant upon networked technologies that allow for this creation and sharing.

VGI emerges as geospatial technologies and Web 2.0 designs and implementations are incorporated. For this reason, VGI can be seen as a special case of Web 2.0 itself (Goodchild 2007). While technology is necessary in order to produce the location-based information, technology is just as necessary for its interpretation, visualization, and sharing. Users are now presented with simplified, easy-to-use tools for the creation, display, and interpretation of geospatial data (Crampton 2009; Tsou 2011). While many of these tools (AJAX, XML, Ruby on Rails, etc.) were developed for Web 2.0 applications, the emphasis on ease of use and flexible structures has benefited geographic Web applications as well. This has resulted in a host of "mash-ups" as various forms of digital information are combined with geographic information over a base layer (Darlin 2005). Application programming interfaces (APIs) like Google Maps and Bing Maps allow users ready access to a large database of satellite images, roads, addresses, and other cartographic information previously unavailable or available only to professionals.

While corporate provision of the most common APIs has raised questions related to the control of visibility, the creation of "god's eye views" of reality, and other critical issues, their present availability provides GeoWeb users with the ability to link almost any type of digital content with geo-temporal information (Zook and Graham 2007; Schuurman 2009; Kingsbury and Jones 2009). As geospatial technologies become mobile and ubiquitous with their inclusion in smartphones, new dimensions of interaction, mobility, and knowledge are created as design, technology, and daily life coalesce in new ways (Dave 2007; Roche et al. 2011). Geospatial information, and particularly the use of mobile geospatially enabled devices, allows individuals to know a place without physical presence (Sutko and de Souza e Silva 2011). As users become the drivers of uses and applications of geospatial information, the definition of creator, presenter, and interpreter shifts along with the emphasis on shared and volunteered information of Web 2.0. Volunteered geographic information, as a field of study, addresses the potential scientific and societal changes of this new relationship between producer and consumer of geographic information, or what Budhathoki et al. (2008) has termed the "*produser*" (Sui 2008; Elwood 2010).

10.3 Inspiration for VGS: VGI and Crisis Response

When considered as any application "in which people, either individually or collectively, voluntarily collect, organize and/or disseminate geographic information" for use by others, VGI has widespread utility (Tulloch 2008). Similarly, VGS – the ability to geospatially link individual users based on possession of or desire for any broadly defined service – has a similar breadth of potential uses. VGS, unlike VGI, is explicitly focused on *action*. Volunteered geographic information is contained in any VGS application, but VGS takes the additional step of linking, through time and space, individual exchange of services. The inspiration for VGS lies in VGI's usage in crisis response and management. This section explores the background of VGS in relation to VGI's use and limitations in crisis response. The section begins by arguing that while VGI may never replace traditional, finite, and verified data sources, it can effectively augment crisis management. Further, while VGI can serve as an asset for traditional crisis responders and managers, the PSUmobile.org implementation of VGS provides a framework in which the ability to respond lies in the hands of groups on the ground. This collapses the conceptual distance between information representation and action while, at the same time, removing the one-to-many, many-to-one bottleneck of traditional crisis responders.

VGI has been used in the management and response of many qualitatively different types of crisis: forest fires (Goodchild and Glennon 2010), hurricanes (Miller 2006), earthquakes (Zook et al. 2010; Morrow et al. 2011), floods (Miller 2006), riots (Presley 2011), and other natural and man-made disasters (Palen et al. 2009; Parry 2011). In many of these situations, VGI provided access to the up-to-date information that responders need in the early stages of crisis management (De Longueville et al. 2010; Roche et al. 2011). Despite successes, research has found crisis responders resistant to incorporating VGI data into their decisions (Zerger and Smith 2003). One potential cause is VGI's perceived problems with validation and verification. Official responders, who are held legally liable for crisis response, are reticent to rely wholly upon user-generated content where erroneous data can lead to loss of life (Ostermann and Spinsanti 2011; Burgener 2004). In fact, NGOs have been found to be far more likely to utilize VGI data than official responders (Roche et al. 2011).

VGS is an attempt to push beyond the limits of VGI with respect to crises in two related ways. First, present crisis management responses that utilize VGI do so from a top-down perspective: Information is gathered from individual users but is then filtered upwards and (re)presented to key decision makers. In many ways, this is consistent with a view of information technology as a tool for enhancing the control and release of information (Zook et al. 2010). Ushahidi, for example, gathered and verified information at distant centers during the Haitian earthquake response.

A similar platform was run out of Tufts University after the Tokyo earthquake (Naone 2011). While this information was predominantly verified through the work of volunteers, it remained centralized and verified far from the immediate locale of action (Morrow et al. 2011). On the one hand, this showcases VGI's ability to allow remote users to offer aid through digital means (Sui and Delyser 2012). OpenStreetMap's rapid increase in editing and updating of digital information during the Haitian Earthquake attests to this ability (Zook et al. 2010). On the other hand, the embedded individual actor experiencing the crisis can become lost in the digital centralization and representation of VGI.

While Ushahidi allows individuals on the ground to upload information about needed services, it provides no means by which individuals may provide those services to each other: A user may request first aid equipment, and his or her neighbor may have such equipment available; however, the Ushahidi platform in its current incarnation connects these local needs only through a centralized representation. Present implementations of VGI lack the ability to link users directly through time and space. Although this ability could easily be added to any given implementation, it represents a concept distinct from present definitions of VGI. As a concept, VGI is concerned with data and information both in its gathering and presentation (Goodchild 2007; Tulloch 2008; Agrios and Mann 2010; Zook et al. 2010). With this focus, a gap emerges between information and action. VGI aids in the accumulation and dissemination of data necessary to make decisions, but the decisions are organized outside of present VGI implementations. For example, in a traditional VGI system like Ushahidi, if multiple requests for first aid or evacuation occur in a given area, an NGO or government agency could decide to allocate resources to that area. Resource allocation and action still occur outside of the VGI system. This gap removes actors on the ground from decision-making processes even as the information they provide is used for these decisions.

VGS attempts to address both the gap between information and action and the top-down nature of present VGI implementations. As a concept, it refers to a system that crowd-sources and coordinates micro-transactions. In a crisis, this allows on-the-ground citizens to organize and allocate available resources directly. It follows from research that suggests that on-the-ground actors often have more invested in immediate response and must be included in crisis management. It is meant to augment traditional crisis management and response through new mobile technology. Pushing VGI beyond its limits, VGS goes from the "rise of local expertise to replace centralization" of information to that of action (Goodchild 2008). In the following section, two potential uses of VGS are presented.

10.4 Volunteered Geographic Services: The General Case for Use and Two Examples

As a concept, VGS is not limited to crisis response. It entails any system where volunteered geographic information is coupled directly to an exchange of services, broadly defined. The particular PSUMobile.org implementation is meant for use in crises where micro-transactions can have a significant positive effect. The following two examples present situations in which the immediate allocation of an available resource by actors on the ground is sufficient. The first example is of an everyday use of the PSUmobile.org implementation of VGS, while the second focuses on use in a more immediate emergency.

10.4.1 Plowing the Driveway During a Snow Storm

In the northeast of the United States, it is not uncommon to encounter severe snowstorms of 12–14 in.of snow in a single day. As anyone living in the area can attest, removing the snow from a driveway is both necessary and time consuming. Traditionally, residents either remove the snow themselves or rely upon neighbors or local plows for aid. The PSUmobile.org implementation of VGS can help residents organize for time-efficient snow removal.

First, a resident in a hurry and without a snowplow can send out a spatially and temporally specific request for assistance (Sect. 10.2). Second, a resident with a snowplow could make an offer of aid to which residents in need could then respond. In each situation, the VGS application used will then display all offers and requests to other users using appropriate (and adjustable) spatial and temporal filters.

While in many ways trivial, this situation demonstrates the ability to facilitate micro-transactions of aid in everyday life. This is done in real time without the need of any professionally verified information or assignments of assistance to locate and match users in time and space.

10.4.2 Coordinating Flood Evacuation

According to the US National Research Council, geospatial data and tools should be, but often are not, used in all aspects of emergency management (NRC 2007). As VGI continues to find its way into crisis response, it remains used to acquire and present up-to-date data from the dynamic situations on the ground (Goodchild and Glennon 2010; Zook et al. 2010; Parry 2011). While this data is critical for coordination and response, the support bottleneck remains: Actions are coordinated through top-down, official channels. VGS systems, like the PSUmobile.org implementation, present a potential solution to this constraint as they allow *ad hoc* coordination of micro-services by individuals on the ground. The following describes a potential use of a VGS in a flood situation.

Regardless of flood severity, those in the path of floodwaters need to acquire or arrange for transportation to safer ground. In such situations, transportation routes and availability may change from moment to moment, making any sort of official decision potentially inadequate to the shifting situation on the ground. Further, while mass evacuation may at times be necessary, individuals may need to relocate as well.

The PSUmobile.org implementation of VGS would allow for individuals to request and offer immediate transportation solutions. For example, if a resident with a truck was leaving his neighborhood and had room for additional passengers, they would post an offer of transportation that would be viewable by users within the immediate vicinity. Likewise, a user in need of transportation can make requests viewable by all other users in the appropriate area (Sect. 10.2). The PSUmobile.org implementation of VGS contains spatial and temporal filters that allow those requesting aid to be appropriately matched with those offering it. The Linked Data model of the PSUMobile.org implementation allows an individual to respond to a request for aid and be automatically linked to multiple users in similar situations (Sect. 10.1).

In the situation of a flood, or any other event where individuals need to rapidly coordinate movement, the removal of top-down coordination allows response to match the rapidly shifting circumstances on the ground. The removal of external verification and organization opens VGS to several potential problems discussed in Sect. 10.6; however, it is clear that VGS, through its ability to connect users at appropriate spatial and temporal scales, offers significant opportunities for augmenting crisis response.

10.5 The PSUmobile.org VGS Implementation

This section describes the specific implementation of VGS by the PSUmobile.org team. The server is described first and represents an open-source platform capable of easy integration with existing applications and data sources. The client is described second and represents one possible implementation of the server and, more broadly, of VGS itself. The descriptions offered here are meant to familiarize the reader with how VGS might work; technical specifications of the implementation are available in the paper by Savelyev et al. (2011).

10.5.1 The Server

As the backbone of the implementation, the server provides a framework for diverse implementations of the VGS concept that address various user communities and purposes. A generic request-offer messaging bus based on a Linked Data model handles all requests, offers, users, and their related elements in an open manner which allows for their combination with other sources (Bizer et al. 2009). The client is just a single implementation that uses the server infrastructure.

Within the implementation, an offer of help may be answered by a request and vice versa; however, in addition to one-to-one matching, a request may attach itself to another request or an offer to another, related, offer. If one user needed transportation during a flood (see Sect. 10.2), another user also in need of transportation may "add" himself or herself to the existing request rather than creating a unique request (Sect. 4.2 explains this process). These requests (or offers) reinforce existing requests (and offers) and are therefore not considered unique requests in their own right. The term "action" is used to generically refer to all requests, offers, and their interlinkages.

In the Linked Data model used by the server, actions contain all of the interactions between requests and users as well as a variety of possible metadata. Actions may contain user identification, the location of the user, the location the action refers to, the time of creation for the action, and the time of expiration of the action, as well as the textual content of the message. The VGS framework ontology does not model this metadata, as diverse implementations of VGS may require all, some, or even additional types of metadata. Avoiding a formalized framework for metadata allows each implementation to create the necessary ontology for its intended use. Each VGS implementation may require its own application-level ontology to parse this metadata, while the implementation framework manages requests, offers, and associated information from the server side.

On the server side, each action receives a unique identifier. This assigned unique identifier is then associated with all related offers, requests, and attached actions following the Linked Data paradigm (Bizer et al. 2009). The server provides a generic, adaptable, and robust means of handling all requests and offers for service as well as connections between these actions. For a complete description of the present server implementation, including server-side ontologies, the Service Bus Implementation, and Web Service model, see the paper by Savelyev et al. (2011) from which this section draws heavily.

10.5.2 The Android Client

While the server implementation and its associated Web Service is the generic backbone for all VGS implementations discussed here, the client is a specific application developed for the Android platform to interface with the VGS server system. The mobile client allows users to browse existing actions, submit new actions, and respond to existing actions through its graphical interface. Here an "action" refers to the user-initiated request for or offer of help. Although, as discussed above, the framework handles all action metadata generically, this specific implementation parses the user name, title of the action, action description, geolocation, and time of creation.

The client implementation has four screen views in the present implementation – the default GUI for browsing existing actions (Fig. 10.1a), a screen to submit a request or offer (Fig. 10.1b), a "detailed action view" that allows users to confirm requests or offer help to the viewed request (Fig. 10.1c), and a screen to confirm an offer to fulfill an existing request or offer (Fig. 10.1d).[1]

---

[1] Image created by Thatcher et al. (2011).

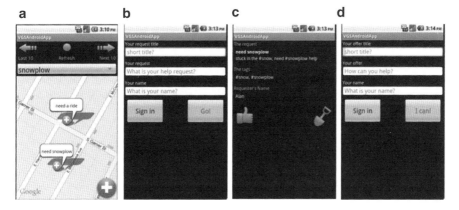

Fig. 10.1 Activity screens of the PSUmobile VGS App

Upon initialization, the client acquired the device's location via GPS and presents the user with the "home" screen (Fig. 10.1a). From this screen, three potential functions of the VGS application (to offer help, to request help, or to confirm an offer or request) diverge.

To send a request, the user presses the "Create Request" button in the lower right-hand corner of the screen (the blue circle with a plus sign in Fig. 10.1a). To offer help, the user selects a request bubble from the map view and is taken to the "detailed action" screen (Fig. 10.1c). In the present build of the client, the ability to offer a service not in response to an existing request is not initialized.

The "home" screen also allows users to filter requests and offers by time and space. Filtering by space is accomplished through panning and zooming the map, while temporal filters are visible at the top, allowing users to step through time by ten requests. For a critical discussion of this design and its ramifications, see the paper by Thatcher et al. (2011).

Creating a request takes the user to Fig. 10.1b. This screen offers three fields – request title, request description, and the user's name. Pressing the "Go!" button sends the request to the server via an HTTP transmission. The user may also "sign in" to his or her user account from this screen. The process for requesting help was intentionally kept simple to allow for rapid publication in times of stress.

Users offer to fulfill requests from Fig. 10.1c. This screen contains all of the information gathered from the request submission and offers the user two choices: A "thumbs-up" hand symbol confirms the request, linking the user to the request and denoting that he or she also requires the same service. This serves both to legitimize the request and to allow those offering aid to rapidly determine how many people in a given area require a service. Pressing the "shovel" button takes the user to the Send Offer screen (Fig. 10.1d), where the user confirms that he or she is able to provide the requested service. When complete, the offer and request are linked together in the server database (Sect. 10.1).

Through the Google Map API and the four screens in Fig. 10.1, users may view offers and requests using temporal and spatial filters. Simple, rapid interactions link offers and requests together on the server. Although the present client implementation is simple and lacks robust features, it demonstrates the strength and versatility of both the VGS concept and the server implementation.

10.6 Summary and Concluding Remarks

This short chapter has introduced the concept of volunteered geographic services and situated it within the larger framework of VGI. While its inspiration and present implementation stem from crisis response, the framework presented can effectively manage a host of potential applications as it provides a generic infrastructure for the storing, retrieving, and linking of human services. Using the Linked Data paradigm and semantic Web technologies, the concept of VGS is supported with a robust server implementation (Savelyev et al. 2011). The software is open source and encourages outside contributions; the latest builds are available at http://vgs.svn.sourceforge.net/viewvc/vgs/.

VGS is distinct from present considerations of VGI as it collapses the conceptual distance between gathering and presenting geospatial information and acting within geographic contexts. The PSUmobile.org implementation supports a framework by which end users can coordinate micro-service transactions in a variety of contexts. The client is one application of VGS and is meant as a proof of concept rather than a definitive tool. The varied uses to which existing VGI applications have been put demonstrate their ability to incorporate new uses like VGS; however, it would be necessary to shift the focus from aggregation and presentation of information to on-the-ground *ad hoc* coordination of action. At present, services like Ushahidi's SwiftRiver platform retain an intrinsic block between information display and crisis response. This separation allows for remote users to aid in crisis response and offers the opportunity to verify information. In the future, the repurposing of existing and proven tools may be necessary as VGS demonstrates its value in augmenting VGI.

Before any large-scale adoption, it is necessary to consider some potential issues created in the use of VGS systems. First, much like VGI, VGS faces potential threats of spam and fraudulent posting. Simple location specificity is not enough to address the "context deficit" of digital information as location may be spoofed using a variety of means (Eysenbach 2008). This becomes an acute issue in situations of crisis response, and although literature has highlighted VGI's successful use in crises, VGS has yet to be examined (Flanagin and Metzger 2008; Goodchild and Glennon 2010; Zook et al. 2010; Roche et al. 2011). Recent work suggests that implementation of automated location-based trust and reputation models can have some success, but testing with VGS applications is necessary (Bishr and Mantelas 2008; Bishr and Janowicz 2010).

In addition to fraud and spam, the VGS system faces the difficulty of illicit, but sincere, uses. Like many frameworks that allow for the creation and exchange of information, the PSUmobile.org framework could easily be used for illegal or harmful purposes. Unlike situations in which only information is exchanged, the concept of VGS itself implies *action* and so further entangles the system in a set of logistic and ethical questions.

Finally, it is critical to reiterate that VGS systems are not meant to replace traditional crisis-response systems but to augment them. In crisis situations, official managers often defer to paper-based maps rather than the more recent, but potentially less accurate, digital ones (Roche et al. 2011). Official responders are legally liable for their actions and so require validated information, something that VGI and VGS may not always provide (Tulloch 2008). VGS systems cannot fully replace official response to disasters; they are not top-down architectures that allow for the type of control and response necessary for legal liability. Further, they inherently function at the level of individuals; while useful (Sect. 10.4), this mode of functionality cannot allocate and mobilize resources on the massive scales that disaster response necessitates.

VGI and VGS can also never fully address issues of unequal socioeconomic access to digital communication or uneven loss of access during crises. VGS removes the bottleneck of decision making for immediate, on-the-ground micro-transactions. It is not meant to coordinate the deployment of thousands of sandbags along a breaking levee but, rather, to enable those individuals most affected by a crisis to organize themselves in small, but significant, ways. Rather than another tool to integrate new information into officially verified, known, and finite data sets, VGS allows local actors, those most often affected by emergencies, to become decision makers (Elwood 2008). The micro-services coordinated by the PSUmobile.org implementation and conceptualized in VGS present a small step towards the bottom-up organization of crisis response.

Acknowledgements The author acknowledges the entire PSUMobile.org team and particularly Krzysztof Janowicz, Alexander Savelyev, Wei Luo, Sen Xu, Christoph Mulligann, and Elaine Guidero, without whom this project could not exist. Madelynn von Baeyer and Courtney Thatcher are thanked for early edits and encouragement.

References

Agrios, B., & Mann, K. (2010). Getting in touch with volunteered geographic information: Use a Javascript API live samples to build a web editing application. http://www.esri.com/news/arcuser/0610/vgi-tutorial.html. Accessed February 2, 2012.

Bishr, M., & Janowicz, K. (2010). Can we trust information? The case of volunteered geographic information (Vol. 640). In *Towards a digital earth search discover and share geospatial data workshop at future internet symposium*, CEUR-WS.

Bishr, M., & Mantelas, L. (2008). A trust and reputation model for filtering and classifying knowledge about urban growth. *GeoJournal, 72*(3–4), 229–237.

Bizer, C., Heath, T., & Berners-Lee, T. (2009). Linked data – The story so far. *International Journal on Semantic Web and Information Systems, 5*(3), 1–22.

Budhathoki, N. R., Bruce, B., & Nedovic-Budic, Z. (2008). Reconceptualizing the role of the user of spatial data infrastructure. *GeoJournal, 72*, 149–160.

Burgener, E. (2004). Assessing the foundation of long distance disaster recovery. *Computer Technology Review, 24*(5), 24–25.

Crampton, J. (2009). Cartography: Maps 2.0. *Progress in Human Geography, 33*(1), 91–100.

Darlin, D. (2005). A journey to a thousand maps begins with an open code. *New York Times Online*. http://www.nytimes.com/2005/10/20/technology/circuits/20maps.html. Accessed February 1, 2012.

Dave, B. (2007). Space, sociality and pervasive computing. *Environment and Planning B, 34*(3), 381–382.

De Longueville, B., Annoni, A., Schade, S., Ostlaender, N., & Whitmore, C. (2010). Digital earth's nervous system for crisis events: Real-time sensor web enablement of volunteered geographic information. *International Journal of Digital Earth, 3*(3), 242–259.

Elwood, S. (2008). Volunteered geographic information: Future research directions motivated by critical, participatory, and feminist GIS. *GeoJournal, 72*, 173–183.

Elwood, S. (2010). Geographic information science: Emerging research on the societal implications of the geospatial web. *Progress in Human Geography, 34*, 349–357.

Eysenbach, G. (2008). Credibility of health information and digital media: New perspectives and implications for youth. In M. J. Metzger & A. J. Flanagin (Eds.), *Digitla media, youth, and credibility*. Cambridge: MIT Press.

Flanagin, A. J., & Metzger, M. J. (2008). The credibility of volunteered geographic information. *GeoJournal, 72*, 137–148.

Francica, J. (2012). Geospatial data content licensing and marketing in the era of data as a service – An interview with James Cutler, CEO, emapsite.com. *Directions Magazine*. http://www.directionsmag.com/articles/geospatial-data-content-licensing-and-marketing-in-the-era-of-data-as-/227410. Accessed February 1, 2012.

Goodchild, M. (2007). Citizens as sensors: The world of volunteered geography. *GeoJournal, 69*(4), 211–221.

Goodchild, M. (2008). Commentary: Whither VGI? *GeoJournal, 72*, 239–244.

Goodchild, M., & Glennon, J. A. (2010). Crowdsourcing geographic information for disaster response: A research frontier. *International Journal of Digital Earth, 3*(3), 231.

Kingsbury, P., & Jones, J. P., III. (2009). Walter Benjamin's Dionysian adventures on Google Earth. *Geoforum, 40*, 502–513.

Kingston, R. (2007). Public participation in local policy decision-making: The role of web-based mapping. *The Cartographic Journal, 44*(2), 138–144.

Lenhart, A. (2006). User-generated content. Pew Internet & American Life Project. http://pewinternet.org/Presentations/2006/UserGenerated-Content.aspx. Accessed February 5, 2012.

Miller, C. (2006). A beast in the field: The Google maps mashup as GIS. *Cartographica, 41*, 1878–1899.

Morrow, N., Mock, N., Papendieck, A., & Kocmich, N. (2011). Independent evaluation of the Ushahidi Haiti project. https://sites.google.com/site/haitiushahidieval/news/finalreportindependentevaluationoftheushahidihaitiproject. Accessed February 1, 2012.

Naone, E. (2011). Internet activists mobilize for Japan. *Technology Review Published by MIT*. http://www.technologyreview.com/communications/35097/. Accessed February 2, 2012.

National Research Council. (2007). *Successful response starts with a map: Improving geospatial support for disaster management*. Washington, DC: The National Academies Press.

O'Reilly, T. (2005). What is web 2.0 design patterns and business models for the next generation of software O'Reilly.com. http://oreilly.com/web2/archive/what-is-web-20.html. Accessed February 1, 2012.

Ostermann, F. O., & Spinsanti, L. (2011). A conceptual workflow for automatically assessing the quality of volunteered geographic information for crisis management. In *Proceedings of Agile 2011*, Utrecht.

Palen, L., Vieweg, S., Liu, S. B., & Hughes, A. L. (2009). Crisis in a networked world: Features of computer-mediated communication in the April 16, 2007, Virginia Tech event. *Social Science Computer Review, 27*(4), 467–480.

Parry, M. (2011). Academics join relief efforts around the world as crisis mappers. *The Chronicle of Higher Education*, March 27, 2011.

Presley, S. (2011). Mapping out #LondonRiots. NFPvoice. http://www.nfpvoice.com/2011/08/mapping-out-londonriots/. Accessed February 1, 2012.

Roche, S., Propeck-Zimmermann, E., & Mericskay, B. (2011). GeoWeb and crisis management: Issues and perspectives of volunteered geographic information. *GeoJournal*. doi:10.1007/s10708-011-9423-9.

Ruiz, M. O., & Remmert, D. (2004). A local department of public health and the geospatial data infrastructure. *Journal of Medical Systems, 28*(4), 385–395.

Savelyev, A., Janowicz, K., Thatcher, J., Xu, S., Mulligann, C., & Luo, W. (2011). Volunteered geographic services: Developing a linked data driven location-based service. In *Proceedings of SigSpatial: International Workshop on Spatial Semantics 2011*.

Schuurman, N. (2009). An interview with Michael Goodchild. *Environment and Planning D: Society and Space, 27*(4), 573–580.

Sui, D. (2008). The wikification of GIS and its consequences: Or Angelina Jolie's new tattoo and the future of GIS. *Computers, Environment and Urban Systems, 32*, 1–5.

Sui, D., & DeLyser, D. (2012). Crossing the qualitative-quantitative chasm I: Hybrid geographies, the spatial turn, and volunteered geographic information (VGI). *Progress in Human Geography, 36*(1), 111–124.

Sutko, D. M., & de Souza e Silva, A. (2011). Location-aware mobile media and urban sustainability. *New Media Society, 13*, 807–823.

Thatcher, J., Mulligann, C., Luo, W., Xu, S., Guidero, E., & Janowicz, K. (2011). Hidden ontologies – How mobile computing affects the conceptualization of geographic space. *Proceedings of Workshop on Cognitive Engineering for Mobile GIS 2011*.

Tsou, M. H. (2011). Revisiting web cartography in the United States: The rise of user-centered design. *Cartography and Geographic Information Science, 38*(3), 249–256.

Tulloch, D. L. (2008). Is VGI participation? From vernal pools to video games. *GeoJournal, 72*, 161–171.

Yang, C., Raskin, R., Goodchild, M., & Gahegan, M. (2010). Geospatial cyberinfrastructure: Past, present and future. *Computers, Environment and Urban Systems, 34*(4), 264–277.

Zerger, A., & Smith, D. I. (2003). Impediments to using GIS for real-time disaster decision support. *Computers, Environment and Urban Systems, 27*, 123–141.

Zook, M., & Graham, M. (2007). Mapping DigiPlace: Geocoded internet data and the representation of place. *Environment and Planning B, 34*, 466–482.

Zook, M., Graham, M., Shelton, T., & Gorman, S. (2010). Volunteered geographic information and crowdsourcing disaster relief: A case study of Haitian Earthquake. *World Medical & Health Policy, 2*(2), 231–341.

Chapter 11
The Geographic Nature of Wikipedia Authorship

Darren Hardy

Abstract The efficacy and use of volunteered geographic information (VGI) is an active research area, but the geography of VGI authorship is largely unknown. Wikipedia is an online collaborative encyclopedia where anyone can edit articles, including those about place. Moreover, Wikipedia's editorial transparency facilitates *in situ* observations of collective authorship. The empirical study described in this chapter collects 32 million contributions to Wikipedia's geographic articles over 7 years. It finds exponential decay in the spatial patterns of Wikipedia's authorship processes, which is consistent with other sociospatial phenomena, like innovation diffusion. As global information infrastructures continue to reduce communication and coordination costs, this study may provide insight into whether geographic distance ultimately matters in information peer production. This chapter begins by discussing core concepts behind collective authorship; then provides an overview of Wikipedia, its contributors, and their production processes; discusses the results and implications from spatial modeling of geotagged Wikipedia article contributions; and concludes with future research issues.

11.1 Introduction

A notable example of a widely popular system with volunteered geographic information (VGI) capabilities is Wikipedia, an online collaborative encyclopedia. Wiki technology provides simple methods for Web-based collective authorship where anyone can contribute. Using this technology, Wikipedia provides a large-scale social computing system in which participants collectively author encyclopedic information.

D. Hardy (✉)
Bren School of Environmental Science and Management, University of California,
Santa Barbara, CA, USA
e-mail: dhardy@bren.ucsb.edu

Since 2001, Wikipedia has 17.5 million articles in 263 languages. Since March 2007, Alexa has ranked Wikipedia in the top 10 Internet sites. As of 23 February 2010, Wikipedia has 15 million articles in 272 languages with 860 million edits from 22 million contributors (Wikimedia 2010). During 2009 alone, Wikipedia had 365 million unique visitors that generated 133.6 billion page views (Zachte 2010a). Its impact on the Web's content is significant. Fifty-one percent of its site visits come from link-based search engine referrals (Alexa Internet, Inc. 2009). Of those page views that were referred to Wikipedia by external sites, 42% were referred by Google search, maps, and other services, and 8% were made by Google's "web-crawling" software GoogleBot (Zachte 2009). Over 1.2 million articles are place-based articles (i.e., "geotagged") (as of April 2011). These geotagged articles span dozens of languages and are accessible through geobrowsers and online mapping services.

As the Internet itself grows, many describe it as place*less*—cyberspace without place. Yet sociological researchers find cultural differences in virtual communities that mimic real-world environments, and a shared understanding of a virtual place is a central determinant in such research. But today, any Internet user can get some sense of place through rich interactive geovisualization technologies. "Slippy maps" depict roads and buildings and other geographical features using simple point-and-drag navigational and informational tools and even 3D imagery. Within these online mapping interfaces, users may access a diverse set of VGI, including geotagged Wikipedia articles and photographs.

Yet, despite the advantages of the Internet for collaborative work, authors are fundamentally engaged in knowledge production processes that are grounded in social structures and norms, and in turn, physical place. Geographic distance, in particular, should be a significant factor in online knowledge production. But the nature of the Internet in a globalized world has led to debate on whether geographic distance matters (cf. Cairncross 1997; Friedman 2005; Goodchild 2004; Marston et al. 2005). That is, the Internet may redefine the role of physical place in our lives due to reduced communication costs and increased ubiquity. Zook (2005, p. 54) summarizes this debate as a new "geography of electronic spaces," as the Internet becomes "a recombinant space for political, cultural, and economic interaction."

This chapter focuses on information production methods and processes behind geographic Wikipedia articles and discusses the nature of these production processes. For example, are contribution patterns similar between VGI and non-VGI content? How do authors geotag articles? What is the geography of Wikipedia's authorship? What is the spatial distribution of articles and contributors, and how does physical proximity influence contributions, either by article topic or language?

11.2 Collective Authorship Processes

Collective authorship is one type of information production process—a mass collective effort by individuals to produce information artifacts within a digital commons. The term "information production" itself has different semantics across disciplines. In the humanities, the term may represent the authoring of a written work or book; in economics, market resources, or commodities, or perception, or even a constitutive force in society (Browne 1997, p. 266); in library science, how we communicate collaborative work to public scientific knowledge (Cronin 2001); and, in social computing, collaborative filtering or recommendation systems (Beenen et al. 2004), blogging as community forums (Nardi et al. 2004), and user-generated tag clouds (Golder and Huberman 2006). For Wikipedia, the terms wikinomics (Tapscott and Williams 2006), collective intelligence (O'Reilly 2005), and crowdsourcing (Brabham 2008) all reflect the user-centric processes that drive information production.

And user-centric it is. Each month, over ten million authors contribute to Wikipedia articles, roughly divided into two classes of contributors—a small, highly productive set, then everyone else. The Web itself has a scale-free, power law distribution in its link structure (Broder et al. 2000) and surfing behavior (Huberman et al. 1998), and Wikipedia has them for both readership (Priedhorsky et al. 2007) and editing (Almeida et al. 2007; Kittur et al. 2007; Voss 2005). For example, the intensity of authorship shows that a small number of Wikipedia articles receive the majority of edits, and the vast majority of articles receive a small number of edits (i.e., the long tail).[1]

Wikipedia's production processes are nontrivial, despite its perception in the popular media as a loose or chaotic system. Wikipedia has many policies and mechanisms to govern contributions, including rule-making, monitoring, conflict resolution, and norms (Forte and Bruckman 2008; Lih 2009; Viégas et al. 2007a, b). Its most well known policy is that contributors must write articles using a neutral point of view, and this is a key discussion point between authors (e.g., Bryant et al. 2005; Viégas et al. 2004). As described by Wikipedia, *neutral point of view* (NPOV) is "a fundamental Wikimedia principle and a cornerstone of Wikipedia," requiring that "all content [be] written from a neutral point of view, representing fairly, proportionately, and as far as possible without bias, all significant views that have been published by reliable sources" (http://en.wikipedia.org/wiki/NPOV).

The term *Wikipedian* does not have a strict definition, other than being a contributor to a Wikipedia article generally.[2] Registered, anonymous, administrative Wikipedians and bots are the four basic types of contributor. *Registered* Wikipedians create an account on Wikipedia, and their contributions are explicitly tagged in the article history using their account. *Anonymous* Wikipedians do not provide any

---

[1] In scientific authorship, Lotska's law predicts an inverse power relationship (e.g., $w \sim n^{-\beta}$) between the number of authors n, the size of their contributions w, and a constant β. Zipf's law is a reformulation of this principle, generalized to individual contributions among group effort — i.e., the rank r of an individual is proportional to the inverse of her contributions n (e.g., $n \sim r^{-\beta}$) (Almeida et al. 2007).

[2] Some have a more narrow definition of highly or consistently active contributors (Zachte 2010b), but in this chapter, "Wikipedian" refers to any contributor to a Wikipedia article, regardless of activity level.

registration information, and their computer's IP address is used in lieu of an account. "Bots" and other *administrative* Wikipedians are both special cases of registered accounts, but they have additional access or permissions to edit articles. The overwhelming majority of Wikipedians do not collaborate with each other in a traditional sense. They do not often discuss their contributions with others (Viégas et al. 2007a) and as such form a loosely collaborative, online collective authorship. The most active segments of the Wikipedian population are 91,817 Wikipedians with at least five contributions *per month* and 1,076,908 Wikipedians with at least ten contributions total (Zachte 2010b). The "long tail" has 21.1 million Wikipedians, each of whom have less than ten contributions total.

Although authorship processes are largely invisible to readers, the authors themselves struggle to control article content around information types, responsibility, perspectives, organization, or provenance and creation (Miller 2005; Sundin and Haider 2007). Wikipedia provides complete article histories for those readers wanting detailed authorship information. *WikiScanner*, for example, is a data-mining tool that extrapolates from article edit histories the location or affiliation of anonymous authors (Griffith 2007). But the utility of explicit authorship information is debatable. As summarized by Viégas (2005, p. 61), on the one hand explicit authorship information may be "an important part of social collaboration in the sense that it adds context to interactions," and on the other hand it may be "irrelevant and sometimes even detrimental to the creation of truly communal repositories of knowledge."

In fact, the success of Wikipedia and other "user-generated content" Web services (O'Reilly 2005) has challenged academic theories of production. Benkler (2002) argues that in terms of economic models of production, when the efficiency gains of "peering" exceeds the costs of organizing human capital into a firm or market, a commons-based peer production system will emerge. Its advantage is based not only on reduced costs of human capital and communications but also on the nonrival aspects of Web-based information artifacts—i.e., many people can read (consume) a webpage simultaneously without degrading its value. This effectively eliminates allocation costs to consumers and increases the pool of potential contributors, which mitigates effects from free riders.

When applied to geographic information production, these factors will likely challenge the "knowledge politics" of spatial data infrastructures (Elwood 2010). For example, they may weaken traditional notions of authoritative sources as the collective social production of spatial information increases (Budhathoki et al. 2008; Coleman et al. 2009; Sieber and Rahemtulla 2010). As Sui (2008, p. 4) argues, the "wikification of GIS is perhaps one of the most exciting, and indeed revolutionary developments since the invention of [GIS] technology in the early 1960s." Moreover, Wikipedia's editorial patterns in the production of VGI content are similar to those for nongeographic content. That is, each of the four types of contributors exhibits editorial patterns that are systematic when contributing to geographic articles, but idiosyncratic across languages (Hardy 2008).

Table 11.1 Example geotag formats for University of California, Santa Barbara (UCSB; approx. 34.41°N, 119.85°W)

(a) Template:Coord and Template:Infobox in Wikipedia

```
UC Santa Barbara {{coord|34|24|35|N|119|50|59|W}}
UC Santa Barbara {{coord|34.41254|-119.84813|display=title|type:edu}}
{{Infobox_University |name=UC Santa Barbara
                    |latd=34  |latm=24  |lats=35  |latNS=N
                    |longd=119 |longm=50 |longs=59 |longEW=W
...}}
```

(b) Geo microformat for HTML (Çelik 2005)

```
<DIV CLASS="geo">UC Santa Barbara
    <SPAN CLASS="latitude">34.41</SPAN>,
    <SPAN CLASS="longitude">-119.85</SPAN>
</DIV>
```

(c) Dublin Core metadata for HTML (Kunze 1999)

```
<META NAME="DC.title"      CONTENT="UC Santa Barbara" />
<META NAME="DC.coverage.x" CONTENT="-119.85"/>
<META NAME="DC.coverage.y" CONTENT="34.41"/>
```

(d) Geo metadata for HTML (Daviel and Kaegi 2007)

```
<META NAME="geo.position"  CONTENT="34.41;-119.85"/>
<META NAME="geo.placename" CONTENT="UC Santa Barbara" />
<META NAME="geo.region"    CONTENT="US-CA" />
```

11.3 Volunteered Geographic Information in Wikipedia

Now, we turn to the specific types of geographic information produced through collective authorship in Wikipedia. Geographic information, in general, informs us about the *where* of things. It is spatial information about a phenomenon's distribution in our geographic world (Goodchild 2000). *Georeferencing* is the set of methods for defining a geographic location on the globe (Hill 2006), and *geotagging* assigns geographic locations to content (Amitay et al. 2004), referring to "tagging" georeferenced metadata to a document or other content. A geotag may contain geographic coordinates, extent, shape, or feature type information. A useful geometry for cataloguing georeferenced content is the *minimum-bounding rectangle*, which is the smallest rectangle aligned with the coordinate axes that spans all coordinates for a given location.

Wikipedia primarily uses single points and bounding rectangles rather than fine-resolution polygons in its geotagging processes. In this case, a geotag contains simple geographic coordinates for latitude and longitude, and this georeferenced information is embedded into articles using one of many microformats and extensions to *Wikitext*, Wikipedia's content markup language. For example, the *Template:Coord* and *Infobox* Wikitext templates accept point coordinates (Wikipedia. org 2008). In fact, there are dozens of ways to include geographic coordinates in an article. There is not a single "geotag" standard or format for Wikipedia, or the Web for that matter (Table 11.1).

The geotagging process itself in Wikipedia is haphazard. Wikipedia started explicitly using structured geotagging in February 2005 when geotags were introduced into

Wikipedia in 2005 by Egil Kvaleberg's *gis* extension to *MediaWiki*. Some authors create geotags manually using a reference digital or paper map to estimate coordinates, while others resolve toponyms based on existing online gazetteers. Alternatively, bots perform a bulk of the automated geotagging based on *GEOnet Names Server*, an online gazetteer, and run periodically. This process also adds geographic feature type (i.e., city, river, mountain, etc.) when it is available from the gazetteer.

The vast majority of geotagging is reportedly done by a variety of bots (Kühn and Alder 2008), and their ad hoc nature ultimately makes it more difficult to extract geotags from articles. For example, a semiautomated bot *Anomebot2* runs periodically to geotag articles or mark those that *may* need a geotag. It cross-references named entities in over 100,000 article titles with online gazetteer services.[3] These bots provide a structural mechanism to integrate existing geographic data sources into articles. But they are not semantic in nature, nor do they generate standardized markup (Table 11.1). In fact, they increase the complexity of extracting structured geographic information from articles because of their chaotic, ad hoc nature and that of the Wikitext markup and templates themselves (Sauer et al. 2007). The end result is that geotag extraction requires ad hoc or data-mining approaches to deal with the nondeterministic, semistructured nature of article templates and ad hoc inclusion of geotags. But, anecdotally, some claim the majority of geotags were created manually and not via automated processes (T. Alder, 22 April 2008, personal communication). This further obscures the lineage of these geographic coordinate data.

To index place-based articles, the *Wikipedia-World* project creates a catalog of geotagged articles (Kühn and Alder 2008). Since geotagging in Wikipedia is chaotic, this process relies on data-mining methods and is largely heuristic (Fig. 11.1). In May 2008, this process found 1,163,797 geotagged articles across 230 languages and 234,474 unique locations (at 1 km resolution). Wikipedia-World uses these data to provide various online mapping services and exports the underlying geographic data as database tables. And the index of place-based articles is growing rapidly. In May 2011, the same process found 1.7 million geotagged articles across 273 languages and 1.1 million unique locations (at 1 km resolution, Fig. 11.2).

11.4 Geography of Authorship

In systems like Flickr and Wikipedia, VGI content itself is spatially clustered (Hecht and Gergle 2010), and Wikipedia articles are also more likely to link to articles about places nearby (Hecht and Moxley 2009). But the literature does not directly address whether

---

[3] These services include *GEOnet Names Server* (GNS) and *Geographic Names Information System* (GNIS) (http://en.wikipedia.org/wiki/User:The_Anomebot2). Using gazetteers as data sources is common for these automated processes, but there are other data sources in use. *Rambot*, for example, uses its own database of 3,141 counties and 33,832 cities to create geographic articles (http://en.wikipedia.org/wiki/User_talk:Rambot).

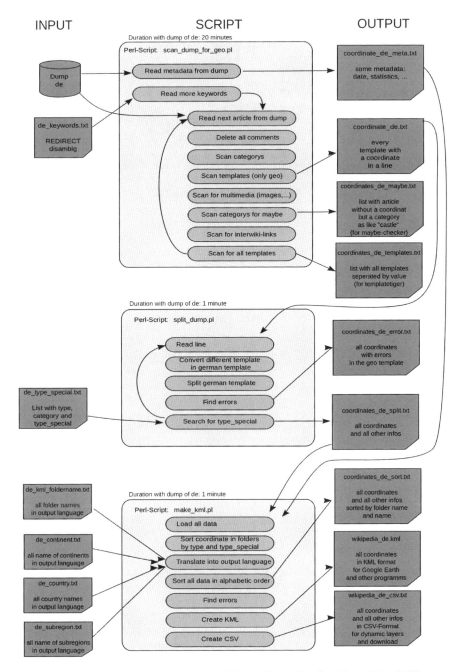

Fig. 11.1 Detailed workflow for geotag data-mining software (Reprinted from Kühn 2008)

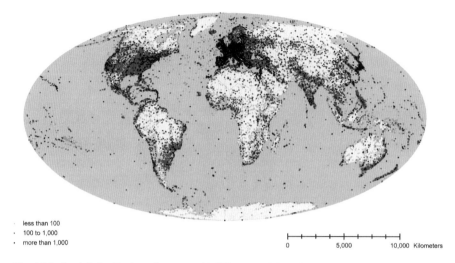

Fig. 11.2 Spatial distribution of geotagged Wikipedia articles, visualized using log-scale density for number of article contributions at 10 km resolution

VGI production processes themselves exhibit regular spatial patterns. This section will discuss a spatial model for contributions, and results that show anonymous contributors exhibit geographic effects that fit an exponential distance decay function.

11.4.1 Data Collection

Wikipedia manages hundreds of individual language-specific databases across three data centers in the United States, Netherlands, and South Korea. Their services use open-source MediaWiki software and data models (MediaWiki 2006). Wikipedia provides article and metadata via periodic dumps of their database and as static HTML files (http://meta.wikimedia.org/wiki/Data_dumps), but historically, these data do not always include complete article contribution records due to their volume and limited operational resources (e.g., the August 2008 dump of the English Wikipedia had 2.5 million articles and 250 million contributions—http://en.wikipedia.org/wiki/Special:Statistics).

The openness of their data lends itself to empirical study by researchers (e.g., Almeida et al. 2007; Priedhorsky et al. 2007; Voss 2005). This study collects data directly via SQL from near real-time replicas of Wikipedia databases, provided by Wikimedia Deutschland's *Toolserver* (http://toolserver.org). These databases use MySQL and the MediaWiki database schema, which organizes articles by revision. Briefly, the *revision* table provides metadata for author contributions and links to the *page* and *text* table for details on the article's contents. For every article, the

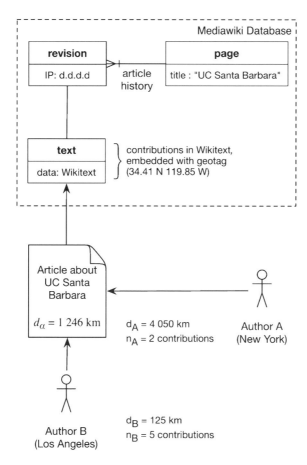

Fig. 11.3 VGI production process in Wikipedia. Authors contribute to place-based articles using Wikitext and embedded geotags that are stored in database tables, including a full history of revisions. For anonymous authors, each revision includes their IP address. In the example, two authors contribute to an article about UC Santa Barbara whose signature distance d_α is 1,246 km, defined as the average distance weighted by contributions, for example, $(2 \cdot 4050 + 5 \cdot 125) / (2 + 5) = 1246$

page table contains a unique identifier and the language-specific title for the article, and the *text* table stores the article's contents. Wikipedians write articles using *Wikitext*, a loosely structured markup language (http://en.wikipedia.org/wiki/Wikipedia:MARKUP), and they embed semistructured metadata *within* the article (Fig. 11.3). The nondeterministic nature of Wikitext's grammar and conventions causes problems for structured data extraction (cf. Sauer et al. 2007). The *WP:GEO* project in Wikipedia governs an infrastructure for adding geographic information to articles (http://en.wikipedia.org/wiki/Wikipedia:GEO). They provide an array of "wiki templates" that have a semistructured syntax for embedding geographic coordinates.

Wikipedia-World's database (Kühn and Alder 2008) from 10 May 2008 uses an extensive data-mining process to extract geotags embedded in Wikitext articles (Fig. 11.1).[4] For each geotagged article, we extract all the authoring history and the most recent version from the replica databases.

To simplify computation across language-specific databases, we migrate the authoring histories into a single shared database, where we modify MediaWiki tables to associate a source language for each record (e.g., *page_id* and a new *page_lang* column comprise the primary key instead of only *page_id*) and to remove data incidental to analysis. This data model provides a multilingual abstraction layer to Wikipedia articles, authors, and their contributions. It has tables for *article*, *author*, and *geotag* data, and *author_article* and *geotag_article* association tuples. It also provides fast access to summary statistics per article and per author. The data extraction from the MediaWiki tables results in *page* and *text* with 990,315 articles, *revision* with 32,141,334 author contributions between 2001 and 2008, and *user* with 578,448 registered author accounts. Since the *user* table contains records only for registered authors, the analysis extracts and parses data from the *revision.rev_user_text* column to identify IP addresses for anonymous users and to integrate them into the data model.

11.4.2 Spatial Model of Authorship

Each author in Wikipedia has a "spatial footprint" comprised of all of the articles to which they have contributed. For anonymous authors, we can estimate their location using IP geolocation (Fig. 11.4). For registered authors and bots (Figs. 11.5 and 11.6), we have no direct estimate of their location, although an indirect estimate based on their spatial footprint is possible (Lieberman and Lin 2009). But are there spatial patterns in these interactions between the authors and the places about which they write?

11.4.2.1 Gravity Models

In regional geography and related disciplines, spatial interaction models form the basis of social theories (Haynes and Fotheringham 1984). These models pertain to flows (interactions) between two or more geographic regions. They have a decades-long history in geography dating back to "social physics" in the early twentieth century (Fotheringham 1981; Wilson 1969, 1971). Distance decay or "gravity" models are one

---

[4] Their software targets a predetermined set of 21 languages: Catalan (ca), Chinese (zh), Czech (cs), Danish (da), Dutch (nl), English (en), Esperanto (eo), Finnish (fi), French (fr), German (de), Icelandic (is), Italian (it), Japanese (ja), Norwegian (no), Polish (pl), Portuguese (pt), Russian (ru), Slovak (sk), Spanish (es), Swedish (sv), and Turkish (tr).

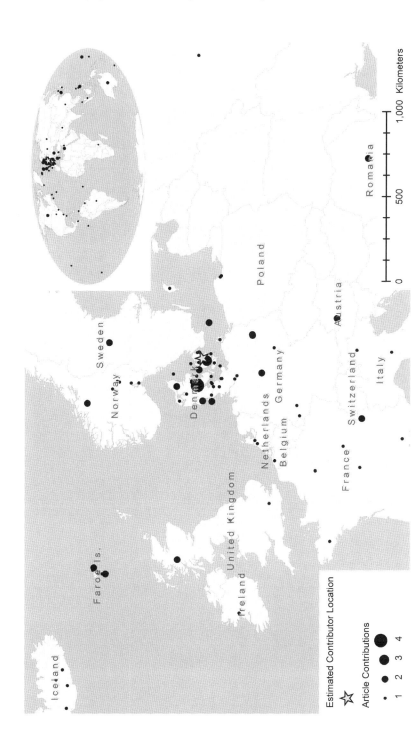

Fig. 11.4 Spatial footprint of an *anonymous* author with 172 contributions to 143 articles in the Danish Wikipedia. The *yellow icon* represents an estimate of the author's location, based in IP geolocation

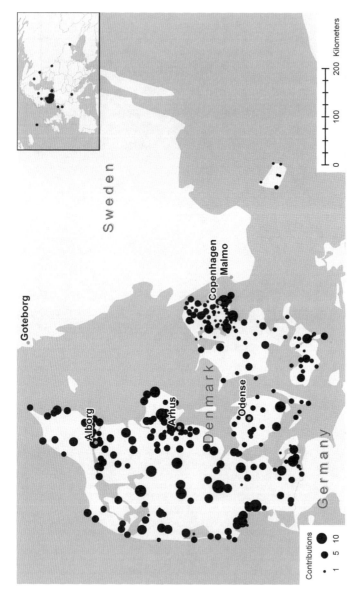

Fig. 11.5 Spatial footprint of a *registered* author with 1,099 contributions to 296 articles in the Danish Wikipedia. *Markers* represent geotagged location of each article edited by author, the vast majority of which are clustered inside Denmark

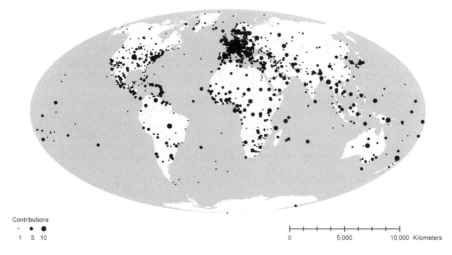

Fig. 11.6 Spatial footprint of a *bot* with 3,006 contributions to 1,601 articles in the Danish Wikipedia

type of spatial interaction model. They use "mass" functions to deal with scale and distance effects. The general gravity model (Sen and Smith 1995, p. 3) is

$$T_{ij} = A_i \cdot B_j \cdot F(d_{ij}), \tag{11.1}$$

where T_{ij} is the interaction between population centers i and j; A_i and B_j are unspecified origin and destination weight (mass) functions; d_{ij} is the spatial factor, or distance between regions i and j; and $F(d_{ij})$ is an unspecified distance decay function, which is commonly a power, exponential, or gamma (combined) function (Sen and Smith 1995, pp. 93–99).

In spatial information theory, an individual's *information field* is the spatial distribution of the "knowledge an individual has of the world" (Morrill and Pitts 1967, p. 406) and is a factor when modeling sociospatial behaviors, like diffusion of innovation or migration (Hägerstrand 1967). An individual's information field decays as the distance from the individual increases. In quantitative geography, gravity models formalize spatial interaction analysis by using this type of distance decay function (Fotheringham and O'Kelly 1989; Sen and Smith 1995). When Wikipedians choose to write about a place, their mean information fields should exhibit distance decay effects found in other sociospatial phenomena, like innovation diffusion. When Wikipedians as a group write more articles, for example, they expand the overall spatial coverage of Wikipedia articles. But when an individual Wikipedian writes an article about a place, that place is likely to be nearby. Thus, our hypotheses for this study are (a) Wikipedians write articles about nearby places more often than distant ones and (b) this likelihood follows an exponential distance decay function.

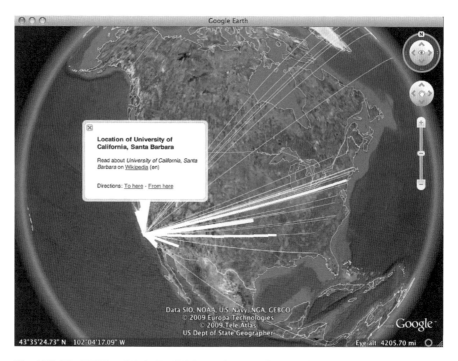

Fig. 11.7 The UCSB article in English has a signature distance of 533 km based on 135 anonymous authors with 719 revisions. Each contribution is shown as a *white line*, with *thicker lines* denoting more contributions

11.4.2.2 Gravity Model for VGI Production

To model VGI production as a spatial process, we define a probabilistic model where the dependent variable is a likelihood for interaction, based on a spatial factor. Specifically, we use a *probabilistic invariant exponential gravity model* (Sen and Smith 1995, p. 102). In terms of Eq. 11.1, T_{ij} is converted to the probability of an interaction based on a spatial factor. The mass terms A_i and B_j are combined into a single invariant constant K to allow for uneven distributions of authors and articles over the Earth's surface. Finally, $F(d_{ij}) = exp(-\beta d_{ij})$, an exponential distance decay function.

$$\Pr(d = d_\alpha) = K \cdot exp(-\beta d_\alpha), \quad \text{where} \quad d = d' \pm \varepsilon. \tag{11.2}$$

Equation 11.2 shows the model using the probability $\Pr(d=d_\alpha)$ as the likelihood that a given article has a signature distance d_α equal to a distance d within a range of $d' \pm \varepsilon$ (K and β are empirically derived constants). For this spatial model, we use a "signature distance" d_α metric to measure the proximity effect for a given article (Hardy et al. 2012). The metric is the average distance between an article and its n authors, weighted by relative number of contributions from each author (Figs. 11.3 and 11.7). That is, each anonymous author has a spatial footprint that is the set of contributions made to any geotagged article by that author. Every author has a single footprint, and every article belongs to its authors' footprints. This model requires a known location for both articles and authors, so we use MaxMind's GeoLite City database, which uses proprietary methods to convert IP addresses into geographic coordinates, to estimate the locations of anonymous Wikipedians whose IP addresses are embedded into their contributions.[5] Location-based services have driven the development of methods to convert IP addresses into geographic coordinates (Muir and Oorschot 2009; Stanger 2008) and to evaluate positional accuracy (Gueye et al. 2006; 2007).

11.4.2.3 Model Results by Article

To fit the model in Eq. 11.2 to the study data, we use an ordinary least squares regression method with a logarithmic transformation to a linear model:

$$\ln\left[\Pr(d=d_\alpha)\right] = \ln K - \beta\, d_\alpha. \tag{11.3}$$

All geographic calculations use ~10-km resolution and great circle distances (where $1'=1.852$ km). We selected the sample from available data to satisfy the methodological requirements that articles have at least one anonymous contribution (for author location estimates) and that articles have one and only one geotag (for signature distance metric). We convert the units of d_α from km to 10^3 km, and use observed relative frequency for $Pr(d=d_\alpha)$. The model fits at $K=0.0022$ and $\beta=0.2842$ ($n=438{,}077$; $R^2=0.9005$; $p<0.01$; $f=17{,}480$; $DF=1{,}930$). When signature distances are relatively low ($d_\alpha<2$), there is no correlation across language databases, suggesting spatial behavior is idiosyncratic across languages.

11.4.2.4 Model Results by Article Category

To test whether signature distances vary by category, we collected categorical data for English articles. Contributors may categorize Wikipedia articles into one or more categories. These categories are not strictly tags but rather registered categories, although anyone may create a new category. These categories are often descriptive of a topic such as "14th-century architecture" or "Art museums and galleries in Paris." They may be editorial, however, and denote workflow items such as "Tokyo railway station stubs," or "All articles needing style editing," or "Articles lacking sources from December 2009." The category space is flat with no consistent nomenclature. Each article's category is displayed at the bottom of the article, and each category has an "article" that lists all articles belonging to that category. From our

---

[5] Wikipedia provides access to IP addresses for anonymous, but not registered, Wikipedians. Reportedly, Wikipedia logs IP addresses for all contributions—from anonymous and registered Wikipedians alike—but they restrict access to those data to authorized administrators.

Table 11.2 Popular topic keywords in English articles, sorted by distance

1,000 km	2,000 km	3,000 km	4,000 km	5,000 km	6,000 km
Carolina	Area	Airports	Areas	Paris	Islands
Channel	California	Architecture	Articles		
County	Census-designated	Building	Australia		
Illinois	Communities	Buildings	Communes		
Indiana	England	Cities	Containing		
Metropolitan	Established	Cleanup	Districts		
Michigan	Establishments	District	Former		
Micropolitan	London	Lacking	Geography		
Missouri	Museums	Municipalities	Language		
Ohio	New	National	Mountains		
Pennsylvania	North	Needing	Prefecture		
Television	Opened	Places	Province		
Texas	Railway	Populated	Region		
TV	States	References	Sites		
Washington	Stations	State	South		
York	United	Structure	Statements		
	Venues	Structures	Stubs		
	Villages	Towns	Text		
		West	Wikipedia		

study, we collected 8,474 unique categories with at least ten English articles, comprising 372,793 articles.[6] We then extracted 4,512 unique keywords (minus common words) from the category title to create an inverted index of category keywords. Each index entry has a unique category keyword, the number of articles that belong to the category, and a mean signature distance for those articles.

For topic keywords with at least 50 articles, Table 11.2 shows the popular topic keywords in English articles by the mean signature distance d_α ($n=372,793$; mean $=3,049$ km). While not conclusive, there is some evidence that signature distances do vary by topic. Topic keywords with lower mean distances are "local" in scope such as cities ("[New] York"), state names, administrative boundary terms ("County" or "Metropolitan"), and buildings ("Museums"). Those with higher mean distances were "regional" in scope such as non-English speaking cities ("Paris"), country names ("Australia"), and regional boundaries ("Islands" or "Province" or "Region").

11.5 Discussion

This section presents some further research issues on architectural, social, and methodological factors, beginning with how both geotagging and geolocation could better support VGI production processes.

---

[6] Other languages also have categories, but this content analysis is restricted to English.

11.5.1 Architectural Factors

The lack of well-structured geotags is problematic. In particular, further research on methods for specifying geotags as *first-class metadata*—rather than as the most basic common denominator of latitude, longitude coordinates—is needed. If collaborative online gazetteers with large-scale global coverage were to emerge, they might serve as a basis for toponym-indexed geotags and thus relieve users from low-level georeferencing tasks. In the meantime, collaborative methods are a possible approach to improving geotag metadata, especially within scientific communities. Currently, geotagging schemes are opaque and inconsistent and are done by automated bots or by users who specify geographic coordinates interactively from a general-purpose mapping service. Neither of these schemes preserve semantic or context information about place and instead leave only precise numerical coordinates of ambiguous intent.

For decades, metadata has been the ever-present, cure-all solution to heterogeneous data integration and use. Yet high-quality, ubiquitous metadata is extremely rare in practice, despite geospatial data infrastructures that are designed to be interoperable and metadata-centric (de By et al. 2009; van Loenen et al. 2009). Current VGI systems may provide insights on how users could produce and manage better metadata for geotags. Metadata is "data about data," intended to facilitate data discovery, integration, and use (or reuse). Practitioners often standardize metadata syntax and semantics, but adherence to metadata standards is extremely rare in distributed systems, especially large or global ones; this is hereafter referred to as the "metadata problem." GIS usually assumes strongly typed spatial data representations, and GIScientists have developed disambiguation methods (e.g., toponym resolution or fuzzy boundaries) for spatial data that do not comply with these structures. These complexities make metadata important for geospatial integration and use. VGI systems, however, successfully integrate heterogeneous data sources on a global scale without solving the metadata problem directly. VGI systems use "best effort" geotagging methods and representations to avoid the complexity of richer GIScience approaches to georeferencing. Moreover, the VGI notion of metadata, and its production and management, is different than in geospatial data infrastructures.

Scientific communities have collaborated on metadata standards and conventions, such as CF (Hankin et al. 2009) and its predecessor (COARDS 1995), but in a study of earth science datasets published via the OPeNDAP protocol (Hardy et al. 2006), they do not accurately follow these conventions. In fact, only a minority of them *claims* their convention (as required), and even of those, only a fraction *accurately* adhere to their stated convention. In practice, scientific data sharing varies by discipline. Ecologists, for example, take idiosyncratic approaches to data sharing and reuse, which depend on disciplinary knowledge and social factors (Zimmerman 2007). This metadata problem forces scientists to use specialized knowledge and manual effort for data reuse.

Wikipedia may provide some lessons for metadata production and management in geospatial data infrastructures (Table 11.3). GIScientists may consider the wiki approach to metadata production and use to address how they might integrate the

Table 11.3 Applying Wikipedia approaches to geotagging

Approach	Benefit
Use trivial geotagging	They avoid the complexity (and implications) of GIScience approaches to georeferencing. Typically, they use decimal degree geographic coordinates in an assumed datum (WGS84) and without enforcing numerical precision. For example, in Wikipedia, the geotag for UCSB is (−119.84813°, 34.41254°), where 0.00001° ≈ 1 m precision[a]. Flickr uses a similar approach but saves context for how users select location (e.g., a zoom level on an interactive map) (Jankowski et al. 2010)
Seemingly trivial metadata structures	Wikipedia uses *Wikitext* (Sauer et al. 2007), a lightweight markup language, for its article content and metadata. Flickr uses perhaps the simplest metadata structure of all: tags which are simply any word or phrase in an uncontrolled vocabulary
Use bots extensively	Wikipedia has hundreds of semiautomated programs to perform a wide variety of editorial functions from removing vandalism to extracting metadata to suggesting work
Promote "refined" and flag problematic content	The community identifies content that is exemplary or meets certain quality standards, and promotes this content. They flag any content that needs further work which is then suggested to those looking for content to work on. They also flag content that is subject to controversy or "edit wars" (Viégas et al. 2007a)
Lazy, but rapid integration (most popular first).	Mashups and other rapid prototyping use service-level, rather than data-level, integration. They focus on APIs and "cookbooks" or working examples rather than formalized specifications. This approach enables rapid integration but also lazy integration since they target specific uses with partial APIs
Complete histories of revisions	The complete context for changes is always available (and easily accessible) when issues arise or for tools to utilize
Data mining	They have tools that search through content looking for ways to improve their service, and they provide APIs for anyone who wishes to mine content programmatically

[a]Typical GPS units report coordinates at ≈1 m resolution in WGS84 datum

increasingly voluminous VGI data into metadata-based geospatial data infrastructures. In particular, the novelty and practicalities of VGI production may benefit the scientific community as they confront increasingly large-scale, heterogeneous data integration problems in metadata-poor environments—a recurrent research area (Hardy 2010; Hardy et al. 2006; Lanter 1991; Rodriguez et al. 2009).

Ideally for analysis, all contributions would have explicit geographic information for the author's location. But these data are not available in most VGI applications, including Wikipedia. Thus, geolocation methods are problematic for VGI contributions due to constraints in data availability and also privacy concerns. This study exploits IP addresses to apply geolocation methods for anonymous contributors. IP geolocation methods, however, are inherently both spatially and temporally dynamic in nature, inaccurate at large scales (i.e., street-level), and relatively easily evaded by savvy users or anonymizing software (Duckham and Kulik 2005; Muir and Oorschot 2009).

Alternatives are similarly constrained. Current survey-based methodologies are limited (e.g., Nov (2007) used email solicitations which yielded about 150 authors) due to the level of anonymity in Wikipedia. Spatial analysis methods based on behavioral patterns, such as the locations of the articles to which an author has contributed, are relatively new in this research area (Lieberman and Lin 2009). Combined approaches (i.e., where quantitative spatial analysis models are calibrated with surveyed locations) may prove useful. Furthermore, VGI is increasingly moving into the mobile domain where users leave (often implicitly) digital traces more conducive to geolocation methods, such as GPS-enabled smart phones, cell phone tower records, or even georeferenced photos (Girardin et al. 2008; González et al. 2008). These trace data can enable spatial data-mining methods for tracking trajectories of individuals or groups (Kisilevich et al. 2010).

Interdisciplinary approaches may also prove useful since geolocation methods are used in other domains. Geographic profiling, for example, is a criminal "investigative methodology that uses the locations of a connected series of crimes to determine the most probable area of offender residence" (Rossmo 2000, p. 1). Geographic profiling systems use spatial distribution and probability distance strategies, such as center of the circle, centroid, median, geometric mean, harmonic mean, and center of minimum distance algorithms (Snook et al. 2005).

11.5.2 Social Factors

How do social factors (such as communication, culture, language, settlement patterns (diaspora), and socioeconomic status) influence VGI contributions? The production and use of VGI will likely shift spatial data infrastructures architecturally to provide for social factors (Budhathoki et al. 2008; Coleman et al. 2009; Elwood 2010; Elwood et al. 2012; Sieber and Rahemtulla 2010). Further modeling of social characteristics in the collaborative authorship process might include spatiotemporal constraints on social networks of Wikipedians or future VGI systems based on increasingly rich social network technologies.

For example, the VGI production model defines work in the signature distance metric in simple terms as an edit count. But the literature has many different definitions for "work," including edit counts (Kittur et al. 2007), edit deltas (Zeng et al. 2006), edit similarity (i.e., information distance) (Voss 2005), edit longevity (i.e., age or survival or persistence) (Adler and de Alfaro 2007; Wöhner and Peters 2009), and edit visibility (Priedhorsky et al. 2007). These metrics may better model social processes and clarify sociospatial factors in collaborative authorship. In particular, edit longevity and edit visibility more directly reflect social phenomena like "edit wars"[7] and herding behaviors, respectively. Similarly, our study had limited comparison of geographic effects across article categories, but further analysis on content-centric dimensions may help study these social processes.

Another question is whether language and population demographics affect spatial patterns in VGI. Ideally, spatial models for collective authorship would include probabilities for how many potential Internet users who speak a given language are available to make contributions for any given location. This study did not normalize authorship by population or potential speakers due to a lack of available data at the needed resolution. Balk and Yetman (2004) provide relatively large-scale data for population estimations but do not include speaker estimates. Moreover, Internet use is spatially variant (Billón et al. 2008; Zook 2005) where large-scale Internet population estimates are not readily available.

Furthermore, at a global scale of the Internet, the concept of "near" is different than in social science research that conducts studies at smaller scales (Graham 1998). For example, in our study, less than 2,000 km is relatively "near" compared to the full scope of available Wikipedia contributors. Notwithstanding global or even virtual travel (Urry 2002), typical scales for nearness are much smaller than 2,000 km, such as walking in urban centers (Turner and Penn 2002) or commuting distance via transportation networks (Weber 2003).

Finally, the notion of collective action through new media is at the core of VGI. VGI and the related phenomena of *neogeography* expand the notion of the "public" from prior work in public participation GIS to include much larger, distributed civic participation (Elwood 2008; Hall et al. 2010; Sieber 2006; Sui 2008).

11.5.3 Methodological Factors

Finally, what methodological advancements are required for future research? The high-volume, highly distributed, real-time, and social nature of VGI is inherently difficult to analyze with simple computational methods. Rather, as shown in our research, significant computational resources and data-mining methods are better suited for empirical studies of VGI. Data-mining methods with a resolution at sub-article levels, such as sections or paragraphs, would improve the sample size. Also, geographic and network visualization methods may enable a visual analytics approach to studying VGI.

---

[7] To address these actions, Wikipedia has a policy that states "Wikipedians should interact in a respectful and civil manner" (http://en.wikipedia.org/wiki/Wikipedia:Five_pillars).

In the coming years, wiki-based VGI systems, where the provenance of information is transparent, may no longer apply as the ephemeral and social nature of VGI rises. Specifically, one of the key challenges in methodology will be to effectively cope with data deluge in an environment where data are filtered through social networks (Watts et al. 2002). If information primitives become based on distance or connectivity through fluctuating social networks, then traditional information science methodologies will not be applicable at large scales. Social network methodologies, which are based on graph theory, are now being used to study online collaborative environments, such as massively multiplayer gaming (Szell and Thurner 2010), and blogging (Liben-Nowell et al. 2005).

11.6 Conclusion

Although the underlying technologies of online geographic services have been in development for many years, the behavioral impacts of VGI production are largely unknown. These services require large-scale data interoperability and collaboration, for example, neither of which has a purely technical solution. VGI production will likely create new knowledge politics, and many of the problematic emerging issues are institutional and sociobehavioral in nature, not technological (Elwood 2008, 2010; Goodchild 2008). For example, the capacity of a ubiquitous Internet to reduce communication costs has raised questions of whether geographic distance matters in information and economic production (Cairncross 1997; Castells 2010).

This chapter addresses two basic questions in VGI production, namely, (1) how individuals contribute place-based information to a digital commons and (2) authorship dynamics of such collective effort. Our approach takes a user-centric perspective of spatial behavior in VGI production. Research on VGI production is a nascent area with many unexplored avenues, in architectural, social, and methodological factors. These factors form a basis of a research agenda that asks (a) how to improve the structure and quality of essential geographic metadata, (b) how language and demographics affect VGI, and (c) how social networks change the nature of VGI.

Acknowledgments This research was supported in part by the National Science Foundation (awards #BCS-0849625 "Collaborative Research: A GIScience Approach for Assessing the Quality, Potential Applications, and Impact of Volunteered Geographic Information" and #IIS-0431166 "Collaborative Research: Integrating Digital Libraries and Earth Science Data Systems") and the US Army Research Office (award #W911NF0910302). Thanks to Wikimedia Deutschland, e.V. in Berlin, Germany, for providing the helpful Toolserver service (http://toolserver.org). They provided database access, Web hosting, and computational resources for this research. Thanks to Tim Alder and Stefan Kühn for comments on geotagging methods in Wikipedia and for sharing their data-mining software and results. Thanks also to reviewer comments and for the many discussions with students and faculty at UCSB's Center for Information Technology and Society and Center for Spatial Studies.

References

Adler, B. T., & de Alfaro, L. (2007). A content-driven reputation system for the Wikipedia. *WWW'07*. doi:10.1145/1242572.1242608.

Alexa Internet, Inc. (2009). Alexa traffic rank. http://www.alexa.com/siteinfo/wikipedia.org. Accessed Dec 2009.

Almeida, R., Mozafari, B., & Cho, J. (2007, March 26–28). *On the evolution of Wikipedia*. Paper presented at the 1st international conference on weblogs and social media, Boulder, CO.

Amitay, E., Har'El, N., Sivan, R., & Soffer, A. (2004). Web-a-where: Geotagging web content. *SIGIR'04*. doi:10.1145/1008992.1009040.

Balk, D., & Yetman, G. (2004). Gridded population of the world (GPWv3). http://sedac.ciesin.columbia.edu/gpw/. Accessed Feb 2010.

Beenen, G., Ling, K., Wang, X., Chang, K., Frankowski, D., Resnick, P. et al. (2004). Using social psychology to motivate contributions to online communities. *CSCW'04*. doi:10.1145/1031607.1031642.

Benkler, Y. (2002). Coase's penguin, or, Linux and the nature of the firm. *The Yale Law Journal, 112*(3), 369–446.

Billón, M., Ezcurra, R., & Lera-López, F. (2008). The spatial distribution of the internet in the European Union: Does geographical proximity matter? *European Planning Studies, 16*(1), 119–142.

Brabham, D. C. (2008). Crowdsourcing as a model for problem solving: An introduction and cases. *Convergence: The International Journal of Research into New Media Technologies, 14*(1), 75–90.

Broder, A., Kumar, R., Maghoul, F., Raghavan, P., Rajagopalan, S., Stata, R., et al. (2000). Graph structure in the web. *Computer Networks, 33*(1–6), 309–320.

Browne, M. (1997). The field of information policy: Fundamental concepts. *Journal of Information Science, 23*(4), 261–275.

Bryant, S. L., Forte, A., & Bruckman, A. (2005). Becoming Wikipedian: Transformation of participation in a collaborative online encyclopedia. *GROUP'05*. doi:10.1145/1099203.1099205.

Budhathoki, N. R., Bruce, B., & Nedovic-Budic, Z. (2008). Reconceptualizing the role of the user of spatial data infrastructure. *GeoJournal, 72*(3), 149–160.

Cairncross, F. (1997). *The death of distance: How the communications revolution will change our lives*. Cambridge, MA: Harvard Business School Press.

Castells, M. (2010). *The rise of the network society* (2nd ed.). West Sussex: Wiley-Blackwell.

Çelik, T. (2005). Geo microformat specification [draft]. http://microformats.org/wiki/geo. Accessed Dec 2009.

COARDS (1995). Conventions for the standardization of NetCDF files. http://ferret.wrc.noaa.gov/noaa_coop/coop_cdf_profile.html. Accessed July 2009.

Coleman, D. J., Georgiadou, Y., & Labonte, J. (2009). Volunteered geographic information: The nature and motivation of produsers. *International Journal of Spatial Data Infrastructures Research, 4*, 332–358.

Cronin, B. (2001). Hyperauthorship: A postmodern perversion or evidence of a structural shift in scholarly communication practices? *Journal of the American Society for Information Science and Technology, 52*(7), 558–569.

Daviel, A., & Kaegi, F. (2007). Geographic registration of HTML documents [draft]. http://tools.ietf.org/pdf/draft-daviel-html-geo-tag-08.pdf. Accessed Dec 2009.

de By, R., Lemmens, R., & Morales, J. (2009). A skeleton design theory for spatial data infrastructure. *Earth Science Informatics, 2*(4), 299–313.

Duckham, M., & Kulik, L. (2005). A formal model of obfuscation and negotiation for location privacy. In H. W. Gellersen et al. (Eds.), *Pervasive 2005* (pp. 152–170, LNCS, Vol. 3468). Berlin: Springer.

Elwood, S. (2008). Volunteered geographic information: Future research directions motivated by critical, participatory, and feminist GIS. *GeoJournal, 72*(3), 173–183.

Elwood, S. (2010). Geographic information science: Emerging research on the societal implications of the geospatial web. *Progress in Human Geography, 34*(3), 349–357.

Elwood, S., Goodchild, M. F., & Sui, D. Z. (2012). Researching volunteered geographic information: Spatial data, geographic research, and new social practice. *Annals of the Association of American Geographers, 102*(3), 571–590. doi:10.1080/00045608.2011.595657.

Forte, A., & Bruckman, A. (2008). Scaling consensus: Increasing decentralization in Wikipedia governance. *HICSS'08*. doi:10.1109/HICSS.2008.383.

Fotheringham, A. S. (1981). Spatial structure and distance-decay parameters. *Annals of the Association of American Geographers, 71*(3), 425–436.

Fotheringham, A. S., & O'Kelly, M. E. (1989). *Spatial interaction models: Formulations and applications*. Dordrecht: Kluwer Academic.

Friedman, T. L. (2005). *The world is flat: A brief history of the twenty-first century*. New York: Farrar, Straus, and Giroux.

Girardin, F., Calabrese, F., Fiore, F., Ratti, C., & Blat, J. (2008). Digital footprinting: Uncovering tourists with user-generated content. *IEEE Pervasive Computing, 7*(4), 36–43.

Golder, S. A., & Huberman, B. A. (2006). Usage patterns of collaborative tagging systems. *Journal of Information Science, 32*(2), 198–208.

González, M. C., Hidalgo, C. A., & Barabási, A.-L. (2008). Understanding individual human mobility patterns. *Nature, 453*, 779–782.

Goodchild, M. F. (2000). Communicating geographic information in a digital age. *Annals of the Association of American Geographers, 90*(2), 344–355.

Goodchild, M. F. (2004). Scales of cybergeography. In E. Sheppard & R. B. McMaster (Eds.), *Scale and geographic inquiry: Nature, society, and method* (pp. 154–169). Malden: Blackwell.

Goodchild, M. F. (2008). Geographic information science: The grand challenges. In J. P. Wilson & A. S. Fotheringham (Eds.), *The handbook of geographic information science* (pp. 596–608). Malden: Blackwell.

Graham, S. (1998). The end of geography or the explosion of place? Conceptualizing space, place and information technology. *Progress in Human Geography, 22*(2), 165–185.

Griffith, V. (2007). WikiScanner. http://wikiscanner.virgil.gr/. Accessed February 2009.

Gueye, B., Ziviani, A., Crovella, M., & Fdida, S. (2006). Constraint-based geolocation of internet hosts. *IEEE/ACM Transactions on Networking, 14*(6), 1219–1232.

Gueye, B., Uhlig, S., & Fdida, S. (2007). Investigating the imprecision of IP block-based geolocation. In S. Uhlig, K. Papagiannaki, & O. Bonaventure (Eds.), *Passive and active network measurement* (pp. 237–240, LNCS, Vol. 4427). Berlin: Springer.

Hägerstrand, T. (1967). *Innovation diffusion as a spatial process*. Chicago: University of Chicago Press.

Hall, G. B., Chipeniuk, R., Feick, R. D., Leahy, M. G., & Deparday, V. (2010). Community-based production of geographic information using open source software and Web 2.0. *International Journal of Geographical Information Science, 24*(5), 761–781.

Hankin, S. C., & 14 co-authors (2009). NetCDF-CF-OPeNDAP: Standards for ocean data interoperability and object lessons for community data standards processes. *OceanObs'09: Sustained ocean observations and information for society*. doi:10.5270/OceanObs09.cwp.41.

Hardy, D. (2008, October 15–19). *Discovering behavioral patterns in collective authorship of place-based information*. Paper presented at the 9th international conference of the association of internet researchers, Copenhagen, Denmark.

Hardy, D. (2010, September 14). *"Title not required": The wikification of geospatial metadata*. Paper presented at the GIScience workshop on the role of volunteer geographic information in advancing science, Zurich, Switzerland.

Hardy, D., Janée, G., Gallagher, J., Frew, J., & Cornillon, P. (2006). Metadata in the wild: An empirical survey of OPeNDAP-accessible metadata and its implications for discovery. *Eos Trans. AGU, 87*(52), Fall Meet. Suppl., Abstract IN54A-04.

Hardy, D., Frew, J., & Goodchild, M. F. (2012). Volunteered geographic information production as a spatial process. *International Journal of Geographical Information Science*. doi:10.1080/13658816.2011.629618.

Haynes, K. E., & Fotheringham, A. S. (1984). *Gravity and spatial interaction models*. Beverly Hills: Sage.

Hecht, B. J., & Gergle, D. (2010). On the "localness" of user-generated content. *CSCW'10*. doi:10.1145/1718918.1718962.

Hecht, B., & Moxley, E. (2009). Terabytes of Tobler: Evaluating the first law in a massive, domain-neutral representation of world knowledge. In K. S. Hornsby (Ed.), *Spatial information theory* (pp. 88–105, LNCS, Vol. 5756). Berlin: Springer.

Hill, L. L. (2006). *Georeferencing: The geographic associations of information*. Cambridge, MA: MIT Press.

Huberman, B. A., Pirolli, P. L., Pitkow, J. E., & Lukose, R. M. (1998). Strong regularities in World Wide Web surfing. *Science, 280*, 95–97.

Jankowski, P., Andrienko, G., Andrienko, N., & Kisilevich, S. (2010). Discovering landmark preferences and movement patterns from photo postings. *Transactions in GIS, 14*(6), 833–852.

Kisilevich, S., Mansmann, F., Nanni, M., & Rinzivillo, S. (2010). Spatio-temporal clustering. In O. Maimon & L. Rokach (Eds.), *Data mining and knowledge discovery handbook* (2nd ed., pp. 855–874). New York: Springer.

Kittur, A., Suh, B., Pendleton, B. A., & Chi, E. H. (2007). He says, she says: Conflict and coordination in Wikipedia. *CHI'07*. doi:10.1145/1240624.1240698.

Kühn, S. (2008). Workflow from Wikipedia-Dump to geodata. http://de.wikipedia.org/wiki/Datei:Wikipedia_Geodata_Workflow.svg. Accessed Oct 2008. Creative Commons license (CC BY-SA 3.0).

Kühn, S., & Alder, T. (2008). Wikipedia-World [in German]. http://de.wikipedia.org/wiki/Wikipedia:WikiProjekt_Georeferenzierung/Wikipedia-World. Accessed Oct 2008.

Kunze, J. (1999). Encoding Dublin core metadata in HTML. http://www.ietf.org/rfc/rfc2731.txt. Accessed Mar 2008.

Lanter, D. P. (1991). Design of a lineage-based meta-data base for GIS. *Cartography and Geographic Information Science, 18*(4), 255–261.

Liben-Nowell, D., Novak, J., Kumar, R., Raghavan, P., & Tomkins, A. (2005). Geographic routing in social networks. *Proceedings of the National Academy of Sciences, 102*(33), 11623–11628.

Lieberman, M., & Lin, J. (2009, May 17–20). *You are where you edit: Locating Wikipedia contributors through edit histories*. Paper presented at the 3rd International AAAI Conference on Weblogs and Social Media, San Jose, CA.

Lih, A. (2009). *The Wikipedia revolution: How a bunch of nobodies created the world's greatest encyclopedia*. New York: Hyperion.

Marston, S. A., Jones, J. P., & Woodward, K. (2005). Human geography without scale. *Transactions of the Institute of British Geographers, 30*(4), 416–432.

MediaWiki (2006). The technical manual for the MediaWiki software: Database layout. http://www.mediawiki.org/wiki/Manual:Database_layout. Accessed Mar 2008.

Miller, N. (2005). Wikipedia and the disappearing "Author". *ETC: A Review of General Semantics, 62*(1), 37–41.

Morrill, R. L., & Pitts, F. R. (1967). Marriage, migration, and the mean information field: A study in uniqueness and generality. *Annals of the Association of American Geographers, 57*(2), 401–422.

Muir, J. A., & Oorschot, P. C. V. (2009). Internet geolocation: Evasion and counterevasion. *ACM Computing Surveys, 42*(1), 1–23.

Nardi, B. A., Schiano, D. J., Gumbrecht, M., & Swartz, L. (2004). Why we blog. *Communications of the ACM, 47*(12), 41–46.

Nov, O. (2007). What motivates Wikipedians? *Communications of the ACM, 50*(11), 60–64.

O'Reilly, T. (2005). What is Web 2.0: Design patterns and business models for the next generation of software. http://oreilly.com/web2/archive/what-is-web-20.html. Accessed Mar 2008.

Priedhorsky, R., Chen, J., Lam, S. T. K., Panciera, K., Terveen, L., & Riedl, J. (2007). Creating, destroying, and restoring value in Wikipedia. *GROUP'07*. doi:10.1145/1316624.1316663.

Rodriguez, M. A., Bollen, J., & Sompel, H. V. D. (2009). Automatic metadata generation using associative networks. *Transactions on Information Systems, 27*(2), 1–20.

Rossmo, D. K. (2000). *Geographic profiling*. Boca Raton: CRC Press.

Sauer, C., Smith, C., & Benz, T. (2007). WikiCreole: A common wiki markup. *International Symposium on Wikis*. doi:10.1145/1296951.1296966.

Sen, A., & Smith, T. E. (1995). *Gravity models of spatial interaction behavior*. Berlin: Springer.

Sieber, R. (2006). Public participation geographic information systems: A literature review and framework. *Annals of the Association of American Geographers, 96*(3), 491–507.

Sieber, R. E., & Rahemtulla, H. (2010). *Model of public participation on the geoweb*. Paper presented at the 6th international conference on GIScience, Zurich, Switzerland, September 14–17, 2010.

Snook, B., Zito, M., Bennell, C., & Taylor, P. J. (2005). On the complexity and accuracy of geographic profiling strategies. *Journal of Quantitative Criminology, 21*(1), 1–26.

Stanger, N. (2008). Scalability of techniques for online geographic visualization of web site hits. In A. Moore & I. Drecki (Eds.), *Geospatial vision: New dimensions in cartography* (pp. 193–217). Berlin: Springer.

Sui, D. Z. (2008). The wikification of GIS and its consequences: Or Angelina Jolie's new tattoo and the future of GIS [editorial]. *Computers, Environment and Urban Systems, 32*(1), 1–5.

Sundin, O., & Haider, J. (2007). Debating information control in Web 2.0: The case of Wikipedia vs. citizendium. *Proceedings of the American Society for Information Science and Technology, 44*(1), 1–7.

Szell, M., & Thurner, S. (2010). Measuring social dynamics in a massive multiplayer online game. *Social Networks, 32*(4), 313–329.

Tapscott, D., & Williams, A. D. (2006). *Wikinomics: How mass collaboration changes everything*. New York: Portfolio.

Turner, A., & Penn, A. (2002). Encoding natural movement as an agent-based system: An investigation into human pedestrian behaviour in the built environment. *Environment and Planning B, 29*(4), 473–490.

Urry, J. (2002). Mobility and proximity. *Sociology, 36*(2), 255–274.

van Loenen, B., Besemer, J. W. J., & Zevenbergen, J. A. (Eds.). (2009). *SDI convergence: Research, emerging trends, and critical assessment*. Delft: Netherlands Geodetic Commission.

Viégas, F. B. (2005). *Revealing individual and collective pasts: Visualizations of online social archives*. Ph.D. dissertation, Massachusetts Institute of Technology, Cambridge, MA.

Viégas, F. B., Wattenberg, M., & Dave, K. (2004). Studying cooperation and conflict between authors with history flow visualizations. *CHI'04*. doi:10.1145/985692.985765.

Viégas, F. B., Wattenberg, M., Kriss, J., & van Ham, F. (2007a). Talk before you type: Coordination in Wikipedia. *HICSS'07*. doi:10.1109/HICSS.2007.511.

Viégas, F. B., Wattenberg, M., & McKeon, M. M. (2007b). The hidden order of Wikipedia. In D. Schuler (Ed.), *Online communities and social computing* (pp. 445–454, LNCS, Vol. 4564). Berlin: Springer.

Voss, J. (2005). *Measuring Wikipedia*. Paper presented at the 10th international conference of the International Society for Scientometrics and Informetrics, Stockholm, Sweden, July 24–28, 2005.

Watts, D. J., Dodds, P. S., & Newman, M. E. J. (2002). Identity and search in social networks. *Science, 296*, 1302–1305.

Weber, J. (2003). Individual accessibility and distance from major employment centers: An examination using space-time measures. *Journal of Geographical Systems, 5*(1), 51–70.

Wikimedia Foundation (2010). List of Wikipedias. http://meta.wikimedia.org/wiki/List_of_Wikipedias. Accessed Sept 2010.

Wikipedia (2008). WikiProject geographical coordinates. http://en.wikipedia.org/wiki/Wikipedia:GEO. Accessed Mar 2008.

Wilson, A. (1969). Notes on some concepts in social physics. *Papers in Regional Science, 22*(1), 159–193.

Wilson, A. (1971). A family of spatial interaction models, and associated developments. *Environment and Planning, 3*(1), 1–32.

Wöhner, T., & Peters, R. (2009). *Assessing the quality of Wikipedia articles with lifecycle based metrics*. 5th international symposium on wikis and open collaboration. doi:10.1145/1641309.1641333.

Zachte, E. (2009). Wikimedia visitor log analysis report: Google requests as daily averages, based on sample period [November 2009]. http://stats.wikimedia.org/wikimedia/squids/SquidReportGoogle.htm. Accessed Feb 2010.

Zachte, E. (2010a). Wikimedia report card [January 2010]. http://stats.wikimedia.org/reportcard/. Accessed Feb 2010.

Zachte, E. (2010b). Wikipedia statistics: Overview of recent months. http://stats.wikimedia.org/EN/Sitemap.htm. Accessed Feb 2010.

Zeng, H., Alhossaini, M. A., Ding, L., Fikes, R., & McGuinness, D. L. (2006). Computing trust from revision history. *International Conference on Privacy, Security and Trust.*. doi:10.1145/1501434.1501445.

Zimmerman, A. (2007). Not by metadata alone: The use of diverse forms of knowledge to locate data for reuse. *International Journal on Digital Libraries, 7*(1), 5–16.

Zook, M. (2005). The geographies of the internet. *Annual Review of Information Science and Technology, 40*(1), 53–78.

Chapter 12
Inferring Thematic Places from Spatially Referenced Natural Language Descriptions

Benjamin Adams and Grant McKenzie

Abstract Places are more than just a location and spatial footprint. A sense of place is the result of subjective experience that a person has from being in a place or from interacting with information about a place. Although it is difficult to directly model a person's conceptualization of sense of place in a computational representation, there exist many natural language data online that describe people's experiences with places and which can be used to learn computational representations. In this paper we evaluate the usage of topic modeling on a set of travel blog entries to identify the themes that are most closely associated with places around the world. Using these representations we can calculate the similarity of places. In addition, by focusing on individual or sets of topics we identify new regions where topics are most salient. Finally we discuss how temporal changes in sense of place can be evaluated using these methods.

12.1 Introduction

J. Nicholas Entrikin (1991) has written that narrative accountings of places are essential resources for understanding the world because they provide "a distinct form of knowing that derives from the redescription of experience in terms of a synthesis of heterogeneous phenomena." A key aspect of these narratives is that they come from an individual point of view and therefore capture qualities of subjective experiences. Much volunteered geographic information (VGI) on the Web comes

---

B. Adams (✉)
Department of Computer Science, University of California, Santa Barbara, CA, USA
e-mail: badams@cs.ucsb.edu

G. McKenzie
Department of Geography, University of California, Santa Barbara, CA, USA

in the form of unstructured, natural language descriptions of places on the Earth. Examples of these kinds of descriptions include Wikipedia articles, travel blog entries, and entries from microblogs such as Twitter. VGI place descriptions form rich datasets for geographic analysis; however, the vast quantity of available information begs for automated approaches to aid analysis. In this chapter, we describe results of using topic modeling, a popular natural language-processing technique, to identify the latent topics in a large corpus of travel blog entries that describe places around the world. We examine how the topics are distributed over space and time and how individual or combinations of topics can be drawn on a map to represent places of thematic distinction.

Geography has traditions in both thematic and regional analysis. Thematic geography examines the commonalities and the differences between geographic structures through the lens of a particular theme, e.g., economics or politics. Regional geography focuses on a particular region of the Earth and takes into account the unique spatial organization of that region. Spatial heterogeneity is the notion that "geographic phenomena do not oscillate around a mean, but drift from one locally average condition to another" (Goodchild 2009). It is an important concept in geographic information science since it means that statistical methods that treat the world as flat will fail to accurately model many geographic scale phenomena. In a large corpus of natural language documents, where the documents are associated with one or more locations, the distribution of topics will be spatially heterogeneous. Some common thematic patterns will be found across the documents that span different geographic regions rather uniformly and other themes will be found that are highly spatially and temporally correlated with particular locations and times. Consequently, in these sorts of documents, there is grist for both thematic and regional geographic analysis.

In this chapter, we describe a method for using topic modeling on georeferenced natural language text to construct regions of thematic saliency. Topic modeling is an automated data-mining technique that has enjoyed popularity as an effective way to discover the latent topics in a large corpus of documents (Blei and Lafferty 2009). Informally, a topic is a semantically coherent grouping of terms that tend to co-occur in a document. For example, a topic might be characterized by the words: *wine, vineyard, tasting,* and *cheese*. A corpus of travel blog entries is used as an exemplar in this chapter, but the techniques presented can be adapted to other texts that are ordered in this manner. The text of each blog entry is modeled as a mixture of topics produced through a random generative process and we generalize those results over all the entries in a location. The result is that we can develop dynamic statistical formal models of places out of highly unstructured volunteered data. We show that some of the resulting topics correspond to specific geographic locations and thus are applicable for predictive analytics of the form "Where (or when) is this text about?" whereas other topics provide a means for thematic and comparative exploration.

One result of applying topic modeling to a large corpus is that it effectively reduces the dimensionality of the topic space allowing researchers to identify and compare geographic contexts based on a fixed number of themes. This reduction of the corpus to an interpretable number of thematic dimensions creates potential for different sorts of analysis. For example, as we show later in the chapter, the topics discovered in the travelogues can be used to discern whether places are generally viewed as natural places or rather dominated by descriptions of human made features. In addition, by looking at the most prominent topics for a given place, researchers can identify the landmarks, features, and associated activities that are most salient in a place from the tourist's perspective, which can be compared and contrasted with other data about the feature distributions at a place or descriptions written by locals. They can also be used to find places that otherwise might be very dissimilar but are analogous with respect to specific thematic dimensions. Finally, when looking at a corpus of travelogues that spans over time, researchers can use this methodology to better understand how the touristic image of a place has changed over time.

The fundamental premise behind our approach for creating thematic regions is that a natural language document that describes a place is an observation of phenomena at a particular location, **x**, and the mixtures of topics that compose these kinds of observations will be spatially autocorrelated. This can be viewed as a rewording of Tobler's first law that near places are more similar than far apart places (Tobler 1970). Note, that while the *mixtures* of topics will show spatial autocorrelation, the individual topics that make up those mixtures will have differing degrees of global spatial autocorrelation. In other words, some topics are more local than others.

The remainder of this chapter is organized as follows. In Sect. 12.2, we present background information on topic modeling, place, and related work on using topic modeling to find regional topics. In Sect. 12.3, we present the data collection and preprocessing process. Section 12.4 shows the results of running latent Dirichlet allocation (LDA) on the data and details the method to describe and analyze places from these results. In Sect. 12.5, we show how the topics generated can be visualized and how the regional extent of topics can be mapped, and in Sect. 12.6, we look at temporal analysis of the themes. Finally, we conclude with future research.

12.2 Background

In this section, we provide background information on place, topic modeling, and related work.

12.2.1 Place

According to Tuan, place is space infused with meaning, i.e., a way of making sense of the world (Tuan 1977). Another commonly cited definition of place by Agnew is that it is the combination of location, locale, and sense of place (Agnew 1987). By this definition, sense of place is subjective and is a product not only of the physical structure of a place but also the phenomenological experiences that an individual has when in a place or when observing a reference to a place (e.g., reading about it) (Cresswell 2004). The conceit of the methods presented in this chapter is that what people choose to write about a place reflects their sense of that place, and by generalizing over many people's writings, we can extrapolate an aggregate view of a place. By doing this kind of analysis, we remain agnostic on the question of whether to approach the study of place from a relatively decentered and objective perspective or relatively subjective perspective (Entrikin 1991). Because topic modeling operates on the level of individual documents, analysis can be performed on the level of aggregations of descriptions (as we do below), or it can be done on the level of individual descriptions reflecting a more individualized notion of place.

Despite the importance of place in geography and related disciplines, there has not been much success in formally modeling sense of place in geographic information systems. The operationalization of place has, however, been identified as an important research agenda; notably, an issue of the journal *Spatial Cognition and Computation* was dedicated to this question (Winter et al. 2009). Multidimensional measures of sense of place have been explored in human geography, but they tend to be tested using psychological experimental studies of very specific geographic settings, e.g., lakeshore properties (Jorgensen and Stedman 2006). The goal in this chapter is to explore unsupervised ways of operationalizing place using a much larger, crowdsourced dataset generated by many different people.

12.2.2 Topic Modeling

Generative topic modeling encompasses a suite of unsupervised data-mining methods for uncovering the semantic structure of textual documents in a large corpus (Steyvers and Griffiths 2007). A generative topic model is a statistical model that explains how the words found in documents are generated as the result of a random process. For the most popular generative topic model, LDA, each word in a document is chosen from one of a set of topics that are shared among all the documents in the corpus (Blei et al. 2003). Each document is modeled as a unique mixture of topics (i.e., a multinomial distribution over topics), and each topic in turn is a multinomial distribution over words. Therefore, to generate each word, one can imagine two weighted dice being tossed. The first die has as many sides as there are topics and is weighted uniquely for each document. It is used to probabilistically sample a topic. Then, given the selected topic, we toss another dice specifically weighted for that topic and with as many sides as there are words in the corpus, drawing the word. This generative process is easily extended and can take into account other information, such as authorship, which has lead to a number of variants of LDA (Steyvers et al. 2004). Topic models are *bag-of-words* models, which means that the order and syntactic context of the words in the text do not factor in the result.

The goal of topic modeling is to infer the latent variables (i.e., the weights on the dice) most likely to have generated the observed words in a corpus of existing documents. This inference is a Bayesian-inferencing problem on a large probabilistic graphical model. The main innovations in topic modeling over the last decade have been to develop efficient algorithms for approximating this inference, given that an exact solution to the problem is computationally intractable. Algorithms that use a Gibbs sampling Markov chain Monte Carlo (MCMC) approach have proved effective to approximate parameters (see Griffiths and Steyvers 2004; Bishop 2006).

When running an LDA inference, the inputs are the α and β hyperparameters, the number of topics, and the observed data. The α hyperparameter determines how many topics are assigned to a given document (a very small α will essentially assign one topic to every document). The β hyperparameter determines whether the words are distributed more or less evenly over the topics. A number of excellent implementations of LDA based on MCMC are freely available. For the work presented in this chapter, we used the topic modeling toolkit contained within the MAchine Learning for LanguagE Toolkit (MALLET) (McCallum 2002).

LDA is very modular, and several extensions to LDA have been developed, including ones that allow for learning the number of topics, supervised labels, and correlations between topics (cf. Blei and Lafferty 2006; Teh et al. 2006; Li and McCallum 2006; Blei and Mcauliffe 2008). Extensions add to the computational complexity of the approximate inferencing, however. In the analysis presented in this chapter, we utilized the original LDA model, though the post hoc spatial analyses of the topic modeling results presented here are fully compatible with any of the many topic modeling variants.

12.2.2.1 Similarity of Documents

The Kullback-Leibler divergence, D_{KL}, (also known as the relative entropy) of the topic distributions of two documents can be used as a similarity measure. Let P and Q be probability distributions of a random discrete variable:

$$D_{KL}(P \mid Q) = \sum_i P(i) \log \frac{P(i)}{Q(i)}$$

Kullback-Leibler is an asymmetric measure; i.e., the distance from P to Q is different than the distance from Q to P. If a symmetric measure is desired, then the Jensen-Shannon divergence, D_{JS}, can be used instead:

$$D_{JS}(P \mid Q) = \frac{1}{2} D_{KL}(P \mid M) + \frac{1}{2} D_{KL}(Q \mid M),$$

where $M = \frac{1}{2}(P + Q)$.

12.2.3 Related Work

A location topic model for travelogues has been developed that explicitly decomposes documents into local and global topics (Hao et al. 2010). There has been some work on extending the LDA model to include information on document labels or links, which can be used to train for topics that are predictive of a variable (e.g., a location label) (Wang et al. 2007; Blei and Mcauliffe 2008; Chang and Blei 2009). In addition, models have been developed to specifically train for spatiotemporal themes (Mei et al. 2006). However, by specifically training for topics that are predictive of location or time, we lose the ability to examine how the spatial and temporal distribution of individual topics *differ* from one another. One goal of the work presented here is to be able to characterize the degree to which different locations share or do not share themes. By focusing on post hoc analyses of topic modeling results, we create techniques that are applicable to a wide variety of source data and are not be overspecified for a specific domain. However, all the above approaches are compatible with the work presented here. In addition, although the methodology presented here focuses on text analysis, it can be augmented by incorporating other kinds of data such as images (Serdyukov et al. 2009).

12.3 Data Preprocessing

There are a number of blogging sites on the Web that allow people to post blog entries about their travel experiences. Since we were interested in exploring a diverse set of entries from around the world that were written by a variety of authors, we looked at some of the larger sites, including travelpod.com, travelblog.org, and travellerspoint.com as sources of data. Travelblog.org[1] was chosen given its relatively simple user interface, and a Web crawler was written to download the text for all the public entries through September 2010. In addition to the text, the date and location, in the form of a geographic hierarchy of the entry, were also saved. Travelblog.org entries often have images and video as well, but since we were only interested in textual analysis, we did not download those; however, such information could be used in the future for a more complex semantic analysis of the entries. In total, 309,683 blog entries were downloaded.

The entries were preprocessed for LDA using the following steps. First, some blog entries consist almost entirely of pictures or video, so entries with fewer than 100 words were removed. Second, during our exploratory analysis, we discovered that the words from blogs written in languages other than English tend to be organized by LDA into their own topics. To mitigate this problem, a language detection script was run on each entry, and those entries labeled as non-English were removed. However, due to the presence of some entries that were written in English as well as in another language, some of this effect was still found. Third, the words in the entries were filtered against a standard list of English stop words, and all punctuation and HTML markups were removed. After preprocessing, the input dataset consisted of 275,468 blog entry documents.

Travelblog.org lets users specify the location associated with an entry within a geographic hierarchy (e.g., North America, United States, California, Los Angeles).

---

[1] http://www.travelblog.org

Users usually select the location from a predefined taxonomy, but they can suggest a new location that will be added to the database dependent on moderator approval. There does not appear to be a standard method for determining how countries are subdivided into regions. Some countries such as the United States and France have regions based on first-level political administrative units, but other countries are subdivided into nonpolitical geographic regions or skip directly to local town/city-level regions. Users have flexibility to specify location at any depth of the geographic hierarchy, which means that some entries are labeled in a coarse-grained manner (e.g., California). In addition, it is difficult to compare regions at the same level because they reference areas that vary greatly in size. For example, Andorra and Russia are both at the same level.

We mapped the user-defined locations to geometric representations to aid the visualization and spatial analysis. One mapping was achieved by geocoding each unique location to a latitude-longitude point using the Google Maps geocoding Web service. The service handled all but about 500 locations, which were hand coded. In total, each entry was mapped to one of 10,496 locations. The coarse granularity of some locations, while problematic, is unavoidable as entries about a person's travel experience will inherently be about fuzzy places or even multiple places rather than a point on a map.

The 10,496 locations are not uniformly distributed around the world, and in some places (e.g., in the London area), several location points exist in close proximity to one another. For the purpose of analysis, we created a one-quarter-degree grid over the Earth and aggregated all locations within a single grid square. A new point location was specified as the centroid of the grid square. Although a grid of that size consists of over a million cells, only 7,227 actually had associated entries. As an alternative method, the locations were also mapped to the United Nations Global Administrative Unit Layers (GAUL), a product of the Food and Agricultural Organization (FAO). That representation has three geographic layers mapped to county, state, and country represented in shapefile format, which allowed us to aggregate entries based on political boundaries.

12.4 Modeling Places from LDA Results

Given a set of georeferenced documents, D, the first step in our approach is to use LDA on the corpus to generate a set of topics, T. Following the LDA training, for each document, d, we have a location $\mathbf{x}^d = <x, y>$ and a T-dimensional vector, θ^d, specifying the multinomial distribution of topics for that document. Because the latitude and longitude specified for a georeferenced article is often an approximation of a vaguely defined region and there may be more than one article for a particular location, we need to relax the location. That is done by generating a fixed-sized grid over the Earth and averaging the topic distributions for each grid square, g_i, by finding the mean topic distribution vector, θ^{gi}, for all the documents spatially included within the square. Given the topic distributions for a set of grid squares, the next step is to spatially interpolate a continuous field representation for each topic.

Fig. 12.1 Sample latent topics from 200-topic LDA run

12.4.1 Topic Modeling Results

During our empirical tests, we ran several Gibbs sampling simulations on the dataset for 20, 50, 100, 200, 300, and 400 topics. For the selection of α, β, and topic number parameter values, we followed suggestions presented in Griffiths and Steyvers (2004). We kept the β value constant at 0.1 and the α value at 50/# of topics.

The topics that are discovered using LDA are often represented as an ordered list of the most commonly generated words for that topic. While these lists do rank the words in a topic from the most commonly generated word on down, they ignore the relative importance of the words. For example, in one topic, the top-ranked word might have probability 0.08 and the next most common word 0.01, which means the first word is eight times as commonly generated as the second. In other topics, the first and second word might have very close probabilities. We found that a word cloud visualization that shows the words in relative sizes is more illustrative and will therefore use that method in lieu of lists.

After observing the results, we found that the topics tend to fall into four broad categories: activity topics, feature topics, locality topics, and miscellaneous topics. Activities and features are distributions of words related to things to do and see, respectively. Locality topics consist of words that are associated with a specific geographic location. Miscellaneous topics are ones that do not appear to have any special relationship to traveling per se but nevertheless reflect semantic structures in the language. Many topics fall into more than one of these categories, which are fuzzy. Figure 12.1 shows some sample topics from a 200-topic run.

12.4.2 Adding Location Information

A trained LDA model results in a topic mixture for each blog entry. The LDA model does not include any spatial or temporal information as a parameter, so we do a post hoc analysis of the topic strengths for entries associated with specific locations. We propose that by combining the topic mixtures for all the entries in a location, an aggregate picture of that location's sense of place can be drawn. The best technique for aggregating the topic mixtures is not immediately apparent, however. The blog entries are not evenly distributed over the locations, and as a result, there are some locations with many more entries than at other locations. For our examples in this and the following section, we will identify the location of places with the one-quarter-degree grid squares as described in the previous section. One simple method of calculating a location topic distribution is to take the average θ values for each topic at the location. Let M be the number of entries for a location, θ_{ij} be the value for the ith topic of the jth entry, and L_θ be the location topic distribution:

$$L_{\theta_i} = \frac{\sum_{j=1}^{M} \theta_{ij}}{M}$$

The location topic distributions can then be used to calculate the similarity between places using the relative entropy measures described in Sect. 12.2 as a semantic distance measure. In order to get a similarity value [0.1], we define the similarity of places as an exponentially decaying function of this distance measure: $e^{-D_{KL}}$ (in this case using the asymmetric relative entropy). Alternately, one could use a linear or Gaussian decay function (Shepard 1987). Figure 12.2 maps out the similarity of places to Santa Barbara, CA. The results show a clear example of Tobler's first law of geography that generally speaking near places are more similar than far places (Tobler 1970), but it also captures that Santa Barbara shares some themes with some far places as well (such as urban areas in New York, for example).

This method treats locations with one or two entries as equal to locations with hundreds of entries. Depending on the goal of the analysis, this may or may not be problematic because locations with many entries (such as London) will tend to have a much smoother topic distribution due to the averaging. Intuitively, this makes sense because it is characteristic of global cities such as London that they are extremely heterogeneous and should reflect many different perspectives and themes. However, it is possible to normalize the distributions using the entropy of L_θ as the normalizing parameter.

An entropy measure, L_e, can be used to determine the degree to which a location topic distribution is about a few topics or many topics:

$$L_e = -\sum_{i=1}^{n} p(L_{\theta_i}) \log p(L_{\theta_i})$$

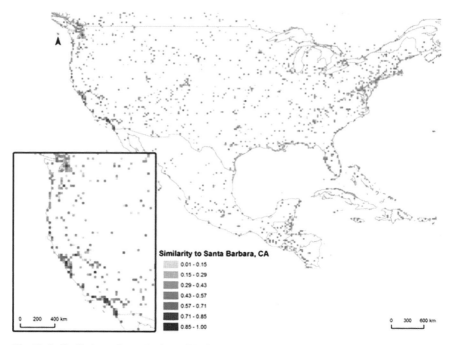

Fig. 12.2 Similarity to Santa Barbara, CA, based on relative entropy measures

The lower the entropy, the fewer the number of topics being written about for that location. Figure 12.3 shows the entropy of all places around Australia where we have five or more entries. This map illustrates that urban areas tend to have more diversity in topics being discussed than rural areas. Presumably, given that we are looking at travel blog entries, this is because certain rural locations are visited by people to undertake specific types of activities. Also, there is less heterogeneity in terms of the human geographic features in those places.

We propose that if an individual topic probability is prominent despite high overall entropy of the topic distribution for a location, it should be considered comparatively more important than an equivalent topic probability in a location with very low entropy. Let L_e be the location entropy, N be the number of topics, and γ be a scalar; the strength s_i of topic i is defined as

$$s_i = \begin{cases} \gamma L_e \theta_i, & \theta_i > \dfrac{1}{N} \\ \theta_i, & \theta_i \geq \dfrac{1}{N} \end{cases}.$$

The conditional is due to the fact that we only want to increase the prominence of topics that already have a probability greater than or equal to $1/N$.

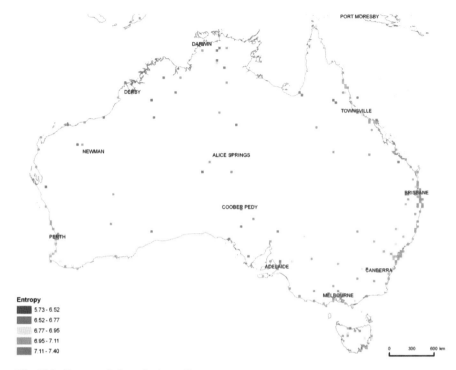

Fig. 12.3 Entropy of places in Australia

12.5 Constructing Regions of Thematic Distinction

Using the s_i as an indicator of the salience of topic i at a specific point location, we can use spatial statistics to generate a field representation of the topic's relevance around the world. Polygonal regions indicating where the topic is most relevant can be constructed by calculating a contour from the field representation based on a threshold value. In our examples, we identify topic strengths for the one-quarter-degree grid square centroids as described in Sect. 12.3. The topic strength at a point is treated as a point count input into an Epanechnikov kernel density estimation function (de Smith et al. 2007).

Figures 12.4 and 12.5 show results of mapping two topics using this method with contour lines at topic strength equal to 0.01.[2] The *wine* topic is shown for Europe. The *temple* topic illustrates the regionality of some topics – temple features are found much more often in South and East Asia, and this is reflected in what people write about in those places. Mapping out topics in this way is a two-way street. Not only does it provide a mechanism for exploratory analysis to find out

---

[2] It should be emphasized that the topic regions for the exemplars in this chapter reflect strong biases from the fact that original data are travel blog entries.

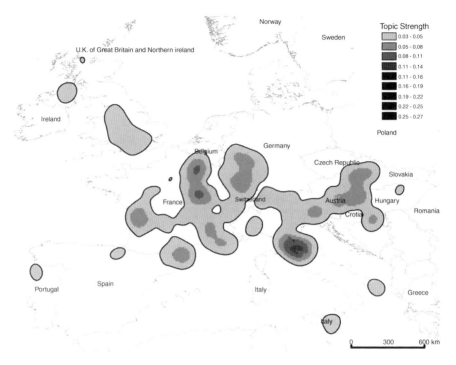

Fig. 12.4 *Wine* topic strengths

where certain topics are being mentioned but also in some cases the visualization can help in the interpretation of the meaning of the topic. For example, in our 400-topic simulation, two distinct topics (nos. 275 and 384) were generated that had *war* as the top-ranked word (see Fig. 12.6). Topic 275 appears to refer to war history, whereas topic 384 appears to be more about current war events. By mapping out the weights, we can begin to confirm those assumptions. Topic 384 is strong in current or recent hotspots (e.g., Iraq, Afghanistan, Sri Lanka), but topic 275 is not.

12.5.1 Visualizing Multiple Topics

Figure 12.7 shows a map of topics grouped into two themes. Topics related to human characteristics of place are shown in contrast to those related to physical characteristics. Through examination of the 200 topics, approximately 12 topics could be characterized as relating exclusively to the physical environment (e.g., mountains, rivers, beaches). Additionally, 13 topics are related specifically to human-constructed features (e.g., churches, cities, markets). Many other topics had both physical and human components. We aggregated the "physical" and "human" topics into two kernel density estimations created from the quarter-degree point values with a 1.5° radius.

12 Inferring Thematic Places from Spatially Referenced Natural Language… 213

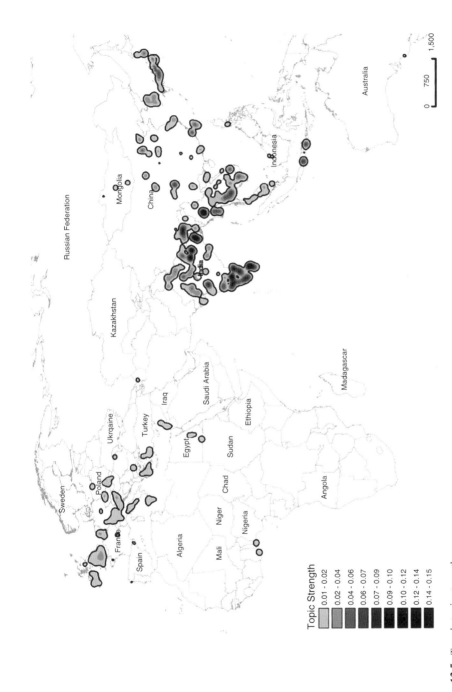

Fig. 12.5 *Temple* topic strengths

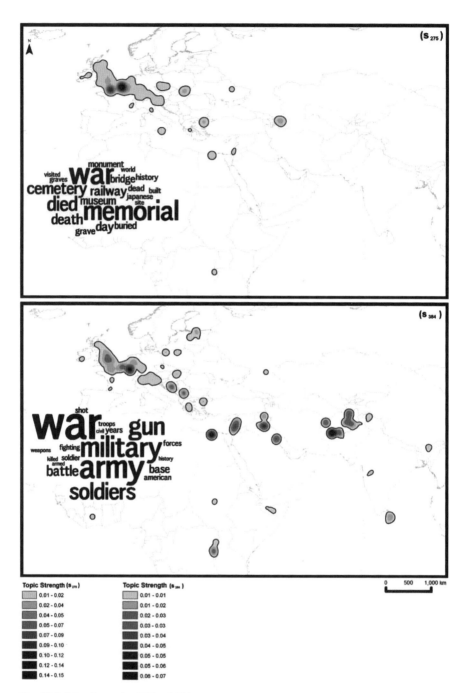

Fig. 12.6 Mapping topics 275 and 384

Fig. 12.7 Comparing regions based on strength of human vs. physical geography topics

The polygons displayed on the map show regions with a kernel density theme value above its mean. For example, the original physical characteristics theme values ranged from approximately 0.203–2.723 with a mean value of 0.457. The polygons shown on the map only contain values above 0.457.

12.5.2 Measuring the Localization of a Topic

As we have mentioned earlier, some of the topics generated by LDA have place names as top words and very clearly correspond to one specific region of the Earth. And as the maps in the previous sections show, there are some topics that show up in more than one part of the world but nevertheless are not evenly distributed everywhere (where *everywhere* in this context means all the locations for which we have blog entries). We would like to be able to measure the degree to which a topic is written about in one area or many or everywhere.

The topic strengths $S_i = \{s_{i_1}, s_{i_2}, \ldots, s_{i_n}\}$ for topic i at a set of n locations (e.g., the 7,277 grid cells with entries) can be interpreted as a kind of categorical probability distribution. The *localization* of topic i can then be evaluated as an inverse function of the entropy measure on that distribution. However, before localization can be evaluated in this manner, the topic strengths must be normalized so that they sum to 1. Let k be a location; the "probability" of k for topic i is

$$p_k = \frac{s_{i_k}}{\sum_{j=1}^{n} s_{i_j}}.$$

Localization of topic i is then defined as follows:

$$\Lambda_i = \gamma e^{\sum_{k=1}^{n} p_k \log p_k},$$

where γ is a constant scalar value.

12.6 Temporal Analysis

In their analysis of scientific topics, Griffiths and Steyvers (2004) showed how a post hoc analysis of the linear trend line of the mean θ value for an LDA topic could be used to infer its "hotness" or "coldness." A similar technique can be used to identify the dynamic change in topics being written about in travel blog entries. To illustrate, we present some results from a 400-topic simulation that was run for 1,000 iterations. The mean theta value for each topic per day was calculated from all entries with at least 100 words that were written between January 1, 2006 and August 31, 2010.

Figures 12.8 and 12.9 show how a *Chinese* locality topic is trending upward over this time period while a *Japanese* locality topic is trending downward. The increase in variance of the point values in 2010 can be explained by the fact that fewer people were blogging on travelblog in 2010 than in previous years. While a linear fit is useful for revealing the overall drift in a topic's popularity, nonlinear fits can illuminate periodic patterns in a topic's popularity. For example, many of the LDA topics show seasonal fluctuations. A *festival* topic peaks during February and March when many festivals (e.g., Carnival and Mardi Gras) happen around the world (see Fig. 12.10). The strength of other topics coincides with specific types of events such as natural disasters. For example, topic 387 shown in Fig. 12.11 peaks after the May 2008 Sichuan earthquake. While it would require a much more in-depth analysis to verify, it is conceivable that some topics act as leading indicators for some types of events, especially ones that are socially constructed.

Combined with geographic information, this kind of analysis shows how some topics trend differently in different locations. A closer look at topic 387 shows that it, in fact, peaks at two different times in 2008 depending on the location of the entry. In China, predictably, it peaks in May 2008 after the Sichuan earthquake. However, in both the United States and India, it peaks when the Mumbai terrorist attacks occurred in October 2008. A review of the entries written at these places and times verifies that these events were referenced using words from the topic 387 distribution.

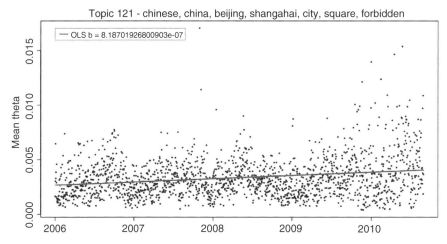

Fig. 12.8 *Chinese* topic trending upward

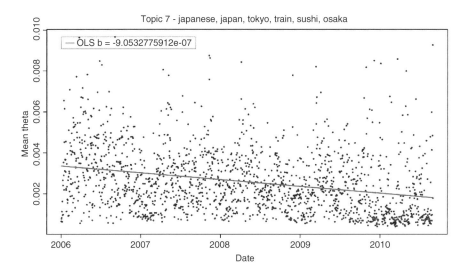

Fig. 12.9 *Japanese* topic trending downward

12.7 Summary and Conclusions

Much of the VGI available on the Web comes in the form of textual descriptions. In this chapter, we presented ways of using topic modeling to identify a place's unique mixture of characteristics directly from natural language observations. These methods allow us to calculate the similarity of places, map out places of thematic

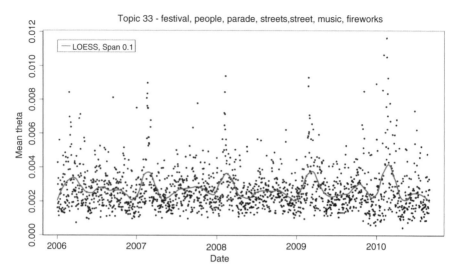

Fig. 12.10 *Festival* topic peaks in February/March

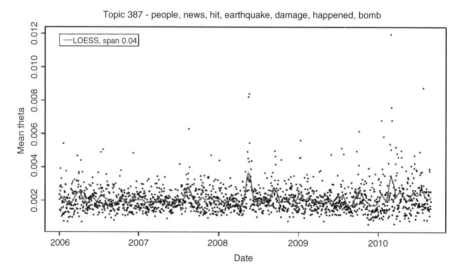

Fig. 12.11 Topic 387 strength corresponds to specific events

distinction, measure the degree to which certain themes are local or global, and evaluate thematic change over time. These results open up new opportunities for understanding the makeup and dynamism of places as described from individual experiences. A future task is to examine how these operationalized representations of place can be incorporated into a more robust framework that affords more indepth reasoning about place, including a context-dependent similarity measure.

Fig. 12.12 Latent Dirichlet allocation model plate notation

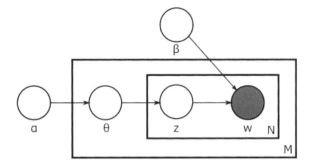

The issue of granularity remains a problem in terms of locating the documents in space, and further work needs to be done to incorporate a granularity metric into the representations generated using our methods.

Acknowledgments The authors wish to thank Mike Goodchild and two anonymous reviewers for their valuable comments.

12.8 Appendix A: Latent Dirichlet Allocation Model

This appendix describes the generative model for LDA (Blei et al. 2003). Let α be the Dirichlet hyperparameter of the per-document topic distributions, β be the Dirichlet hyperparameter for the per-topic word distributions, θ_i be the multinomial topic distribution for document i, z_{ij} be the topic for the jth word in document i, and w_{ij} be the jth word. The generative model for LDA is then defined as follows:

- Choose θ_i proportional to Dirichlet(α), where $i \in \{1, \ldots, M\}$.
- For each of the words w_{ij}, where $j \in \{1, \ldots, N\}$:
  - Choose a topic $z_{i,j}$ proportional to multinomial(θ_i).
  - Choose a word $w_{i,j}$ proportional to multinomial($\beta z_{i,j}$).

Figure 12.12 shows the plate notation representation of the LDA model. Plate notation is a shorthand representation of graphical probabilistic models that have many repeating variables. Each circle represents a variable in the model and the number in the lower right corner indicates the number of times the variable is repeated. For example, θ is repeated M times in the model shown. The shaded variable w is the only observed variable (i.e., the words in the documents).

References

Agnew, J. (1987). *The United States in the world economy*. Cambridge, MA: Cambridge University Press.
Bishop, C. (2006). *Pattern recognition and machine learning* (Vol. 4). New York: Springer.

Blei, D. M., & Lafferty, J. D. (2006). Correlated topic models. In Y. Weiss, B. Schölkopf, & J. Platt (Eds.), *Advances in neural information processing systems (NIPS) 18* (pp. 147–154). Cambridge, MA: MIT Press.

Blei, D. M., & Lafferty, J. D. (2009). Topic models. In A. N. Srivastava & M. Sahami (Eds.), *Text mining: Classification, clustering, and applications* (pp. 71–94). Boca Raton: CRC Press.

Blei, D., & McAuliffe, J. (2008). Supervised topic models. In J. C. Platt, D. Koller, Y. Singer, & S. Roweis (Eds.), *Advances in neural information processing systems (NIPS) 20* (pp. 121–128). Cambridge, MA: MIT Press.

Blei, D. M., Ng, A. Y., & Jordan, M. I. (2003). Latent Dirichlet allocation. *Journal of Machine Learning Research, 3*, 993–1022.

Chang, J., & Blei, D. (2009). Relational topic models for document networks. In D. van Dyk & M. Welling (Eds.), *Proceedings of the 12th international conference on artificial intelligence and statistics (AISTATS)* (pp. 81–88). Clearwater Beach: Journal of Machine Learning Research.

Cresswell, T. (2004). *Place: A short introduction*. Malden: Blackwell Publishing Ltd.

de Smith, M., Goodchild, M., & Longley, P. (2007). *Geospatial analysis: A comprehensive guide to principles, techniques and software tools* (2nd ed.). Leicester: Winchelsea Press.

Entrikin, N. (1991). *The betweenness of place: Toward a geography of modernity*. Baltimore: Johns Hopkins University Press.

Goodchild, M. F. (2009). What problem? Spatial autocorrelation and geographic information science. *Geographical Analysis, 41*(4), 411–417.

Griffiths, T. L., & Steyvers, M. (2004). Finding scientific topics. *Proceedings of the National Academy of Sciences, 101*(Suppl. 1), 5228–5235.

Hao, Q., Cai, R., Wang, C., Xiao, R., Yang, J. M., Pang, Y., & Zhang, L. (2010). Equip tourists with knowledge mined from travelogues. In M. Rappa, P. Jones, J. Freire, & S. Chakrabarti (Eds.), *Proceedings of the 19th international conference on world wide web (WWW'10)* (pp. 401–410). New York: ACM Press.

Jorgensen, B. S., & Stedman, R. C. (2006). A comparative analysis of predictors of sense of place dimensions: Attachment to, dependence on, and identification with lakeshore properties. *Journal of Environmental Management, 79*, 316–327.

Li, W., & McCallum, A. (2006). Pachinko allocation: DAG-structured mixture models of topic correlations. In *ICML'06: Proceedings of the 23rd international conference on machine learning* (pp. 577–584). New York: ACM Press.

McCallum, A. (2002). *MALLET: A machine learning for language toolkit*. http://mallet.cs.umass.edu. Accessed October 8, 2011.

Mei, Q., Liu, C., Su, H., & Zhai, C. (2006). A probabilistic approach to spatiotemporal theme pattern mining on weblogs. In L. Carr, D. D. Roure, A. Iyengar, C. A. Goble, & M. Dahlin (Eds.), *Proceedings of the 15th international conference on world wide web* (pp. 533–542). New York: ACM Press.

Serdyukov, P., Murdock, V., & van Zwol, R. (2009). Placing flickr photos on a map. In J. Allan, J. A. Aslam, M. Sanderson, C. Zhai, & J. Zobel (Eds.), *Proceedings of the 32nd international ACM SIGIR conference on research and development in information retrieval* (pp. 484–491). New York: ACM Press.

Shepard, R. N. (1987). Toward a universal law of generalization for psychological science. *Science, 237*(4820), 1317–1323.

Steyvers, M., & Griffiths, T. (2007). Probabilistic topic models. In T. Landauer, D. Mcnamara, S. Dennis, & W. Kintsch (Eds.), *Handbook of latent semantic analysis* (pp. 424–440). Hillsdale: Lawrence Erlbaum Associates.

Steyvers, M., Smyth, P., Rosen-Zvi, M., & Griffiths, T. (2004). Probabilistic author-topic models for information discovery. In *KDD'04: Proceedings of the tenth ACM SIGKDD international conference on knowledge discovery and data mining* (pp. 306–315). New York: ACM Press.

Teh, Y. W., Jordan, M. I., Beal, M. J., & Blei, D. M. (2006). Hierarchical Dirichlet processes. *Journal of the American Statistical Association, 101*, 1–30.

Tobler, W. (1970). A computer movie simulating urban growth in the Detroit region. *Economic Geography, 46*(2), 234–240.

Tuan, Y. F. (1977). *Space and place: The perspective of experience*. Minneapolis: The Regents of the University of Minnesota.

Wang, C., Wang, J., Xie, X., & Ma, W. Y. (2007). Mining geographic knowledge using location aware topic model. In R. Purves & C. Jones (Eds.), *Proceedings of the 4th ACM workshop on geographic information retrieval* (pp. 65–70). New York: ACM Press.

Winter, S., Kuhn, W., & Krüger, A. (2009). Guest editorial: Does place have a place in geographic information science? *Spatial Cognition and Computation, 9*, 171–173.

Chapter 13
"I Don't Come from Anywhere": Exploring the Role of the Geoweb and Volunteered Geographic Information in Rediscovering a Sense of Place in a Dispersed Aboriginal Community

Jon Corbett

Abstract This chapter explores the role of participatory mapping, the geoweb, and volunteered geographic information in rediscovering a sense of place within a physically dispersed Aboriginal community, the Tlowitsis Nation from Northern Vancouver Island. Centered on a community-based research project, this chapter examines how the participatory geoweb might be used by Tlowitsis members to better understand and reconnect with their land-related knowledge, as well as examine the ways in which these technologies serve to re-present place-based memories and facilitate dialogue amongst community members located in different geographic settings.

13.1 Background

The territory of the Tlowitsis Nation spans the coastal area of Northern Vancouver Island, British Columbia. Seasonal travel routes, food processing spots, burial and cultural sites, and other named places extend across the entire territory. Karlukwees, located on remote Turnour Island, has been a central settlement for the Tlowitsis Nation since the turn of the twentieth century. In the early 1960s, the British Columbia provincial government halted essential services to the island. With little prospect of schooling and access to health care, the Tlowitsis community began to leave the island. In the ensuing diaspora, community members have become culturally, as well as physically, removed from their traditional territories. Many Tlowitsis members lack a deep sense of their identity and are most often poorly acquainted

---

J. Corbett (✉)
Department of Community, Culture and Global Studies,
University of British Columbia Okanagan, Kelowna, BC, Canada
e-mail: jon.corbett@ubc.ca

with their relatives and other members of the Nation. One member expressed these feelings, the essence of which makes up the title of this chapter, in a group discussion during a Tlowitsis Nation meeting in 2006:

> It's kind of hard to say where I come from because I don't come from anywhere. To say that, being First Nations is important, but to say that I'm Tlowitsis doesn't really have any significance for my family ... I went there as a child – but for me to pass anything on to my children, its really hard to explain to them where our extended family came from because there's nothing, there's no land, there's nothing to go to.

A rising urban population with little attachment to these lands has reduced the opportunity and ability for members to take an active and informed role in their community. The Nation's governing body has had difficulty in maintaining communication and participation of its members in Tlowitsis activities. Despite these issues, the Tlowitsis involvement in land claims negotiations have compelled them to engage their members in treaty-related decision-making. There remains a core group of elders in the community with knowledge of the lands, resource use, and language, as well as a number of community leaders and youth with a desire to participate in planning and decision-making activities.

This chapter details a community-university research project that investigates the role of the participatory mapping activities, focusing on the geoweb, to support the contribution and sharing of community members' spatial knowledge. The research investigates to what extent online mapping tools and processes can be used by the Tlowitsis to understand and reconnect with their land-related knowledge and examine the ways in which these technologies serve to re-present place-based memories and facilitate dialogue among community members located in dispersed geographic settings.

13.1.1 Place

There is a fundamental connection between Aboriginal people and place (Cajete 1994, 2000; Deloria and Wildcat 2001). From Aboriginal perspectives, places are complex entities not simply defined by spatial location or physical structure. Places carry symbolic meaning as moral guides for individual behavior and social norms (Deloria 2001). A traditional Aboriginal sense of place develops over time through generations of interaction with and practical experience on the land. It stresses a process of relationship with place that supports the production and transfer of knowledge and the maintenance of cultural identity (Cajete 2000). In other words, an Aboriginal sense of place connects people to the land through both practical and symbolic relationships.

Interpretations and understandings of places by individuals and social groups are never uniform; places are continuously contested and defended (Cresswell 2004; Till 2003). While distinct understandings of place can coexist, they may also become sources of conflict as individuals and groups struggle to control or transform places and human relationships with those places (Gieryn 2000). This reflects an understanding that places, and people's relationships with places, are fraught with tension and never static. Places are defined not only by the people who live and interact with them but also through the power of others. This leads Escobar (2001, 140) to conclude that we should "understand by place the experience of a particular location with some measure of groundedness (however, unstable), sense of boundaries (however, permeable), and connection to everyday life, even if is constructed, traversed by power, and never fixed."

Basso contends there is an inextricable link between place and community as demonstrated through his work with the Western Apache of Cibecue. According to Basso, "people's sense of place, their sense of their tribal past, and their vibrant sense of themselves are inseparably intertwined" (Basso 1996, 35). Other authors echo this view asserting that meaningful places provide stability and security (Brown and Perkins 1992), function as "anchors" (Marcus 1992), and become "symbolic life lines" (Hummon 1992) and "fields of care" (Relph 1976) for individuals and communities. These human attachments to place "facilitate a sense of security and well-being, define group boundaries and stabilize memories against the passage of time" (Gieryn 2000, 481).

While many Aboriginal groups have maintained a continuous and contiguous physical connection to their ancestral homelands, for others the narrative of place is one of displacement – of individuals, families, and entire communities compelled, by force or might or circumstance, to move from their ancestral territories. It is in this context of displacement and dispersion that place has "become an important object of struggle in the strategies of social movements…and a critical rallying point for indigenous communities" (Escobar 2001, 139).

13.1.2 Loss of Place

Given the significance of place to people's lives, it follows that the loss of place has a profound effect on both a personal level as well as for the social fabric of communities (Gieryn 2000). Relph (1976) contends that the experience of displacement, the destruction of places, and other disruptions to human-environmental relationships can result in placelessness, in other words having no sense of place. He further notes that mobility can weaken a sense of place, which, in turn, can erode psychological and emotional ties to a sense of belonging and to community (Relph 1976, 66). Because places "reflect and shape peoples' understandings of who they are as individuals and members of groups" (Brown and Perkins 1992, 280), the narratives of those who have experienced the loss of place express a sense of loss for more than merely the physical location itself, as places are intrinsically tied to social relationships that occur there (Low 1992, 180). Furthermore, territorial displacement can result in "disagreement about the nature, origins, and remedies for the disruption." Brown and Perkins (1992, 299) explain that this "dissensus impedes community action ... [and] ... is more likely to occur in geographically dispersed communities."

In addition, territorial displacement has been attributed to a weakening of the kinship, political, and spiritual aspects of community life while also disrupting the intergenerational transfer of cultural knowledge between elders and youth (Canada 1996, vol. 1, 469; White et al. 2003, 26). In some cases, displacement has been found to contribute to social and cultural discontinuity that can persist for multiple generations (Canada 1996, vol. 1, 469). Although the histories, impacts, and symptoms of displacement are diverse, there is consensus that territorial displacement ruptures connectivity between people and places and contributes to cultural disintegration within Aboriginal communities. The Tlowitsis Nation is an example of an Aboriginal group that has been physically as well as culturally removed from their lands. This loss has had, and continues to have, a profound effect on the functioning of the community and the sense of cohesion between its members.

The concept of memory is integral to theorizations of both displacement and placelessness, especially when a sense of place exists only in memory or imagination for members of dispersed communities. Fentress and Wickham (1992, 24) emphasize that memory "is not a passive receptacle, but instead a process of active restructuring ... [it] represents the past and the present as connected to each other." Memory can also become "a social activity and an active binding force of group identity" (Hoelscher 2003, 658). Similarly, Said (2000, 179) notes that social groups rely on memory, "especially in its collective forms...to give themselves a coherent identity, a national narrative, a place in the world...endowed with political meaning." Collective memories deployed to restore, reestablish, repatriate territory, and reconnect a people with its original homeland reveal the symbolic significance embedded within place, as well as the value of collective memory as a strategy of resistance and viable political tool. Said (2000, 182–184) further argues that memory is increasingly called on to support small groups' struggles to reestablish identity, culture, and language and, above all, reappropriate historically expropriated territories and contemporarily contested places.

The challenge is to find ways that enable Aboriginal communities to document, share, and reflect on their place-based memories and knowledge and in doing so reestablish identity, culture, and language, which in turn will facilitate the reappropriation of contested places. Participatory mapping has been used to great effect to support these goals, and emergent participatory geoweb applications show further potential to pursue these aims.

13.1.3 Maps and Participation

Maps have been used in both a historical and contemporary context to normalize and reinforce the colonial endeavor (Harley 1988) and to perpetuate both intentional and unintentional lies (Monmonier 1991). Yet, within the discipline, there has been a growing recognition that cartographers need to explore whether maps should "be an inherent mirror of majority values or can they play a wider role in the struggle for social improvement" (Harley 1988). Recent trends in GI Science have been towards the democratization, as well as decolonization, of geographic information and its associated tools (Dunn 2007). It is increasingly recognized that the process of map creation can be shifted away from the realm of the professional cartographer and that this trend is beginning to reach a point where "maps are no longer imparted to us by a trained cadre of experts, but along with most other information we create them as needed ourselves" (Crampton and Krygier 2006, 15). In an Aboriginal context, this has partially come about through a resounding critique of the colonial, toxic, underlying nature of mapping and GIS (Eades 2006; Rundstrom 1995; Wainwright and Bryan 2009), but it is also being facilitated by the radical changes in access to and the usability of new cartographic and geographic information software, in particular the increasing significance of the geoweb.

The past 25 years have witnessed an explosion of participatory mapping initiatives throughout the world, in both developing and developed countries (see Alcorn 2001; Di Gessa 2008; Poole 1995). Participatory mapping is, in its broadest sense, the creation of maps by local communities – often with the involvement of supporting organizations including governments (at various levels), nongovernmental organizations, universities, and other actors engaged in development and land-related planning (IFAD 2009). Participatory maps provide a valuable visual representation of what a community perceives as its place and the significant features within it. These include depictions of natural physical features and resources and sociocultural features known by the community. What makes it significantly different from traditional cartography is the process by which the maps are created (i.e., by the layperson as identified above by Crampton and Krygier) and the uses to which they are subsequently put. Ideally participatory mapping focuses on providing the skills and expertise for community members to contribute the knowledge required to create the maps themselves, to represent the spatial knowledge of community members, and to ensure that the map creators determine the ownership and communication for the maps.

The participatory mapping process can influence the internal dynamics of a community (Aberley 1993; Chapin et al. 2005); it can contribute to building community cohesion (Corbett and Keller 2005), help stimulate community members to engage in land-related decision-making, raise awareness about pressing land-related issues (Peluso 1995), and ultimately contribute to empowering local communities and their members (Craig et al. 2002). Participatory mapping projects also can take on an advocacy role and actively seek recognition for community spaces through identifying traditional lands and resources, demarcating ancestral domain (Brody 1981) and, in some cases, be used as a mechanism to secure tenure (Flavelle 1996; Tobias 2009). Participatory maps play an important role in helping Aboriginal communities work towards legal recognition of customary land rights. Although NGOs, from small local to large international ones, often play a crucial role as interlocutors, trainers, advocates, and facilitators in community-mapping initiatives (IFAD 2009), it should be noted that maps increasingly are being created by Aboriginal communities with their own initiative and without impetus from outsiders

(Tobias 2009). This is especially the case with First Nations communities in Western Canada who see the potential for participatory maps to document their historical and cultural association with the land in order to influence land claims and stimulate interest in local spatial knowledge among their communities' youth.

Participatory mapping uses a range of tools that are commonly associated with participatory learning and action (PLA) initiatives (Corbett et al. 2006). These tools include mental mapping, ground mapping, participatory sketch mapping, transect mapping, and participatory three-dimensional modeling. More recently, participatory mapping initiatives have begun to use more technically advanced geographic information technologies including Global Positioning Systems (GPS), aerial photos and remote-sensed images, Geographic Information Systems (GIS), and increasingly the Geospatial Web (geoweb).

The geoweb is the geographic platform for Web 2.0 digital social networking applications. Web 2.0 refers to what many perceive to be a second generation of the Internet whereby it has become more interactive, allowing users to contribute their own content, seamlessly communicate and collaborate with one another in real time, and share or display a variety of qualitative data using a range of media with a constantly evolving range of web-based of applications. Web 2.0 differs from earlier models of the Internet, where most users would primarily use the web to retrieve information, seldom contributing content of their own. The geoweb incorporates applications such as Google Earth, Google Maps, Microsoft's Bing Maps, and other location-based Internet technologies. The geoweb is beginning to have a profound impact on the way that spatial knowledge is being organized and the extent to which it is communicated (Cisler 2007; Scharl and Tochtermann 2007). In the geoweb model, everyone is potentially a contributor, producer, and consumer of geographic content (Haklay et al. 2008; Sui 2008). Geoweb applications are thus considered highly democratic due to their ability to enhance citizen access and participation (Crampton 2009; Tulloch 2008). The geoweb has achieved broad acceptance, thanks to its widespread availability on the Internet, its platform independence, and because it is superficially "free" to use (although there are associated costs, e.g., some services claim ownership over data collected through their systems and preserve the right to reuse it, see Zook and Graham 2007). Another reason for the geoweb's popularity is its ability to aggregate and present user-generated digital content – referred to as crowdsourcing (Howe 2006; Hudson-Smith et al. 2009). Within the field of geography, this process of citizen-contributed locational information is increasingly referred to as volunteered geographic information (VGI) (Goodchild 2007). This interactive functionality allows for a range of community voices and opinions to be shared, potentially reflecting the inherent heterogeneity within communities and supporting "many-to-many communication" (Ruesch and Bateson 1987).

Within the emergent field of the geoweb and VGI, there appears to be a tension emerging around whether VGI presents a mechanism to exploit the geoweb's potential to capitalize on "citizens as sensors" (Goodchild 2007) or whether the geoweb continues to support the precedent set by participatory mapping practice, in other words using maps to support and bring about social and political change (IFAD 2009). Of course these two approaches are not necessarily oppositional; indeed, the practice of collecting and utilizing VGI allows for a range of uses and applications. However, the term is susceptible to misuse; we therefore need to explore how we use the term in greater detail. For example, much of the initial literature on VGI, as well as the current practice, focuses on collection but not dissemination of data, the reduction of knowledge to information (and often times simply data), as well as the focus of this information being provided to experts (or even corporate entities) and not the compilation of data to support communication and ultimately the "social improvements" that Harley refers to in the beginning of this section.

The practice of Aboriginal mapping, which almost universally aims to support social change and community empowerment, helps frame some important issues related to using the geoweb as a medium for managing and communicating locational knowledge, but through a lens of colonization, dispossession, and conflicting interests. To date most of the Internet, as well as geoweb application, reflects non-Aboriginal culture and values, even though a number of Aboriginal scholars have recognized the potential for it to be appropriated to support an Aboriginal mindset (Wemigwans 2008). There is a need for the VGI debate to take into consideration the linking of science and Aboriginal ways of thought and expression (e.g., through the geoweb's ability to incorporate multiple layers and multiple media) and in doing so consider the issue of privacy, intellectual property, and knowledge transfer in order to ensure that VGI does not become another exploitative and colonizing undertaking whereby community members are reduced to sensors and their knowledge becomes repositioned as mere data.

13.2 Research Project

The next section of this chapter introduces a research project that began in October 2010 and remains ongoing at the time of writing this chapter. The project seeks to explore two areas of research interest: firstly, how the geoweb represents, mediates, and transforms the flow of knowledge related to these traditional lands and, secondly, how the notion of authoritative knowledge plays out in the documentation of Aboriginal VGI on the geoweb. The project is still in its initial stages; it is premature to report definitive results. However, a number of interesting outcomes have begun to emerge.

This project, as with most research undertaken by non-Aboriginal scholars in Aboriginal contexts, is tainted by a legacy of appropriation and marginalization. In the past, Aboriginal people have not been consulted about what information should be collected, who should gather that information, who should maintain it, or who should have access to it. The information gathered may or may not have been relevant to the questions, priorities, or concerns to the people being studied (Canada, vol. 5, 1996 in Schnarch 2004, 82). There is a clear danger that a research project focused on examining the role of the geoweb and VGI in recording and storing community knowledge might fall into this classification. Therefore, in setting up this research project, we have attempted to ensure that the Tlowitsis maintain the exclusive jurisdiction over their knowledge and for this to be explicitly acknowledged and incorporated into project design (both the geoweb component and the research component) in order to ensure that research proceeds in a manner that is respectful, relevant, reciprocal, and responsible (Kirkness and Barnhardt 1991).

The methodological approach adopted by this project is community-based research (CBR). CBR is not so much a set of methods as it is a research philosophy that emphasizes collaboration over positivist notions of objectivity and neutrality and the idea that science is apolitical (Hall 1992; Strand et al. 2003). The active engagement of community members as co-researchers helps to ensure that research products are more accessible, accountable, and relevant to people's lives (Israel et al. 1998; Wallerstein and Duran 2008).

13.2.1 Research Collaborators

The initial phase of this project has been developed by the Treaty Advisory Team, which included members of the Tlowitsis governing body and researchers from the University of British Columbia Okanagan. The Treaty Advisory Team has worked directly with two groups within the Tlowitsis Nation – the Tlowitsis Citizens Advisory Group (TCAG) and the Tlowitsis Elders Advisory Group (TEAG). The two community groups are comprised of around 20 Tlowitsis members who act as a representative decision-making and advisory body for the Nation. The TCAG, formed in June 2008, provides advice and insight related to land and resource decision-making. The TEAG provides guidance and direction for the TCAG. The elders are the knowledge holders of historical and cultural information, as well as of the community's land-based memories. TCAG members also act as facilitators and liaisons in community outreach activities; in other words, they deliver the materials discussed in their group and in the TEAG to other dispersed community members in locations throughout British Columbia.

13.2.2 Mapping Process

Since the formation of the TCAG and TEAG, the Treaty Advisory Team has produced a large quantity of digital video, photographic, and text-based materials (Tlowitsis Nation 2009, 2010). These have all been produced, edited, and disseminated through a collaborative process involving the Treaty Advisory Team and the two community groups. The video products cover a range of issues, but perhaps most important for this chapter are the recordings of community elders sharing their land-based knowledge of their territories. These sessions were convened in community members' houses. They began with a series of sketch-mapping activities and moved on to using topographic maps of the Tlowitsis territories to orientate the elders. This was followed by conversations about memories that directly related to points on the map; these stories were recorded on video. The map became the medium through which the elders engaged and reached a level of comfort whereby they were prepared to share their memories. When elders told their stories, their words were punctuated by vigorous pointing and sweeping arcs drawn over the map as if the map itself was telling the story and the elder was just the narrator.

Our process began with sketch maps where community members drew a map from memory onto paper. They are not to scale but represent the relative position of features to one another and made an effective icebreaker. The elders love to talk about Karlukwees, their ancestral village, and sketch mapping provided them with a way to resituate themselves in that place. They discussed the location of friends and families' houses and shared short stories of particular memories located in specific places (e.g., "remember the time we built a toboggan and came right down the middle of the village" or "…we would pick clams in this spot").

Sketch mapping helped elders remember and begin the process of community reengagement. However, the impact of the process was limited to the group involved in the map's production. Developing a sketch map was also contingent on having some form of lived experience in the place being mapped; as a result, this tool was less relevant for the majority of the membership, most of whom have never visited the Tlowitsis territories.

The Treaty Advisory Team together with elders also used scale maps, in the form of marine charts and topographic maps, to locate named features in the landscape and important resource spots, as well as identify family summer settlements, smoke houses, staging grounds, and travel routes throughout the territory. These scale maps were covered in transparent plastic, and the elders' knowledge was added using colored nonpermanent marker pens. The material added to these scale maps has provided vital substance for the creation of a series of Tlowitsis-centric maps of the territory. They provided precise spatial information that was initially intended to inform and pique the interest of the urban membership. However, as it turned out, they became a highly sensitive set of documents that members of the Treaty Advisory Team determined should be used selectively and strategically at the treaty table because they were too sensitive to be released into the public where the information might be used against the nation in the land claims process.

Much of the information from the sketch and scale mapping exercises has been incorporated into the Tlowitsis Geographic Information System (GIS). The community's GIS has been used to better understand competing claims over current land use throughout the region. The GIS also contains data brokered through sharing agreements with the provincial government and resource extraction industries. Using the GIS, select maps have been created to produce a Tlowitsis atlas for distribution to government and community members. However, the database remains in the hands of experts and its contents only made accessible within carefully constructed messaging that supports the treaty process.

In the most recent mapping project, which is still in the nascent stage, the Treaty Advisory Team together with members of the TCAG has begun to explore the use of the geoweb as a tool for not only one way information flow of the elders spatial memories to other community members but also as a tool to support a map-mediated dialogue on land-related issues among community members. We have used the geoweb's ability to aggregate, or mashup, qualitative information and incorporate video recordings of elders, photographs taken by community members, and other relevant text documents, and display them directly through a Google Map's type interface.

The Treaty Advisory Team has taken into consideration the requirements for privacy and the protection of the Tlowitsis knowledge. The University of British Columbia Okanagan has developed a geoweb tool, called Geolive, to specifically address this issue. Geolive is a participatory mapping tool that combines Google Maps and Joomla! – an open source content management system. The application allows administrators to set varying levels of user access. Registered users can create and share their own spatial information using a single dynamic map-based interface; they can drop information markers onto a map; link these markers to videos, photos, or text; and turn different data layers on and off as well as take part in "instant messenger" type discussions. All information added to the map is stored directly on a server that is housed at the university; the information does not reside on the cloud (i.e., web-based digital storage services). This gives the administrators of the Tlowitsis Nation the ability to determine and control exactly who is accessing what specific information.

The Tlowitsis geoweb site is now established. We are in the process of seeding the online map with multimedia information gathered from the earlier TCAG and TEAG participatory mapping activities, as well as other contributed information from community members.

13.3 Discussion

This next section will address two questions raised in the research project section above: firstly, how the geoweb represents, mediates, and transforms the flow of knowledge related to Tlowitsis traditional lands and, secondly, how the notion of authoritative knowledge plays out in the documentation of Aboriginal VGI on the geoweb. This discussion is informed by initial findings from the implementation and impact of the Tlowitsis geoweb project. This research project is ongoing at the time of writing this chapter; therefore, the results presented in this chapter are only preliminary.

13.3.1 Exploring the Flow of Knowledge Using Geoweb

Broadly, the application of VGI can be classified into two areas. The first uses VGI to update, augment, or complement existing spatial databases. Goodchild (2007) refers to these augmented databases as patchworks and notes that the associated data have been increasingly used to fill gaps in governmental databases after it became too cost prohibitive to collect rapidly changing spatial data themselves. Academic interest in patchworks concerns debate related to "citizens as sensors," data accuracy, and validity as well as issues of interoperability through the combining of "official" data with publicly contributed data. The second area of VGI relates to the creation of novel forms of knowledge production that can be used to foster new social and political practices (Elwood 2008). It is this second set of applications that has caused academics to speculate that VGI is changing the way that society views and interacts with spatial data, the knowledge politics of geographic information, and relations to these data (Elwood 2010). It is this second area of VGI that closely relates to the focus of the Tlowitsis geoweb project.

13.3.2 Volunteered Information vs. Aboriginal Knowledge

There remains a fundamental tension that is closely tied to the focus of VGI, and more broadly Web 2.0 and the geoweb, and the nature of Aboriginal knowledge. This tension relates to the fact that many applications are specifically designed to support the uninhibited sharing of volunteered information, however whimsical (Keen 2007), whether this is information about a user's location, their photographs and videos, or simply broadcasting their mood and feelings. The focus of the majority of these applications is on ease of use and unfettered access to other users' information. Aboriginal knowledge, however, is not whimsical, nor is it a commodity to be shared indiscriminately. Aboriginal knowledge is "a key element of the social capital of the poor and constitutes their main asset in their efforts to gain control of their own lives" (World 2002, 1), and the process of knowledge acquisition is in itself of intrinsic value, which in turn shapes the long-term likelihood of its survival.

The words "information" and "knowledge," although often used interchangeably, are distinguishable. Information "takes the shape of structured and formatted datasets that remain passive and inert until used by those with the knowledge needed to interpret and process them" (David and Foray 2002, 12). In this sense information remains close to the concept of citizens volunteering locational data as sensors (Goodchild 2007; Coleman and Georgiadou 2009). Knowledge is the understanding that people derive from that information. It is neither objective nor static, but is ever changing and infused with the principles and the daily realities faced by those who have it. Knowledge used in this context is not just cerebral but includes values, beliefs, skills, attitudes, and practices (UNDP 1999). Furthermore, Aboriginal knowledge is owned and shared collectively within a local community (Greaves 1996). This knowledge has been accumulated over time by successive generations. These communities have used this knowledge to sustain themselves and to maintain their cultural identity (Johnson 1992). In other words, it is considerably different in both nature and custom from the types of VGI gathered to augment government databases, as well as the types of qualitative and highly personal information that are commonly shared using locationally aware social networking applications. For many Tlowitsis members, social networking applications are tools that are used for entertainment and communication by the youth. They are not considered serious tools; for this to happen, there needs to be a shift in the users' perception. For the geoweb to become an effective tool for dispersed community members to learn land-related knowledge about the Tlowitsis territory, the users will need to perceive the tool, as well as the contained knowledge, sincerely as well as interact with it in a serious way.

13.3.3 Aboriginal Knowledge and the Public Domain

Knowledge is often given meaning and value through its cultural setting and interpretation (Brodnig and Mayer-Schönberger 2000). Stevenson (1997) stated that it is the spiritual dimension that determines how Aboriginal knowledge is collected, managed, and transmitted and it is this dimension that sets it apart from basic information that anyone can acquire through observing and experiencing their environment. As a result, this knowledge at times is "providing a world view of which outsiders are rarely aware, and at best can only incompletely grasp" (Greaves 1996). This complicates the codification and recording and therefore communication of this knowledge. Presenting sensitive land-based knowledge through the geoweb, especially if it becomes understood as simply volunteered information, might serve to render it irrelevant, inappropriate, or even harmful (especially given the significance and sensitivity of information and its use and/or misuse in the British Columbia Treaty context). Following a legacy of external domination over the Tlowitsis lands and natural resources, the community is concerned that this is not followed by the appropriation of knowledge without proper acknowledgment of the communities from where it originates (Brush 1996). There has been a persistent fear within the Treaty Advisory Team that any information presented on the web, in particular sensitive land-related knowledge, cannot be controlled. The information might be transformed, used selectively, and reused in ways that at this point in time cannot even be imagined. Despite the attention given to restrict the information's availability in the public domain and to maintain privacy through the design of various levels of access using Geolive, this fear has remained throughout the duration of the project.

Trust is a major theme affecting peoples' willingness to participate in many social networks. In general, the more trust users have of an application, the more they will participate and the more they will be willing to share (Boyd and Ellison 2008). Although there remains a clear desire to implement the geoweb tool by the Tlowitsis (particularly by the youth), and although it is recognized that community wide collaboration and the collective geo-wikification of the community's knowledge is desirable, there remains a fundamental distrust of the Internet as a whole and by association the geoweb. This distrust is directly related to the fear of losing control over their land-related knowledge. Elwood (2008) notes that not all users are fully aware of the potential uses of their volunteered geographic data. However, many Aboriginal communities in Canada are already highly sensitized to the potential that their data will be appropriated and misused because this has occurred numerous times in their colonized past. As a result many Tlowitsis community members are deeply cynical about the potential application of tools that facilitate the sharing of this knowledge, especially when organizations such as universities, who are often perceived as expropriators of knowledge, are involved in facilitating its acquisition. In consideration of this, most Tlowitsis community members involved in the project would rather apply the precautionary principle and carefully restrict the types of spatial knowledge communicated through the geoweb.

13.3.4 The Role of Authoritative Aboriginal Knowledge and the Geoweb

Goodchild (2007, 29) contends "the world of VGI is chaotic, with little in the way of formal structures. Information is constantly being created and cross-referenced, and flows in all directions, since producers and consumers are no longer distinguishable." This raises the question of the extent to which VGI can ever be considered authoritative in documenting spatial data, as well as representing qualitative spatial information. This issue is particularly important when users are unrelated and anonymous or assume "virtual" personas. However, our research has begun to show that the issues of authoritative information and the contested nature of knowledge are also significant to Tlowitsis members who, despite physical separation, still retain a strong sense of shared culture and history.

Commonly, VGI is viewed as being nonauthoritative and therefore something to use, but not rely on. As Goodchild (2007, 31) asserts, "missing at this point are the mechanisms needed to ensure quality, to detect and remove errors." Grira and Bédard (2009) further note that every Internet user has different requirements and expectations of quality; this relates to both the material that they obtain through the Internet as well as information that they contribute. In the same way that Goodchild and Grira and Bédard recognize a significant variety in data quality, the Tlowitsis members also identify that the content and details of their own memories and local spatial knowledge vary considerably between different families and individuals; since the diaspora, these differences have become even more marked.

Cresswell (2004) stresses that interpretations and understandings of places by individuals and social groups are never uniform; places are continuously contested and positions defended. Both elders and members of the TCAG expressed similar concerns about contested memories of place and the Tlowitsis geoweb project. Some felt that openly sharing spatial memories using the geoweb has the potential to expose contested recollections and exacerbate discordance in a way that would not emerge using nondigital, face-to-face forms of participatory mapping techniques. This is because of the inability of the web medium to fully support a running dialogue, to be able to explain the purpose and intent behind information shared through the web interface, as well as talk one's way through disagreements and misunderstandings, all core elements of nondigital participatory mapping activities. Although Geolive is designed to enable online "instant messenger" type conversations, it clearly lacks the subtlety and nuance of face-to-face discussion. Given the sensitivity around memories that reflect such a painful past, also taking into account that there is a clear disruption in the intergenerational transfer of cultural knowledge between elders and youth, there is a danger that the sharing of information in an open and unmediated manner will serve to alienate and perhaps even anger some community members. The geoweb therefore has the potential to exacerbate the same rift between community members that the project attempts to overcome.

Within the VGI literature, there is considerable debate related to spatial data accuracy being diminished when collected and submitted by amateurs (Grira and Bédard 2009), as well as discussion related to evaluating its quality in an environment where "standard conventions of determining credibility break down" (Flanagin and Metzger 2008, 140). Web 2.0 applications seek to enable a less vertical assessment of information quality; in other words, they allow the collective wisdom of all users to play a significant role in the arbitration of data credibility. In many VGI applications, individual user credibility builds around the perceived and relative trustworthiness as identified by the majority of users. Credibility is determined through rating systems where reliability of volunteered data is openly assessed. The content and tone of this form of evaluation, although most often serious, can be humorous, glib, confrontational, and at times demeaning. In many Aboriginal communities this form of evaluation is far removed from a past that deeply respected the position of knowledge holders and in particular elders. This form of open judgment has also served to silence a number of Tlowitsis community members, including elders, not only because they fear being judged to be wrong but because they recognize the variance in memories held by different individuals and families and are concerned about presenting their stories, histories, and memories in a manner that might be construed as being authoritative and/or definitive, and thus in conflict with other views.

In conclusion, the Treaty Advisory Team have recognized that it is important for the project to proceed with extreme caution and to explicitly acknowledge that the geoweb might be best used as a tool to reconnect community members and to start to develop relationships and an active online community, but not yet to treat the geoweb as a tool to transfer land-related knowledge – especially if that knowledge is in any way related to the ongoing Treaty process.

13.4 Conclusions

When members of the Treaty Advisory Team examined the project from a short term "project scope" perspective, we feel that both the earlier participatory mapping activities and the geoweb project have been superficially successful in engaging elders and generating interest in land-related knowledge among members of the TCAG and more broadly in the community. In doing so, the project has helped move the dispersed community towards, as Cajete (2000) notes, a relationship with place that supports the production and transfer of knowledge and the maintenance of cultural identity. Elders have been willing to share limited place-based memories and other spatial knowledge from the territory. That information in turn has been incorporated into the geoweb component of the Tlowitsis web site and thus communicated to other community members through both direct viewing of the web site and facilitated training workshops. During a series of training workshops, community members have used the geoweb interface to comment on how they have appreciated learning more about the lands that they have never visited. So, at least outwardly, the tool has become "a social activity and an active binding force of group identity" (Hoelscher 2003, 658).

However, in regard to longer-term outcomes associated with using the geoweb to "rediscover a sense of place," it is too early to comment given the newness and current reach of the project. We recognize that presently the Tlowitsis geoweb project has limited ability to bring about systemic social and/or political change within the community. The information presented on the map and the ability to interact with this information and other community members through the geoweb medium will never substitute for the need for community members to physically interact with their lands and each other. The geoweb's strength lies in its capacity to offer an intermediary, albeit superficial, step in the process towards reestablishing a deeper connection to the land through the memories, stories, and multimedia displayed on the geoweb map. However, the impact of these stories and collective memories on their own to "foster a sense of community, identity and belonging" (Basso 1996, 31) remains limited.

The Tlowitsis geoweb project has proven an effective way to repurpose existing qualitative multimedia material, especially digital videos and photographs gathered by members of the Treaty Advisory Team and other community members over the past 5 years. This multimedia information is made available to the dispersed membership through a geoweb interface. However, to date there has been little contribution to the map from community members that are not a part of the TCAG and TEAG. This is partially because the project is currently in its infancy; furthermore, the Treaty Advisory Team has not yet widely promoted the geoweb component of the web site through a community outreach campaign. We also recognize that the Tlowitsis geoweb portal illustrates a disconnect that exists within the community in regard to sharing of experiences related to the land – notably that most dispersed community members have no background or practical experience on the land and thus no memories to share. Beyond contributing their personal insights and experiences related to being removed from the land, which they are already doing on the existing web site forum, there is little spatial information and few memories to share. Thus, the geoweb becomes a tool for them to learn directly from the elders' knowledge that the Treaty Advisory Team has seeded on the site. At this stage of the project's development, the geoweb site reflects a unidirectional flow of information, rather than the participatory "many-to-many communication" (Ruesch and Bateson 1987) that we had initially envisaged. As the project develops, a core research objective will be to examine the types of comments, stories, and experiences that the general community membership posts on the site. However, at this stage the geoweb project more closely emulates how knowledge was traditionally passed between generations, in other words from elders to younger generations.

In conclusion, this chapter stresses that within Aboriginal contexts, there is a fundamental difference between information and knowledge. This difference suggests a need for extreme caution in the deployment of the geoweb to record and communicate Aboriginal land-based memories. It is essential that the Tlowitsis geoweb project, as well as similarly motivated projects, does not serve to further disempower the community through inadvertently (and often unintentionally) facilitating the loss of control over their own knowledge. Researchers and practitioners alike need to be conscious that these tools can potentially cause further erosion of community cohesion through memories being construed in contentious ways that ultimately become uncontrollable. There is a critical need for the geoweb to incorporate feedback mechanisms that facilitate multiple pathways of communication in order to introduce this new technology in a culturally appropriate and meaningful manner.

Acknowledgments This chapter could not have been written without both the commitment to research and the financial support of the Tlowitsis Treaty Office, in particular the Chief Negotiator Ken Smith and Treaty researcher Zach Romano. Furthermore, it reflects the views of many Tlowitsis members who have been, and continue to be, overwhelmingly gracious in sharing their time, experiences, and involvement. The Social Sciences and Humanities Research Council of Canada funds the geoweb project described in this chapter. Geolive has been developed using funding from the GEOIDE network Project 41.

References

Aberley, D. (1993). *Boundaries of home: Mapping for local empowerment*. Gabriola Island: New Society Publishers.
Alcorn, J. B. (2001). *Borders, rules and governance: Mapping to catalyse changes in policy and management*. London: International Institute for Environment and Development.
Basso, K. (1996). *Wisdom sits in places: Landscape and language among the western apache*. Albuquerque: University of New Mexico Press.
Boyd, D. M., & Ellison, N. B. (2008). Social network sites: Definition, history, and scholarship. *Journal of Computer-Mediated Communication, 13*(1), 210–230.
Brodnig, G., & Mayer-Schönberger, V. (2000). Bridging the gap: The role of spatial information technologies in the integration of traditional environmental knowledge and western science. *The Electronic Journal on Information Systems in Developing Countries, 1*, 1–16.
Brody, H. (1981). *Maps and dreams: Indians and the British Columbia frontier*. New York: Pantheon Books.
Brown, B., & Perkins, D. (1992). Disruptions in place attachment. In I. Altman & S. Low (Eds.), *Place attachment* (pp. 279–304). New York: Plenum Press.
Brush, S. B. (1996). Whose knowledge, whose genes, whose rights? In S. B. Brush & D. Stabinsky (Eds.), *Valuing local knowledge: Indigenous people and intellectual property rights*. Washington, DC: Island Press.
Cajete, G. (1994). *Look to the mountain: An ecology of indigenous education*. Skyland: Kivaki Press.
Cajete, G. (2000). *Native science: Natural laws of interdependence*. Skyland: Kivaki Press.
Canada (1996). *Royal commission on aboriginal peoples. Volume 1: Looking forward, looking back*. Ottawa: Canada Communication Group. http://www.collectionscanada.gc.ca/webarchives/20071115053257/http://www.ainc-inac.gc.ca/ch/rcap/sg/sgmm_e.html. Accessed July 2011.
Chapin, M., Lamb, M., & Threlkeld, B. (2005). Mapping indigenous land. *Annual Review of Anthropology, 34*, 619–638.

Cisler, S. (2007). *Open geography: New tools and new initiatives*. Santa Clara: Center for Science Technology and Society, Santa Clara University.

Coleman, D. J., Georgiadou, Y., & Labonte, J. (2009). Volunteered geographic information: The nature and motivation of producers. *International Journal of Spatial Data Infrastructures Research, 4*, 332–358.

Corbett, J. M., & Keller, C. P. (2005). An analytical framework to examine empowerment associated with participatory geographic information systems (PGIS). *Cartographica: The International Journal for Geographic Information and Geovisualization, 40*(4), 91–102.

Corbett, J. M., Rambaldi, G., Kyem, P., Weiner, D., Olsen, R., Muchemi, J., & Chambers, R. (2006). Overview – Mapping for change the emergence of a new practice. *Participatory Learning and Action, 54*, 13–20.

Craig, W. J., Harris, T. M., & Weiner, D. (2002). *Community participation and geographic information systems*. London/New York: Taylor and Francis.

Crampton, J. (2009). Cartography maps 2.0. *Progress in Human Geography, 3*(1), 91–100.

Crampton, J., & Krygier, J. (2006). An introduction to critical cartography. *ACME, 4*(1), 11–33.

Cresswell, T. (2004). *Place: A short introduction*. Malden: Blackwell.

David, P. A., & Foray, D. (2002). An introduction to the economy of the knowledge society. *International Social Science Journal, 54*(March), 9–23.

Deloria, V. (2001). American Indian metaphysics. In V. Deloria & D. Wildcat (Eds.), *Power and place: Indian education in America* (pp. 1–6). Golden: Fulcrum Publishing.

Deloria, V., & Wildcat, D. (2001). *Power and place: Indian education in America*. Golden: Fulcrum Publishing.

Di Gessa, S. (2008). *Participatory mapping as a tool for empowerment: Experiences and lessons learned from the ILC network*. Rome: International Land Coalition. http://www.landcoalition.org/pdf/08_ILC_Participatory_Mapping_Low.pdf. Accessed July 2011.

Dunn, C. (2007). Participatory GIS: A people's GIS? *Progress in Human Geography, 31*(5), 617–638.

Eades, G. L. (2006). *Decolonizing geographic information systems*. MA thesis, Carleton University, Ottawa, Ontario, Canada.

Elwood, S. (2008). Volunteered geographic information: Key questions, concepts and methods to guide emerging research and practice. *GeoJournal, 72*, 133–135.

Elwood, S. (2010). Geographic information science: Emerging research on the societal implications of the geospatial web. *Progress in Human Geography, 34*(3), 349–357.

Escobar, A. (2001). Culture sits in places: Reflections on globalism and subaltern strategies of localization. *Political Geography, 20*(2), 139–174.

Fentress, J., & Wickham, C. (1992). *Social memory*. Oxford: Blackwell.

Flanagin, A. J., & Metzger, M. J. (2008). The credibility of volunteered geographic information. *GeoJournal, 72*(3), 137–148.

Flavelle, A. (1996). *Community mapping handbook*. Vancouver: Lone Pine Foundation.

Gieryn, T. (2000). A space for place in sociology. *Annual Review of Sociology, 26*(1), 463–496.

Goodchild, M. F. (2007). Citizens as voluntary censors: Spatial data infrastructure in the world of web 2.0. *International Journal of Spatial Data Infrastructures Research, 2*, 24–32.

Greaves, T. (1996). Tribal rights. In S. B. Brush & D. Stabinsky (Eds.), *Valuing local knowledge: Indigenous people and intellectual property rights*. Washington, DC: Island Press.

Grira, J., & Bédard, Y. (2009). Spatial data uncertainty in the VGI world: Going from consumer to producer. *Geomatica, 64*(1), 61–71.

Haklay, M., Singleton, A., & Parker, C. (2008). Web mapping 2.0: The neogeography of the GeoWeb. *Geography Compass, 2*(6), 2011–2039.

Hall, B. (1992). From margins to center? The development and purpose of participatory research. *The American Sociologist, 23*(4), 15–28.

Harley, J. B. (1988). Maps, knowledge and power. In D. Cosgrove (Ed.), *The iconography of landscape* (pp. 277–312). Cambridge, MA: Cambridge University Press.

Hoelscher, S. (2003). Making place, making race: Performances of whiteness in the Jim Crow South. *Annals of the Association of American Geographers, 93*(3), 657–686.

Howe, J. (2006). The rise of crowdsourcing. *Wired, 14*(6), 176–183.

Hudson-Smith, A., Batty, M., Crooks, A., & Milton, R. (2009). Mapping for the masses: Accessing web 2.0 through crowdsourcing. *Social Science Computer Review, 27*(4), 1–15.

Hummon, D. (1992). Community attachment: Local sentiment and sense of place. In I. Altman & S. Low (Eds.), *Place attachment* (pp. 279–304). New York: Plenum Press.

IFAD. (2009). *Good practices in participatory mapping*. Rome: The International Fund for Agricultural Development (Prepared by J. M. Corbett, 2009).

Israel, B., Schulz, A., Parker, E., & Becker, A. (1998). Review of community based research: Assessing partnership approaches to improve public health. *Annual Review of Public Health, 19*, 173–202.

Johnson, M. (1992). *Lore: Capturing traditional environmental knowledge*. Ottawa: International Development Research Centre.

Keen, A. (2007). *The cult of the amateur: How today's internet is killing our culture*. New York: Doubleday.

Kirkness, V. J., & Barnhardt, R. (1991). First nations and higher education: The 4 Rs – respect, relevance, reciprocity, responsibility. *Journal of American Indian Education, 30*(3), 1–15.

Low, S. (1992). Symbolic ties that bind: Place attachment in the plaza. In I. Altman & S. Low (Eds.), *Place attachment* (pp. 279–304). New York: Plenum Press.

Marcus, C. (1992). Environmental memories. In I. Altman & S. Low (Eds.), *Place attachment* (pp. 279–304). New York: Plenum Press.

Monmonier, M. (1991). *How to lie with maps*. Chicago: The University of Chicago Press.

Peluso, N. L. (1995). Whose woods are these? Counter-mapping forest territories in Kalimantan, Indonesia. *Antipode, 27*(4), 383–406.

Poole, P. (1995). Geomatics, who needs it? *Cultural Survival Quarterly, 18*(4), 1–77.

Relph, E. (1976). *Place and placelessness*. London: Pion.

Ruesch, J., & Bateson, G. (1987). *Communication: The social matrix of psychiatry*. New York: W. W. Norton and Company.

Rundstrom, R. A. (1995). GIS, indigenous peoples, and epistemological diversity. *Cartography and Geographic Information Systems, 22*(1), 45–57.

Said, E. W. (2000). Invention, memory, and place. *Critical Inquiry, 26*(2), 175–192.

Scharl, A., & Tochtermann, K. (2007). *The geospatial web: How geobrowsers, social software and the Web 2.0 are shaping the network society*. New York: Springer.

Schnarch, B. (2004). Ownership, control, access, and possession (OCAP) or self-determination applied to research: A critical analysis of contemporary first nations research and some options for first nations communities. *Journal of Aboriginal Health, 1*(1), 80–95.

Stevenson, M. G. (1997). Ignorance and prejudice threaten environmental assessment. *Policy Options, 18*(2), 25–28.

Strand, K., Donohue, P., & Stoecker, R. (2003). *Community-based research and higher education: Principles and practices*. San Francisco: Jossey-Bass.

Sui, D. (2008). The wikification of GIS and its consequences: Or Angelina Jolie's new tattoo and the future of GIS. *Computers, Environment and Urban Systems, 32*, 1–5.

Till, K. (2003). Places of memory. In J. Agnew, K. Mitchell, & G. Toal (Eds.), *A companion to political geography* (pp. 230–251). New York: Wiley-Blackwell.

Tlowitsis Nation. (2009). *Tlowitsis nation lands and resources: Planning for the future*. Kelowna: The Centre for Social, Spatial and Economic Justice.

Tlowitsis Nation. (2010). *Tlowitsis governance: Values from the past, vision for the future*. Kelowna: The Centre for Social, Spatial and Economic Justice.

Tobias, T. (2009). *Living proof: The essential data-collection guide for indigenous use-and-occupancy map surveys*. Vancouver: Union of BC Indian Chiefs and Ecotrust Canada.

Tulloch, D. (2008). Is volunteered geographic information participation? *GeoJournal, 72*(3/4), 161–171.

UNDP. (1999). *A guidebook for field projects: Participatory research for sustainable livelihood*. New York: United Nations Development Program.

Wainwright, J., & Bryan, J. (2009). Cartography, territory, property: postcolonial reflections on indigenous counter-mapping in Nicaragua and Belize. *Cultural Geographies, 16*, 153–178.

Wallerstein, N., & Duran, B. (2008). The theoretical, historical and practical roots of CBPR. In M. Minkler & N. Wallerstein (Eds.), *Community based participatory research for health* (pp. 25–46). San Francisco: Jossey Bass.

Wemigwans, J. (2008). Indigenous worldviews: Cultural expression on the world wide web. *Canadian Woman Studies, 26*(3/4), 31.

White, J., Beavon, K., & Maxim, P. (2003). *Aboriginal conditions: Research as a foundation for public policy*. Vancouver: University of British Columbia Press.

World Bank. (2002). *The exchange of indigenous knowledge*. New York: World Bank. http://www.worldbank.org/html/afr/ik/exchange.htm. Accessed May 2005.

Zook, M., & Graham, M. (2007). The creative reconstruction of the Internet: Google and the privatization of cyberspace and DigiPlace. *Geoforum, 38*, 1322–1343.

Part III
Emerging Applications and New Challenges

Chapter 14
Potential Contributions and Challenges of VGI for Conventional Topographic Base-Mapping Programs

David J. Coleman

Abstract This chapter introduces the context and characteristics implicit in conventional digital topographic mapping programs and then contrasts them to important underlying assumptions regarding volunteered geographic information. It defines the term "authoritative data" and challenges its use in the context of comprehensive topographic base-mapping programs. After examining prevailing cultures and assumptions that must be adjusted and workflows that must be modified to manage risk and make the best use of VGI in this role, case studies from the state of Victoria, Australia; the United States Geological Survey; and TomTom describe the early experiences of conventional mapping organizations in this regard. The author contends that VGI is *not* the ultimate solution to all geospatial data updating and maintenance challenges now faced by mapping organizations. However, it does represent an important potential channel of such updates that needs to be investigated seriously and implemented responsibly.

14.1 Introduction

Use of volunteered geographic information (VGI) by public and private comprehensive mapping organizations is now either under way or under consideration. As of summer 2011, Google Map Maker provided citizens in 188 jurisdictions with the ability to help populate and update Google Maps' graphical and attribute data (Google 2011). *OpenStreetMap*, *TomTom*, and *NAVTEQ* all routinely use volunteer contributions to maintain their databases (Coleman et al. 2010). In Australia, the Victoria State

---

D.J. Coleman (✉)
Department of Geodesy and Geomatics Engineering,
University of New Brunswick, Fredericton NB, Canada
e-mail: dcoleman@unb.ca

Government now permits (registered) individual government employees to update state-level mapping features and attributes.

Volunteered geographic information or VGI and its related terms have been discussed at length in other chapters of this book. The more general concepts of "user-generated content," "user-created content," and "crowdsourcing" are well documented (OECD 2007), and Cook (2008) offers a taxonomy of both passive and active "user contribution systems" in the consumer market. In addition to the better-referenced works defining neogeography and VGI by Turner (2007) and Goodchild (2007), respectively, more recent articles by Coote and Rackham (2008), Grira et al. (2010), and Heipke (2010) also do an excellent job of examining VGI contributors and their contributions.

Coote and Rackham (2008) describe neogeographic datasets as possessing the following characteristics:

- Creation has been stimulated by a lack of available data or by frustrations with costs, restrictions, and limitations of existing conventional data sources.
- They involve the capture, processing, and dissemination of geographic information provided voluntarily by individuals.
- Approaches to creation and management are neither intuitive nor necessarily tied to accepted standards or methods.
- Data is licensed using some open-source approach, which allows for users to consume the data without charge provided the original creator is acknowledged and any other user can do the same with anything you produce.

Web-enabled VGI has been used extensively over the past 3 years to support emergency operations by mapping the extent of affected areas, highlighting important incidents, and documenting disaster-recovery operations (e.g., Zook et al. 2010; Heinzelman and Waters 2010; Roche et al. 2011). As companies like Google, TomTom, and NAVTEQ have already discovered (Coleman et al. 2010), the potential exists for *government* mapping agencies to harness the power of Web 2.0, new media, and voluntarism in order to improve their own change-detection and geospatial data-updating processes.

There has been no shortage of online discussion regarding whether and how public-sector mapping and charting organizations might employ VGI in their map production, updating and even enriching the attributes of selected features (e.g., Casey 2009; Dobson 2010b; Ball 2010). There has also been interest from national government organizations in examining the role and potential of employing VGI in their map updating and enrichment of attributes (e.g., Guélat 2009). However, such efforts are still in their early stages.

This chapter will review the potential advantages and challenges of using volunteered geographic information as a tool in the updating and elaboration of features contained in government and commercial map databases. After examining prevailing cultures and assumptions that must be adjusted, as well as workflows that must be modified before making the best use of VGI in this role, the author discusses how existing developments are already answering important questions posed by conventional mapping organizations.

14.2 Challenges Faced by Professional Mapping Organizations

The missions, mandates, accomplishments, and perceived shortcomings of national mapping organizations have been well documented (e.g., Andrews 1970; Hardy and Johnston 1982; Cowen et al. 2003). Since the products maintained by these organizations are regarded as important information assets, most of these organizations have had to be responsive to users regarding, for example, how best to improve the content and currency of their products, update data structures to allow for more extensive geographical analyses, and modify pricing policies and distribution infrastructure to facilitate online access and increase downloads—all to meet a broad range of evolving requirements and technologies.

14.2.1 What Is "Authoritative Data"?

The term "authoritative data" has been used to describe products produced by professional mapping organizations (Goodchild 2009; Coleman et al. 2010; Ball 2010). However, no definitions of the word "authoritative" are offered in those articles.

One possibility is offered by Van der Molen and Wubbe (2007) in discussing government policy in the Netherlands. The authors describe the creation and designation of six key datasets as official "authentic" national registers, each of which was defined as "… a high quality database accompanied by explicit guarantees ensuring for its quality assurance that, in view of the entirety of statutory duties, contains essential and/or frequently-used data pertaining to persons, institutions, issues, activities or occurrences and which is designated by law as the sole officially recognised register of the relevant data to be used by all government agencies and, if possible, by private organisation's [*sic*] throughout the entire country, unless important reasons such as the protection of privacy explicitly preclude the use of the register."

In this context, two of the first six authentic registers were in fact geographically related—the cadastral registers and maps, and the 1:10,000-scale topographic base mapping (Kadaster International 2007).

Nautical charts in some countries are recognized to be "authoritative" documents (Fisheries and Oceans Canada 2011; LINZ 2011). They are updated regularly through "notices to mariners," and in some countries, producers or value-added repackagers of electronic nautical chart data may assume some liability in the event charts contain erroneous or out-of-date information (Obloy and Sharetts-Sullivan 1994). Similarly, aeronautical charts are updated regularly, and their updating has long been conducted by specialists possessing a "…comprehensive and authoritative personal knowledge" of reliable source materials and cartographic activities in a given area (UNECA 1966).

It can be appreciated how regularly maintained cadastral maps, nautical charts, aeronautical charts, and even local zoning maps may be seen to be "authoritative" sources of public information within a jurisdiction. Beyond the Netherlands, however, there is little mention in other countries of formally recognized "authentic" or "authoritative" *topographic* map series that would adhere to a similar definition. With the exception of the Ordnance Survey of Great Britain, most national government mapping agencies have neither the funding nor the mandate to keep their mapping databases current within specified time frames. Budgets for topographic digital map maintenance activities in some jurisdictions are either declining or nonexistent. The larger the country's size, the older some of its base mapping is likely to be.

With more up-to-date mapped information now available online from other sources, referring to national or regional government topographic mapping databases as being "authoritative" has become misleading in some cases. The practice is also becoming a source of division and controversy within the geospatial user community when examining the relative merits and uses of volunteered geographic information (Ball 2010; van der Vlugt 2011).

Coote and Rackham (2008) offer the term "conventional" as an alternative to "authoritative" and suggest the following characteristics of conventional datasets:

- Created for a specific and defined set of requirements whether for legal, administrative, or commercial purposes.
- Depending on the context, these may or may not be freely available, but usually there is at least some dissemination charge and, most likely, restrictions on access and use.
- Managed by organizations established for the purpose, whether as public or commercial bodies. There may be collaboration between organizations but on the basis of legal agreements including commercial contracts.
- Collected by professional staff who are paid to do so.
- Based on established methods, standards, specifications, and practices.
- Quality assured to varying degrees during the production of the data and supplied with some information, however basic, on the quality of the data.
- Protected by some form of copyright and governed by formal agreements or licenses.
- Access limited, in many cases, to only certain organizations or individuals for reasons of security, data protection, or commercial advantage.

While some of these points are UK specific—and more conventional data may be "freely" available now than when their paper was presented—these characteristics can be seen in government topographic mapping products across Europe, North America, and Australasia. Accordingly, the adjective "conventional" rather than "authoritative" will be used through the remainder of this chapter to describe comprehensive base-mapping programs.

14.2.2 Implications of Aerial Mapping on the Characteristics of Conventional Base Mapping

Since 1945—and well before in developed countries—most national topographical map series have been produced by photogrammetric means using aerial surveys.

14 Potential Contributions and Challenges of VGI for Conventional Topographic... 249

As a result, some important characteristics of the products involved must be kept in mind:

1. Data compilation has been done remotely by trained mapping technicians who may possess only a limited knowledge of the features in the area being mapped.
2. Mapped features are classified into relatively broad categories, with their corresponding range of attributes often limited to what can be determined from photo interpretation and a limited set of support documents.
3. Prior to any data structuring and logical consistency considerations, preliminary quality assurance concerns in map compilation center on (a) proper rectification of imagery or restitution of stereo models in advance of the mapping, (b) geometrically accurate representation of the center or selected edges of a given feature (e.g., roof lines rather than building footprints), and (c) correct classification and coding of features in accordance with the given classification scheme.
4. Field verification and completion of the mapping content is labor intensive, depends on available program funding, and may vary widely within and between programs depending on prevailing budget considerations.
5. Production and subsequent updating are organized on the basis of geographic coverage, where mapping of one or more adjacent map sheets, files, or tiles is undertaken, completed, and distributed to users within a fixed period of time. Attention and budgets then shift to a different geographic area. It may be years or even decades before attention returns to a given geographic area. (An exception is the United Kingdom, where revision cycles are much shorter.) When updating does occur on such maps or tiles, all features in the given area associated with that particular mapping product are typically updated unless otherwise specified.

Contrast this list with Coote and Rackham's proposed characteristics of neogeographic information offered in Sect. 14.1. Further, Bruns (2008) identifies four important characteristics of information "produsage" in a Web 2.0 environment as distinct from more traditional information production:

- Collection and review operations are community based rather than relying on "… a narrow elite of knowledge workers."
- Roles of producers will be fluid, alternating between collector, reviewer, arbitrator, and user at different times.
- A given product will be never be finished—it will be under continuous review, and different aspects or portions will be updated at different times.
- Producers favor more permissive approaches to rights in intellectual property than those found in traditional content production.

To employees within a mapping organization, then, the implications of incorporating VGI into its processes include rethinking (1) entire aspects of production workflow, (2) who should be involved, and even (3) what constitutes a "product" they are prepared to offer to users. Important cultural hurdles may include the following:

- Accepting that untrained "outsiders"—even trusted ones—may be willing and able to make reliable contributions

- Assessing whether or not some larger community of users has the willingness and capability to collectively offer some level of editing and quality assessment of individual contributions
- Moving from a coverage-based to a feature-based updating model
- Accepting that such volunteered information will be "perpetually unfinished"
- Accounting for and balancing the respective rights of individual contributors, the VGI community, and the mapping organization itself
- Accommodating practical, political, social, and possibly even legal implications implicit in the characteristics mentioned above

Allowing even trusted "outsiders" to collect and/or modify internally collected mapping can be a difficult culture shift. In Canada, it took at least 3 years in the late 1970s before internal government mapping inspectors agreed the quality of national topographic series mapping compiled by professional private-sector firms was sufficient to shift contracting out from pilot-project efforts to standard practice. Even exchanges of digital map data between different levels of government mapping agencies were undertaken with caution and only after considerable negotiation (Pearson and Gareau 1986).

The cultural and processual changes involved in shifting the planning and production focus from a "coverage-based" to a "feature-based" orientation cannot be underestimated. Road-network firms like TomTom and NAVTEQ have already made this shift and realized quicker turnaround times of updates and improved customer service (TomTom 2008), but many government topographic mapping organizations have not.

14.3 Workflow, Quality Assurance, and Risk Management Considerations

Despite the challenges mentioned in Sect. 14.2, the idea of using volunteered geographic information nevertheless remains of considerable interest to mapping organizations. The prospects of more descriptive and up-to-date information in "high-usage" geographic areas where changes may occur frequently are attractive to base-mapping organizations interested in exploring more cost-effective ways in which to improve their products. This section describes some key considerations mapping program managers should take into account when designing or reworking production processes.

14.3.1 Attracting and Retaining Volunteer Contributors

Will individuals want to contribute to government in the same way they contribute to social networks and even to commercial databases from TomTom, NAVTEQ, and others? What questions should an organization ask in determining how, if at all, it should employ VGI? How does an organization assess the credibility of a new contributor and the degree of trust it can place in that person's contributions? How do organizations attract new volunteer contributors, and how do they keep existing volunteers engaged—or is it assumed they will cycle in and out?

There is ample evidence that interested volunteers do exist—at least in the early stages of a program. Results of early research into the nature and motivation of contributors—and the types of contributions they make—are discussed in depth by Coleman et al. (2009), Budhathoki et al. (2010), Dobson (2010b), and Cooper et al. (2011) among others. Coleman et al. (2010) further examined how three different spatial data organizations employed VGI in the updating of their map databases—and summarized in each case the respective motivator(s) the program was directly or indirectly offering.

14.3.2 Quality Assurance Considerations

If and as volunteer contributions are solicited, how will they be integrated into conventional production workflows? Given the challenges discussed in Sect. 14.2 of this chapter, how will quality assurance considerations be addressed? Who, if anyone, will assume the risks associated with the introduction and use of contributions from different sources?

As discussed in Sect. 14.2, there are fundamental differences between how quality assurance is viewed by VGI contributors versus individuals in professional mapping organizations. Conventional mapping is produced in accordance with mature, well-documented specifications and is assessed by individuals who are trained in interpreting those specifications and understanding the products themselves and well versed in the inherent errors or blunders encountered in data compilation.

The positional accuracy of volunteered geographic information in comparison with data from conventional programs is well documented. Rigorous investigations conducted by Haklay (2010), Coleman et al. (2010), and Girres and Touya (2010), among others, all attest to the acceptability and accuracy of VGI contributions to (e.g.) OpenStreetMap in relation to other well-documented map series. Moreover, the repeated capture of the same feature's location by multiple contributors through active and (especially) passive means has proven to greatly improve the accuracy of positioning and representation (Haklay et al. 2010; Dobson 2010a). Finally, related research (e.g., Zandbergen 2009; Gakstatter 2010) has isolated constraints on existing cell phones as positioning devices *and* led to technology breakthroughs that will further improve cell phone-based positioning in the near future.

Positional accuracy is only one aspect of data quality, though. Coote and Rackham (2008) point out that "quality" in the context of VGI is more subjective in nature and depends on:

1. A user's requirement and his or her expectations
2. The benefits the user wants to derive
3. What the user or contributor means by "data quality"

In this regard, currency of data and reliability of feature attributes within a given area may be far more important elements of quality to a given user or users than positional accuracy or the completeness of coverage over an entire county or map tile.

Also, in some VGI initiatives, there may be no clear line of authority regarding who is ultimately responsible for assessing the quality of positioning and representation of a given feature, nor who possesses the rights to modify those things.

This degree of subjectivity, the possible existence of multiple contexts, and the lack of clear lines of authority are possibly baffling and certainly unacceptable to a practitioner accustomed to conventional mapping workflow and practices.

Finally, from a user perspective, the potential lack of consistency in terms of up-to-date content, interpretation, and structuring may constrain a dataset's use for analysis and ultimately lead to only guarded acceptance of the product (Coote and Rackham 2008).

14.3.3 Assessing the Credibility of Contributors

One of the major concerns of using VGI as a source of input to authoritative databases is how to assess the credibility of contributors and the reliability of their contributions. The success of reputation-based services like *eBay.com* holds one key for building trust for handling VGI. *eBay* users who log in to purchase items online may leave feedback for the sellers and future purchasers based on the success of the transactions. *eBay* then uses a centralized user reputation system that drives its inputs from buyer ratings of the sellers. Social networking sites which make use of VGI contributions of point- and route-based data have adopted similar approaches and, in some cases, automated the ways in which improvements can be noted and incorporated.

Different lessons can be learned from leading wikis such as *Wikipedia.org*. Wikipedia originally relied solely upon the "wisdom of the crowds" to evaluate, assess, and if necessary, improve upon entries from individual contributors, usually with great success. However, beginning in December 2009, it has relied on teams of editors to adjudicate certain "flagged entries" before deciding whether or not to incorporate a volunteered revision (Beaumont 2009).

Theoretical approaches to characterizing VGI contributions and/or their contributors are now being formulated by the international research community. For example, Lenders et al. (2008) theorize an automated approach to establishing the level of trust inherent in different user-generated contributions to local-based services. The proposed architecture of their "secure localization and certification service" maintains user privacy by tagging volunteered content with the location and time of the contribution rather than the identity of the contributor. The level of trust in a given change increases in direct proportion to how recently the contribution was made and how close the contributor was geographically to the proposed change. Other examples of research to better categorize contributions and automate processes include work by Maué and Schade (2008), Poser et al. (2009), and Brando et al. (2011), among others.

14.3.4 Practical Examples of Managing Risk

Coleman et al. (2010) investigated how three different public and private organizations incorporated volunteered contributions into their production workflow: the state of Victoria's Notification for Edit Service in Australia, the National Map Corps Initiative of the United States Geological Survey, and TomTom's MapShare™ service. These are summarized in the following subsections.

14.3.4.1 Notification for Edit Service, Victoria Department of Sustainability and Environment, Australia

Victoria DSE's Notification for Edit Service (Thompson 2011; NES 2008) employs internal contributions of volunteered geographic information by internal government staff outside the formal mapping agencies to update widely used base-mapping databases. A network of registered "knowledgeable notifiers"—state and local government users of Victoria's Corporate Spatial Data Library (CSDL)—use a password-protected Web-based system to either correct or update selected mapped features based on field evidence found in routine government operations. Suggested updates to a given feature are routed automatically to the designated custodian agency responsible for that feature type, and then the custodian is given the option to either confirm or refute the amendment. An update tracking system provides regular reports on the status of each update, where it currently sits in the process and, if complete, whether or not it was accepted (Fig. 14.1).

14.3.4.2 National Map Corps, United States Geological Survey

The National Map Corps was a pioneering effort in using VGI to update and supplement government mapping in North America. The NMC's "Adopt-a-Map" program had by 2001 over 3,000 volunteers identifying and annotating topographic map corrections and updates to hardcopy United States Geological Survey (USGS) map sheets (Fig. 14.2) and National Map data files (Bearden 2009). Later aspects of the program included incorporation of updates and additions based on hand-held GPS observations. Still later, a Web-based map and image viewer (Fig. 14.3) enabled volunteer users to easily identify and label buildings and other structures requiring annotation (Bearden 2007b).

Although the volunteer response was very impressive, the USGS simply did not possess the internal resources necessary to act on these notifications. Traditional coverage-based map revision workflows and long updating cycles meant that the feature-based annotation work completed by the volunteers was rarely used. Volunteer numbers diminished, and the map annotation aspect of the program was ultimately stopped altogether in 2005. Further, the large number of GPS updates submitted overwhelmed the limited staff resources assigned to assessing and using the volunteered input. By 2007, there was a 16-month backlog of GPS-collected points increasing almost daily (Bearden 2007a). Due to program budget cuts, issues

1 2 3 4

Change Request ID	Description	Notifier	Status	WF	Date Last Modified	View Details
341	Apiry Point Editing with UFI=346	Local Government Authority (Towong Shire)	SUBMITTED		8/09/2008 10:16:51 AM	View Details
340	Apiry Point Editing with UFI=346	Local Government Authority (Towong Shire)	SUBMITTED		8/09/2008 10:16:42 AM	View Details
339	Apiry Point Editing with UFI=707	Local Government Authority (Towong Shire)	SUBMITTED		8/09/2008 10:16:37 AM	View Details
338	Apiry Point Editing with UFI=707	Local Government Authority (Towong Shire)	SUBMITTED		8/09/2008 10:16:31 AM	View Details
337	Apiry Point Editing with UFI=2633	Local Government Authority (Towong Shire)	SUBMITTED		8/09/2008 10:16:22 AM	View Details
336	Apiry Point Editing with UFI=1563	Local Government Authority (Towong Shire)	SUBMITTED		8/09/2008 10:16:08 AM	View Details
334	road demo	NES General Public	CHANGE ACCEPTED		5/09/2008 3:03:00 PM	View Details
A 332	Sept M1	Local Government Authority (Towong Shire)	SUBMITTED		3/09/2008 3:26:59 PM	View Details
330	Road name demo 1	NES General Public	CHANGE ACCEPTED		3/09/2008 3:21:41 PM	View Details
325	test change parcel	NES General Public	CHANGE ACCEPTED		3/09/2008 12:30:45 PM	View Details
324	test change	NES General Public	CHANGE ACCEPTED		3/09/2008 11:30:24 AM	View Details
298	Deleted a property	NES Sample State Org	SUBMITTED		1/09/2008 9:05:17 AM	View Details
293	name to both	NES General Public	SENT TO ROAD & ADDRESS MAINTAINERS		29/08/2008 4:39:53 PM	View Details
292	name to 2	NES General Public	SENT TO ROAD MAINTAINER		29/08/2008 4:39:41 PM	View Details
291	name to 1	NES General Public	SENT TO ADDRESS MAINTAINER		29/08/2008 4:39:27 PM	View Details
288	road name dual -2	NES General Public	CHANGE ACCEPTED		29/08/2008 3:35:30 PM	View Details
287	road name dual -1	NES General Public	CHANGE ACCEPTED		29/08/2008 3:34:37 PM	View Details
285	Add Address (Testing Add Multi-Feature Function)	Local Government Authority (Wangaratta Rural City)	SUBMITTED		29/08/2008 12:22:27 PM	View Details
284	this is road name extent - other was road name	NES General Public	CHANGE ACCEPTED		29/08/2008 3:39:21 PM	View Details
283	road name extent	NES General Public	CHANGE ACCEPTED		29/08/2008 12:32:14 PM	View Details
280	Road 5	NES General Public	DECLINED		29/08/2008 10:49:34 AM	View Details
279	Road 4	NES General Public	CHANGE ACCEPTED		29/08/2008 3:42:30 PM	View Details
278	Road 3	NES General Public	CHANGE ACCEPTED		29/08/2008 12:34:25 PM	View Details
268	m1 for Augyst	Local Government Authority (Towong Shire)	SENT TO MAINTAINER		28/08/2008 2:11:06 PM	View Details

Fig. 14.1 Tracking updates within NES (NES 2008)

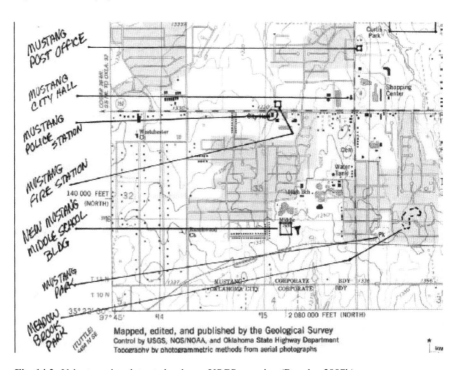

Fig. 14.2 Volunteered updates to hardcopy USGS mapping (Bearden 2007b)

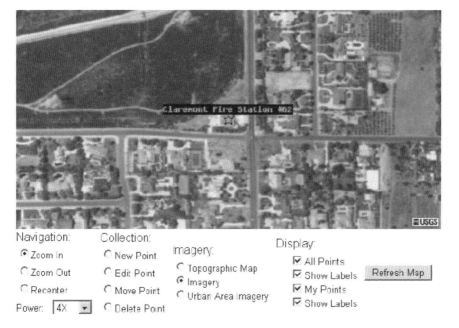

Fig. 14.3 National Map Corps image and map viewer (Bearden 2007b)

over continued resourcing, and internal disagreements over the validity of volunteered content, National Map Corps activities were suspended in fall 2008 with the exception of its online viewer and labeler website (National Map Corps 2008).

Although the program itself had been suspended, interest remained strong in the concept itself. A USGS-sponsored workshop on VGI was held in 2010 (CEGIS 2011), and a collaborative pilot project with the OpenStreetMap organization, started in 2011 to have volunteers digitize new road information, is now under evaluation (Wolf et al. 2011). A follow-on pilot project to collect data on thirty different types of structures in the greater Denver, Colorado, area is now under way (National Map Corps 2011).

14.3.4.3 TomTom's MapShare™ Service

TomTom's online MapShare™ service is a popular operational example of how one large commercial data supplier manages risk in terms of assessing volunteered contributions and disseminating such noncertified updates to its customers (Club TomTom 2007). The company employs a graduated approach to sharing, assessing, and using the volunteer-provided updates. First, MapShare contributors have the choice of only using their updates on their own TomTom units, of sharing within their own group, or of sharing them with the general TomTom community. Second,

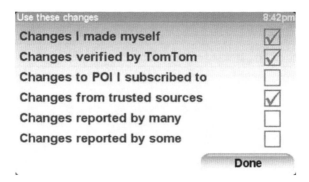

Fig. 14.4 Customer choices in selecting MapShare updates to be used on own TomTom unit (Club TomTom 2007)

TomTom itself assigns a progressively higher level of credibility to a given update through its independent confirmation by (1) more than two independent contributors, (2) many independent contributors, (3) a "trusted partner" or corporate user, and finally (4) its own crews or contractors in the field. Finally, it allows its customers to interactively select the "level of trust" they desire for the data used on their navigation unit (Fig. 14.4). Customers may elect to use only updates reported by TomTom/Tele Atlas field crews, by trusted commercial partners and many customers, by only a few customers, or even only by themselves.

Chapter 17 by Dobson in this volume offers a much more detailed description of TomTom's MapShare service and the data collection production workflow it employs. As well, it provides valuable comparisons with TomTom's competitor NAVTEQ and with the hybrid approach to collection and updating adopted by Google.

14.4 Discussion

Coleman et al. (2009) suggested important questions that conventional public and private mapping organizations should ask themselves when considering the opportunities and risks posed by introducing and employing VGI in their production processes. *What problems or objectives are we trying to address? To what extent should we initially adopt VGI? How may credible contributors be distinguished from those who are mischievous or malicious? How do we cultivate the volunteers— whether they are one-time or regular contributors? Who makes the final decisions regarding the reliability and integrity of a given update?*

These are legitimate questions, and early lessons can already be drawn from reviewing these volunteered contributions as well as those to other online communities.

Regarding the Programs Themselves
- Experience has shown that, once the request goes out, the number and extent of volunteered contributions can be significantly underestimated. Organizations that

underresourced the VGI acceptance and verification aspects of their production workflows have found themselves overwhelmed and unable to use all the input. Unless the rationale and goals of a VGI initiative—be it pilot project or mainstream activity—are clearly defined, communicated, and measured, the initiative risks being curtailed when the next round of budget cuts occur. Program managers need to be clear on their rationale, their goals, and the metrics used to assess progress towards those goals. "Shortening updating cycles," "verifying selected attributes," and "adding new attributes" for specific features are all examples of legitimate and measurable goals.

Regarding Contributors
- The size and scope of the "contributor pool" may be controlled. Initially, it may be restricted to employees of the organization and knowledgeable, long-standing users of the data. Access may be opened up gradually as confidence grows in the contributions received and as the organization provides additional resources to accept and process the growing number of contributions.
- Volunteer contributors value some recognition of their contribution. This may range from prompt recognition of the contribution by a return e-mail message (Tele Atlas and Google both acknowledge such contributions to their Map Insight and My Map websites) to more formal inclusion of contributors in metadata or tags associated with the feature.
- Contributors want to see their contribution used—and quickly. Case studies cited from both the Wikipedia and the open-source software communities identified the importance of contributions being acted upon and either incorporated or refuted quickly. The falloff of National Map Corps volunteers after their updates were not acted upon quickly is just one example of this. It may take a while to verify a contribution, but early acknowledgement of its receipt and follow-on communication that it is being reviewed are both signs of good service that contributors value.
- As a program matures, failing to retain all volunteer contributors is not necessarily a sign of failure. Experience has shown that the majority of contributors of new information to such databases may make only one or two contributions—a new road, an update or correction to a given feature in their neighborhood, etc. A limited group of dedicated and long-serving members of the volunteer community then assess that contribution and refine it to fully meet existing specifications.
- This, in turn, implies that mature programs must have separate but integrated interfaces. A simple, easy-to-use interface with limited functionality—but perhaps more extensive postprocessing—is required to satisfy occasional or one-time-only contributors. With an appropriate hierarchy of access privileges in place, a more sophisticated and multifunctional Web interface would be employed to accommodate more extensive edits by internal production staff and external "power users."

Regarding Their Contributions
- The majority of contributions—especially to road networks—deal with amendments or additions to a feature's *attributes* rather than to its location or representation.

- Major private-sector initiatives like TomTom's have already recognized that they must develop hybrid systems that accommodate different channels of volunteered input. Following the examples discussed in this chapter, a conventional mapping organization may start by accepting only amendments to attribute data and active contributions of positional data collected using GPS. They may subsequently add the capability to incorporate active contributions of features digitized from online satellite or aerial imagery. As technologies and attitudes evolve, their system may grow to include approved passive contributions of a person's position in real time as he or she travels in a car or ATV, on a bicycle, or even on foot.
- There *are* established and tested ways to validate the reliability of VGI contributions and the credibility of their contributors. Certain spatial and temporal considerations make VGI contributions unique, and these may be used to support or refute the credibility of a given contributor. Tools built atop early technologies like WikiScanner (Borland 2007) may help identify the rough geographic location from which a contribution originated. Geotagged cell phone images may be used to provide supporting evidence of a given update.

Some VGI activities—for example, digitizing features from satellite imagery in areas with limited vector mapping or where a natural disaster has occurred—may not lend themselves to this type of validation. However, this is where engagement of other volunteers helps to validate or refute a contribution, or assess the reliability of competing or contradictory contributions.

14.5 On the Future of Conventional Public-Sector Mapping Programs

The significant growth and notoriety of *private-sector* mapping services and their more rapid adoption of VGI beg questions as to the future of *public*-sector mapping programs. Certainly government budgets are shrinking. Funding for both in-house *and* subcontract production has declined in some jurisdictions. Where new or updated production *is* under way in national- and state-level programs, funds are more likely to be directed towards creation of image-based mapping products and higher-accuracy digital elevation models rather than towards updates to vector mapping.

Moreover, rather than compete, many governments have opted to provide a base upon which *others* can develop and provide value-added applications. Even in cases where the basic road centerline and cultural information may come from government base mapping, that information is now perceived to have originated from the Web service provider itself. This places program executives in the position of having to explain to their public-service and political masters the reasons why they should continue to be funded at all.

Why indeed? While they may form the vast majority of operations conducted, can we assume that *all* users will be satisfied solely by the map display, address-finding, routing, area-based query, and point-mapping capabilities of services like Google and Bing?

The largest and most intensive groups of users of multipurpose government mapping are usually found in other government departments. Clearly, a full investigation of this larger question is beyond the scope of this chapter. However, the following focused questions would help drive such an investigation:

1. By virtue of its mission, to what extent must a government or private organization commit to certain norms or expectations of areal coverage and consistency of reliability or currency of a geospatial dataset in a given jurisdiction? Have such expectations become negotiable?
2. What assumptions are taken into account when using conventional base mapping for specific applications? What operations (if any) must be undertaken to convert or prepare this conventional mapping for use in internal applications?
3. Under what conditions can geospatial data provided by commercial suppliers like Google, Bing (Microsoft), and others be used by organizations to meet their own unique mission-driven requirements? (The act of "importing" per se may be a misnomer in this era of cloud computing.) What are the legal, financial, technological, and labor implications involved?
4. To what extent do the geospatial data content, coverage, accuracy, attribute correctness, currency, and structure of commercial datasets meet the requirements of specific applications to be found in such customer organizations? Does the potential heterogeneity of such data across a jurisdiction represent a concern? Under what circumstances?
5. Under what conditions are commercial providers prepared to upgrade the quality or consistency of their data to meet the needs of specific customers? Are these upgrades subsequently available to all users and under what conditions?
6. To what extent are user organizations now prepared either to independently update such commercial information, to update their own datasets, or even to collect and use their own geospatial data from scratch in their GIS applications as a project requires?
7. To what extent is a government organization prepared to tie itself and its data holdings to any one particular service provider? What conditions and provisions should be considered by that organization in order to protect its investment in its own data holdings?

Responses to these and other questions should have a real bearing on the nature of enabling infrastructure, products, and services provided by public-sector mapping programs in the future.

14.6 Concluding Remarks

This chapter has introduced the context and characteristics implicit in conventional digital topographic mapping programs and contrasts them to important underlying assumptions regarding volunteered geographic information. Despite claims to the contrary, government topographic mapping products in most countries are not "authoritative" by any practical definition. Far from it. They are typically out of date, possibly inconsistent, and usually the victims of diminishing maintenance budgets.

At the same time, claims that VGI in itself—or VGI-supported private data suppliers—will replace the overall role of conventional mapping organizations in developed countries in the near term are likely overblown. Research has demonstrated that there remain remote areas in which change may have occurred but which are not being mapped by volunteers. As well, further research is required to determine the conditions under which such suppliers would have the interest and resources required to satisfy the varied requirements of such conventional programs.

VGI is *not* the ultimate solution to all geospatial data updating and maintenance challenges now faced by mapping organizations. However, there is growing agreement that it represents *one important channel* of such updates—one that needs to be investigated, prototyped, and introduced in a reasonable, informed manner.

Acknowledgments The author would like to recognize the Earth Sciences Sector of Natural Resources Canada, the Natural Sciences and Engineering Research Council of Canada (NSERC), and the GEOIDE Network of Centres of Excellence for their financial support of this research over the past 3 years. As well, he would like to thank UNB graduate students Krista Amolins, Andriy Rak, and Titus Tienaah for their constructive suggestions during the review of this chapter.

References

Andrews, G. (1970). *Administration of surveys and mapping in Canada, 1968*. Report of the National Advisory Committee on Control Surveys and Mapping to the Surveys and Mapping Branch, Dept. of Energy, Mines and Resources Canada, Ottawa, Canada.

Ball, M. (2010). What's the distinction between crowdsourcing, volunteered geographic information, and authoritative data? Editorial in *v1 Magazine*. http://www.vector1media.com/dialog/perspectives/16068-whats-the-distinction-between-crowdsourcing-volunteered-geographic-information-and-authoritative-data.html. Accessed 1 Aug 2011.

Bearden, M. J. (2007a, Dec 13–14). *The national map corps*. Paper presented at workshop on volunteered geographic information, University of California at Santa Barbara, Santa Barbara, CA. http://www.ncgia.ucsb.edu/projects/vgi/docs/position/Bearden_paper.pdf. Accessed 10 Feb 2009.

Bearden, M. J. (2007b, Dec 13–14). *The national map corps*. Presentation from workshop on volunteered geographic information, University of California at Santa Barbara, Santa Barbara, CA. http://www.ncgia.ucsb.edu/projects/vgi/docs/present/Bearden_USGS_MapCorps19.pdf. Accessed 10 Feb 2009.

Bearden, M. J. (2009). Personal communication with national map corps program coordinator, U.S. Geological Survey National Geospatial Technical Operations Center, Rolla, Missouri, USA, 21 April.

Beaumont, C. (2009, Aug 26). Wikipedia ends unrestricted editing of articles. *The Telegraph*. http://www.telegraph.co.uk/technology/wikipedia/6088833/Wikipedia-endsunrestricted-editing-of-articles.html. Accessed 11 Aug 2011.

Borland, J. (2007). See who's editing Wikipedia—Diebold, the CIA, a Campaign. Politics/Online Rights Blog. *Wired*. http://www.wired.com/politics/onlinerights/news/2007/08/wiki_tracker. Accessed 14 Jan 2012.

Brando, C., Bucher, B., & Abadie, N. (2011). Specifications for user generated spatial content. In S. Geertman, W. Reinhardt, & F. Toppen (Eds.), *Advancing geoinformation science for a changing world* (pp. 479–495). New York: Springer.

Bruns, A. (2008). *Blogs, Wikipedia, second life, and beyond: From production to produsage.* New York: Peter Lang.

Budhathoki, N., Nedovic-Budic, Z., & Bruce, B. (2010). An interdisciplinary frame for understanding volunteered geographic information. *Geomatica, 64*(1), 11–26.

Casey, M. (2009, May). *Citizen mapping and charting: How crowdsourcing is helping to revolutionize mapping & charting.* Proceedings of the 2009 U.S. Hydrographic Conference, Norfolk, VA. http://www.thsoa.org/hy09/0512A_02.pdf. Accessed 15 Aug 2011.

CEGIS. (2011). *U.S. Geological Survey volunteered geographic information workshop.* http://cegis.usgs.gov/vgi/index.html. Accessed 30 June 2011.

Club TomTom. (2007). *Get to know MapShare™. The official blog for TomTom in North America.* http://www.clubtomtom.com/general/get-to-know-tomtom-mapshare™/. Accessed 15 Apr 2009.

Coleman, D. J., Georgiadou, Y., & Labonte, J. (2009). Volunteered geographic information: The nature and motivation of producers. *International Journal of Spatial Data Infrastructures Research, 4*, 332–358. http://ijsdir.jrc.ec.europa.eu/index.php/ijsdir/article/view/140/198. Accessed 14 Jan 2012.

Coleman, D. J., Sabone, B., & Nkhwanana, N. (2010). Volunteering geographic information to authoritative databases: Linking contributor motivations to program effectiveness. *Geomatica, 64*(1), 383–396.

Cook, S. (2008, Oct). Why contributors contribute. *Harvard Business Review.* http://usercontribution.intuit.com/The+Contribution+Revolution+linked+version. Accessed 3 Aug 2011.

Cooper, A., Coetzee, S., Kaczmarek, I., Kourie, D., Iwaniak, A., & Kubik, T. (2011, May 31–June 2). *Challenges for quality in volunteered geographical information.* Proceedings of the AfricaGEO 2011 Conference, Cape Town, South Africa. http://researchspace.csir.co.za/dspace/handle/10204/5057. Accessed 20 Aug 2010.

Coote, A., & Rackham, L. (2008, Sept). *Neogeographic data quality—Is it an issue?* Paper delivered at the 2008 AGI Geocommunity Conference 2008, ConsultingWhere Ltd., Stratford-upon-Avon, UK.

Cowen, D. et al. (2003). *Weaving a national map: Review of the U.S. Geological Survey concept of the national map.* Report of the National Research Council (U.S.). Committee to Review the U.S. Geological Survey Concept of the National Map. Washington, DC: National Academies Press.

Dobson, M. (2010a, Aug 29). TomTom, TeleAtlas and MapShare. *Telemapics Blog.* http://blog.telemapics.com/?p=323. Accessed 2 Aug 2010.

Dobson, M. (2010b, Sept 22). Crowdsourcing—How much is too much? *Telemapics Blog.* http://blog.telemapics.com/?p=328. Accessed 2 Aug 2010.

Fisheries and Oceans Canada. (2011). About CHS: What we do. http://www.charts.gc.ca/about-apropos/wwd-qfn-eng.asp. Accessed 30 Aug 2011.

Gakstatter, E. (2010, June 1). The dawn of a new era in GPS accuracy. *GPS World Column.* http://www.gpsworld.com/gis/gss-weekly/the-dawn-a-new-era-gps-accuracy-10016. Accessed 3 June 2010.

Girres, J.-F., & Touya, G. (2010). Quality assessment of the French OpenStreetMap dataset. *Transactions in GIS, 14*(4), 435–459.

Goodchild, M. F. (2007). Citizens as voluntary sensors: Spatial data infrastructure in the world of Web 2.0. *International Journal of Spatial Data Infrastructures Research, 2*, 24–32. http://ijsdir.jrc.ec.europa.eu/index.php/ijsdir/article/view/28/22. Accessed 25 Aug 2011.

Goodchild, M. F. (2009). NeoGeography and the nature of geographic expertise. *Journal of Location Based Services, 3*(2), 82–96.

Google. (2011). Countries editable in Google map maker. http://www.google.com/mapmaker/mapfiles/s/launched.html. Accessed 10 Aug 2011.

Grira, J., Bédard, Y., & Roche, S. (2010). Spatial data uncertainty in the VGI world: Going from consumer to producer. *Geomatica, 64*(1), 61–71.

Guélat, J-C. (2009, Aug). *Integration of user generated content into national databases—Revision workflow at Swisstopo.* 1st EuroSDR Workshop on Crowdsourcing, Federal Office of Topography swisstopo, Wabern, Switzerland. http://www.eurosdr.net/workshops/crowdsourcing_2009/presentations/c-4.pdf. Accessed 8 Dec 2011.

Haklay, M. (2010). How good is volunteered geographical information? A comparative study of OpenStreetMap and ordnance survey datasets. *Environment and Planning B: Planning and Design, 37*(4), 682–703.

Haklay, M., Basiouka, S., Antoniou, V., & Ather, A. (2010). How many volunteers does it take to map an area well? The validity of Linus' law to volunteered geographic information. *The Cartographic Journal, 47*(4), 315–322.

Hardy, G. A., & Johnston, W. F. (1982). The future of the ordnance survey? *The Geographical Journal, 148*(2), 155–172.

Heinzelman, J., & Waters, C. (2010). *Crowdsourcing crisis information in disaster-affected Haiti* (Special Report 252). Washington, DC: United States Institute of Peace. http://www.usip.org/files/resources/SR252%20-%20Crowdsourcing%20Crisis%20Information%20in%20Disaster-Affected%20Haiti.pdf. Accessed 15 Aug 2011.

Heipke, C. (2010). Crowdsourcing geospatial data. *ISPRS Journal of Photogrammetry and Remote Sensing, 65*(6), 550–557.

Kadaster International. (2007, Sept). Abroad. Special issue on how e-government and land information align. *Periodical newsletter of Kadaster International*. http://www.kadaster.nl/pdf/abroad_092007.pdf. Accessed 25 July 2011.

Lenders, V., Koukoumidis, E., Zhang, P., & Martonosi, M. (2008). Location-based trust for mobile user-generated content: Applications, challenges and implementations. In *Proceedings of the 9th workshop on mobile computing systems and applications* (pp. 60–64). New York: ACM.

LINZ. (2011). About notices to mariners. Land information New Zealand. http://www.linz.govt.nz/hydro/ntms/about-ntms. Accessed 20 Aug 2011.

Maué, P., & Schade, S. (2008, May). *Quality of geographic information patchworks*. Proceedings of the 11th AGILE International Conference on Geographic Information Science, Girona, Spain. http://plone.itc.nl/agile_old/Conference/2008-Girona/PDF/111_DOC.pdf. Accessed 30 May 2009.

National Map Corps. (2008). Important notice concerning national map corps. http://nationalmap.gov/tnm_corps.html. Accessed 29 May 2009.

National Map Corps. (2011). The national map corps pilot projects. http://nationalmap.gov/TheNationalMapCorps/pilot.html. Accessed 20 Aug 2011.

NES. (2008). NES help module 3 change request/easy editor, p. 31. http://www.land.vic.gov.au/CA256F310024B628/0/DA7C602F526419D8CA25750D00053B7B/$File/NES+Quick+Module3.pdf. Accessed 29 July 2011.

Obloy, E. J., & Sharetts-Sullivan, B. H. (1994, Oct). *Exploitation of intellectual property by electronic chartmakers: Liability, retrenchment and a proposal for change*. Proceedings of the conference on law and information policy for spatial databases, Tempe, AZ. http://www.spatial.maine.edu/~onsrud/tempe/obloy.html. Accessed 20 Aug 2011.

OECD. (2007). *Participative web: User-created content: Web 2.0, wikis and social networking*. Report of the Working Party on the Information Economy, Committee for Information, Computer and Communications Policy, Directorate for Science, Technology and Industry, Organization for Economic Co-Operation And Development. 12 April. Paris: OECD Publishing. ISBN 9264037462, 9789264037465.

Pearson, M., & Gareau, R. (1986). Exchanging digital base mapping data. In *Proceedings of the second international symposium on spatial data handling* (pp. 611–615). Williamsville: International Geographical Union.

Poser, K., Kreibich, H., & Dransch, D. (2009). *Assessing volunteered geographic information for rapid flood damage estimation*. Proceedings of the 12th AGILE International Conference on Geographic Information Science, Leibniz Universität, Hannover, Germany, pp. 1–9.

Roche, S., Propeck-Zimmermann, E., & Mericskay, B. (2011). GeoWeb and crisis management: Issues and perspectives of volunteered geographic information. *GeoJournal*. doi:10.1007/s10708-011-9423-9.

Thompson, Y. (2011, July 20). Notification for edit service (NES). *Mygeoplace Blog*. http://mygeoplace.com/2011/07/30/notification-for-edit-service-nes/. Accessed 10 Aug 2011.

TomTom. (2008, Dec 16). TomTom announces five millionth map share™ improvement. *TomTom News*. http://www.tomtom.com/news/category.php?ID=4&NID=660&Lid=4. Accessed 1 May 2009.

Turner, A. (2007, Dec 6). *Neogeography—Towards a definition*. A weblog posted on *High Earth Orbit*. http://highearthorbit.com/neogeography-towards-a-definition/. Accessed 15 Aug 2011.

UNECA. (1966, Sept 12–24). *Maintaining current base information for aeronautical charts*. Paper submitted by the United States Government to the 2nd United Nationals Cartographic Conference for Africa, Tunis, Tunisia. United Nations Economic and Social Council. http://repository.uneca.org/bitstream/handle/123456789/15884/Bib-64431.pdf?sequence=1. Accessed 25 Aug 2011.

Van der Molen, P., & Wubbe, M. (2007). *E-government and E-land administration. As an example: The Netherlands*. Proceedings of the 6th Regional FIG Conference 'Coastal Areas and Land Administration—Building the Capacity', San Jose, Costa Rica. http://www.fig.net/pub/costarica_1/papers/ts10/ts10_02_wubbe_vandermolen_2480.pdf. Accessed 15 Aug 2011.

van der Vlugt, M. (2011, Jan 4). PSMA, sensis or OpenStreetMap: What makes spatial data "authoritative"? *Spatial Information in the 21st Century* Blog. http://spatial21.blogspot.com/2011/01/psma-sensis-or-openstreetmap-what-makes.html. Accessed 1 Aug 2011.

Wolf, E. B., Matthews, K. M., & Poore, B. S. (2011, June). *OpenStreetMap collaborative prototype, phase one*. Open-file report 2011–1136. http://pubs.usgs.gov/of/2011/1136/pdf/OF11–1136.pdf. Accessed 31 Aug 2011.

Zandbergen, P. A. (2009). Accuracy of iPhone locations: A comparison of assisted GPS, WiFi and cellular positioning. *Transactions in GIS, 13*(1), 5–26.

Zook, M., Graham, M., Shelton, T., & Gorman, S. (2010). Volunteered geographic information and crowdsourcing disaster relief: A case study of the Haitian earthquake. *World Medical & Health Policy, 2*(2), 7–33.

Chapter 15
"We Know Who You Are and We Know Where You Live": A Research Agenda for Web Demographics

T. Edwin Chow

Abstract In the digital era, personal data, such as full name, address, age, phone number, and household members, may scatter across various administrative records, databases of private companies, and social networks, as well as through information contributed as volunteered geographic information (VGI). People-finder sites gather such data and provide user interfaces for Internet users to query web demographics. The emergence of web technologies such as mash-ups, web mapping, and webscraping presents opportunities to capitalize on the availability of web demographics and opens up new frontiers of research. The main objectives of this chapter are to (1) examine web demographics as an example of VGI and (2) explore the research agenda of web demographics. More research and development are needed to enhance extraction rules, identify and remove erroneous enumeration (e.g., duplicate, fictitious, and incomplete records), validate the coverage and accuracy of web demographics, and explore potential applications. Web demographics must be used cautiously in light of the uncertainty of web demographics (e.g., digital divide), privacy issues, and other societal impacts.

This title is partially adapted from Goss (1995) as a tribute to his early vision of "data merchants" acquiring massive personal demographic data and the instrumentation prospects and potential for resistance to geodemographic marketing systems.

T.E. Chow (✉)
Texas Center for Geographic Information Science, Department of Geography,
Texas State University – San Marcos, San Marcos, TX 78666, USA
e-mail: chow@txstate.edu

15.1 Introduction

The mission of a census is to determine where people live and their demographic attributes at a specific time. The effort to achieve this goal at the national level is no small task. Common enumeration strategies include mailings, phone calls, and personal interviews. These conventional enumeration instruments, however, remain labor intensive, time-consuming, and costly. The US Government Accountability Office estimated the cost of Census 2010 to be between $13.7 billion and $14.5 billion, doubling the $6.5 billion expense of Census 2000 and four times the cost of Census 1990 (US GAO 2001, 2008).

In addition to these challenges, the low response rate and other complicating factors (e.g., mobile population and illegal migrants) result in misrepresentation of some population groups (Anderson and Fienberg 1999). The US Census Bureau assessed the accuracy of Census 2000 population count and estimated the error to be between −0.48% and 0.12% (an underestimation of 1.3 million people to an overestimation of 0.34 million people, respectively) (Anderson and Fienberg 2002; Robinson and Adlakha 2002). These results suggest that the task to accurately enumerate every person, especially ethnic minorities, females, and/or people younger than 18, remains problematic (Robinson and Adlakha 2002).

Aside from official censuses, in the digital era, personal data such as full name, address, age, phone number, etc. may scatter across various administrative records, databases of private companies, and social networks, as well as through information contributed as volunteered geographic information (VGI). Compiling such valuable information into a demographic database is useful not only for marketing but also for urban planning, resource allocation, disaster response, and various social studies.

Moving forward from the conventional web architecture in which a multiplicity of one-way transactions between client and server take place, Web 2.0 represents a collection of services that enable bidirectional communication for dynamic content delivery. By proliferating the interconnectivity and interactivity of web content across the Internet, Web 2.0 empowers smart applications to access the distributed databases and transform Internet users from web content consumers to web content producers. For example, a people-finder site (a phone-book equivalent) gathers demographic data from multiple sources and provides an interface for Internet users to query web demographics online. As the Internet connects these remote demographic databases onto a common platform, it redefines the societal implications that Goss (1995) discussed earlier: the prevalence of electronic surveillance, an erosion of privacy, as well as the discourse of geodemographic analysis. The emergence of web technologies such as mash-up, web mapping, and web scraping opens up new frontiers of research (Chow 2008, 2011). The main objectives of this chapter are to

1. Examine web demographics as an example of VGI (Sect. 15.2)
2. Explore the research agenda of web demographics for monitoring and modeling population dynamics (Sect. 15.3)

The sections below are organized in the same order as the objectives listed above and are followed by a concluding remark at the chapter's end.

15.2 Web Demographics as VGI?

Demographic data have been stored, maintained, and updated by various public and private institutions. In general, these databases are scattered across geographic spaces and disconnected from each other. People-finder sites such as Zabasearch (www.zabasearch.com) compile these diverse sources of demographics into a centralized repository and make the web demographics searchable on the Internet. In addition to the conventional databases, some web demographics also include information solicited from social networking websites such as Facebook (www.facebook.com) and Myspace (www.myspace.com). Moreover, some demographic data providers allow users to update their own but also others' listings.

Given its origin, web demographics may seem to be different from "typical" VGI in which individuals volunteer their spatially referenced observations through active participation on a specific project, such as the OpenStreetMap (www.openstreetmap.org) and Wikimapia (www.wikimapia.org). However, similar to VGI, web demographics are generated from multiple data sources and are enabled by the advancement of the Web 2.0 phenomenon. In order to label web demographics either as VGI or merely a gigantic demographic database available on the web, it is perhaps useful to distinguish the VGI community from traditional mapping agencies (Goodchild 2007):

> [t]he world of VGI is chaotic, with little in the way of formal structures. Information is constantly being created and cross-referenced, and flows in all directions, since producers and consumers are no longer distinguishable. Timescales are enormously compressed... tens of thousands of citizens are willing to spend large amounts of time contributing, without any hope of financial reward, and often without any assurance that anyone will ever make use of their contributions. (p. 29)

According to Goodchild's (2007) framework, web demographics resemble and differ from VGI in the following ways:

- *Chaotic structures*: While a valid record of web demographics must include a name, other demographic data (e.g., address, phone number) do not have a formal structure or clear definition. Names can be composed of first and family names that are spelled out in full or initialized. Addresses can range from full address at the street level to a city location. Some people-finder sites include relatives or associated members (e.g., roommates). There are no metadata about the producer, timestamp, accuracy, and coverage with regard to any of the fields in web demographics.
- *Instant and cross-referenced information*: As web demographics are typically integrated from multiple sources, most records, if not all, have multiple contributors and are cross-referenced instantaneously. For example, it is common for an individual record in web demographics to have hyperlinks connected to services such as credit reports and genealogy. Interestingly, some people-finder sites share the same databases with other websites. The WhitePages (www.whitepages.com) database empowers Switchboard (www.switchboard.com), MSN WhitePages (msn.whitepages.com), 411 (www.411.com), and Address (www.address.com).

Similarly, Addresses (www.addresses.com) also shares the same database with AnyWho (www.anywho.com) and People Finder (www.peoplefinder.com).

- *Ambiguous roles of contributor*: The producers of web demographics are arguably the public and private sectors who acquire the personal demographic data. The people-finder sites acquire the data from multiple sources and often sell the compiled records as value-added products for marketing purposes. It is not uncommon that the consumers of such web demographics are the producers. Some people-finder sites allow Internet users, who may be subjects of the web demographics themselves, to update the listing or disable it from being searched (i.e., opt-out). It is not exactly clear which party is the sole contributor or who "owns" the right to these records.
- *Compressed timescales*: It is not known how quickly the web demographics can be acquired, compiled, and consolidated from conventional data sources. Theoretically, web demographics can be created and updated instantaneously in near real time. Preliminary findings and reasonable estimation suggest that the timescale ranges from weeks to months and possibly to years (Chow et al. 2010). Nevertheless, web demographics can be gathered and updated at a much shorter and flexible timescale than conventional enumeration strategies, such as the labor-intensive mail, telephone, or face-to-face surveys involved in a population census.
- *Motivation of volunteers*: In web demographics, the primary method of contributing VGI is submitting/editing a personal profile in the people-finder sites. The subjects of web demographics might not actively surrender their personal information due to information privacy, disclosure rights, and other related issues. However, positive and negative motivations common to the production of VGI might affect the quality of web demographics (Coleman et al. 2009). For example, positive motivations (e.g., professional or personal interest, pride of place) might drive users to correct and update the personal profiles, whereas negative motivators (e.g., privacy concerns) might cause users to "opt-out" or tolerate errors in web demographics. In addition to voluntary input, people-finder sites also solicit web demographics from social networking sites in which individuals willingly allow their personal information to be searchable by friends and acquaintances. Thus, the network of web demographics producers consists of a network of citizen sensors, as well as a centralized hierarchy of many web crawlers over dozens of public/private contributors who are connected to numerous unaware, "volunteering" citizens.

Whether or not web demographics are an example of VGI, a critical but unresolved question is how volunteerism should be defined and interpreted. The key distinction central to this dialogue is "whether the volunteers have actually volunteered the information or whether it is being volunteered largely unbeknownst to them" (Tulloch 2008). The former group of volunteers is a classic example of VGI contributors who understand the call of a particular VGI application and make active contributions to achieve that goal by crowdsourcing. In reality, the participants of VGI often have varying degrees of understanding about the stated goal(s) and little consensus, if any, about the procedures to be used to review and utilize VGI. On the other hand, the VGI production from the latter group of volunteers can be perceived as passive contribution in that their GI was acquired unknowingly. The nature of passive acquisition may be similar to the collection of GI through GPS-enabled mobile devices – a phenomenon referred as geoslavery (Dobson and Fisher 2003; Tulloch 2008). Therefore, web demographics can be perceived as VGI collected by using a mixed mode of active contribution and passive acquisition. It would be interesting to examine how various channels of VGI contribution can affect the quality, for better or worse, of web demographics.

From these perspectives, web demographics have elements that are common but different from "conventional" VGI. Using the OpenStreetMap (OSM) project as an analogy, any voluntary submission of personal profile information in web demographics is an active contribution of VGI similar to the volunteered editing of geographic features in OSM. However, the evolving database of web demographics is constantly being compiled and updated by the passive acquisition of information from unaware individuals, as opposed to the raw TIGER file used as a "baseline" database in OSM, where there is no channel of passive acquisition. As such, it is perhaps more accurate to describe web demographics as VGI in which individuals volunteer both directly and indirectly through third party data suppliers without full realization.

The multifaceted nature and taxonomy of VGI has yet to be fully explored and acknowledged. After the term, VGI was coined by Goodchild (2007), a surge of research followed, first appearing in published outlets in 2007. Similar to VGI, web demographics have great potential for geospatial application and share similar challenges, such as the digital divide and privacy issues (Elwood and Leszczynski 2011). Thus, it is appropriate to perceive web demographics as a special type of VGI and to examine the research agenda of web demographics within the larger umbrella of VGI research.

15.3 Research Agenda of Web Demographics

As the research of web demographics is still in its infancy, there is a great need to understand the nature of web demographics as an alternative data source, discover its potential applications, examine conceivable shortcomings, and explore possible solutions to those shortcomings. Studies in recent years highlight the development of a web-based demographic data extraction tool and offer preliminary comparisons between the web demographics and census enumeration of Vietnamese-Americans in Texas (Chow et al. 2010, 2011). This section explores the procedure of utilizing web demographics for population monitoring, including acquisition, processing, applications, validation, privacy, and other related issues.

15.3.1 Acquisition

The acquisition of web demographics is typically conducted by web scraping – the act of extracting (or "scraping off") the demographic content from an existing website at a fixed temporal interval. A web-scraping program consists of a web crawler that can identify targeted content and a web wrapper that extracts the data of interest in specific formats. The basic idea is to parse the HyperText Markup Language (HTML) document and analyze the hierarchical tags embedded within the content structured in various degrees. Chow et al. (2011) outlined a five-step framework to automate the extraction of web demographics based on surname analysis. The process can be extended from a predefined HTML schema to statistical clustering, machine-learning algorithms, data-mining techniques, or ontology-based relationships to better refine the extraction and navigation rules (Chang et al. 2006).

To improve population monitoring, researchers should also investigate the use of the semantic web, an ongoing attempt to transform existing web infrastructure from a collection of meaningless documents to a web of data with well-defined relationships (Berners-Lee et al. 2001). In the context of web demographics, the semantic web may empower "smart" agents that can discover, navigate, and extract relevant demographics from distributed sources regardless of their diverse formatting and frequent design updates.

Another option is the deep web approach that focuses on the development of novel crawler algorithms to discover the "hidden files" (e.g., database) that are typically buried behind the web interface and dynamically created content. A deep search for web demographics would return any information related to the keywords searched by the query string, whether it is about a restaurant, a person, or an object. More work is needed to screen out the vast volume of irrelevant and redundant information (e.g., multimedia files) from the desired content. An ideal hybrid of the two approaches, in which the acquisition program can crawl through the deep web and understand the meaning of the search, is a topic to be further explored.

While the acquisition of web demographics can be conducted with a more flexible schedule than conventional demographics surveying, there is a lack of understanding on how often the procedure should be repeated for continuous population monitoring. How frequently web demographics are acquired largely depends on the currency of data (the time difference between actual event and database update) relative to the temporal requirement for a specific application. For example, it may be desirable to study the short-term and long-term impacts of Hurricane Katrina on population redistribution at various scales. Processing a change of address from the US post office takes approximately seven to ten postal business days. It is uncertain how much time it would take for the web demographics to be updated through this bureaucratic process (Landsbergen 2004).

As mentioned, the web demographics are compiled from multiple sources. In this regard, one may refer to the web demographics accessible through the people-finder sites as secondary data sources compiled from the primary data sources

Table 15.1 A preliminary assessment of common people-finder sites on their uniqueness of database, spatial precision of address, and minimum search criteria

	Uniqueness[a]	Address[b]	Minimum search criteria
Address.com	Same as WhitePages.com	Full	Last name and zip code
Addresses.com	Unique	Full	Last name and zip code
AnyWho.com	Same as Addresses.com	Full	Last name and zip code
Classmates.com	Unique	–	Last name and graduation information
Intelius.com	Unique	City	Last name and state
Facebook.com	Unique	City	Anything
Myspace.com	Unique	City	Anything
PeopleFinder.com	Same as Addresses.com	Full	Last name and zip code
PeopleFinders.com	Unique	City	Last name and city
Pipl.com	Integrated	Full	Last name and zip code
PublicRecordsNow.com	Unique	City	Last name and city
MyLife.com	Unique	Full	Full name
Spock.com	Same as Intelius.com	City	Last name and state
Switchboard.com	Same as Addresses.com	Full	Last name and zip code
USA-peoplesearch.com	Unique	City	Last name and zip code
Ussearch.com	Unique	City	Full name and city
People.yahoo.com	Integrated	–	Last name and city/zip code
WhitePages.com	Unique	Full	Last name and zip code
Wink.com	Integrated	City	Last name and city/zip code
Zabasearch.com	Unique	Full	Last name and city

[a]The category "integrated" indicates "deep" search that integrates findings from multiple secondary data sources
[b]The category "full" indicates full address at the street level

of various public and private institutions in which people "volunteered" their personal information. A limitation on tracing the data genealogy is the unwillingness of secondary data vendors to reveal their distributed sources of demographic information. In the era of information explosion, it is not uncommon to resell, or share under agreement, value-added information among the data vendors (Phillip 2005). Thus, it may be difficult to identify the exact origin(s) of certain demographic attributes and distinguish the primary data sources. This would be an interesting subject to examine using informetric laws (e.g., Zipf's law) with the number of data sources to demographic records (Egghe 2005). On the other hand, people-finder sites are more straightforward for assessment because they serve as gateways for accessing web demographics. Future research in this area should evaluate the performance of secondary data sources in terms of database coverage, uniqueness, precision, accuracy and currency, query requirement, and demographics available (some of these topics will be discussed in the "validation" section to come). For example, a secondary data source may provide street-level addresses while others may include only city references, along with additional demographic attributes (Table 15.1).

Table 15.2 A subset of web demographics with hypothetical data

ID	Name	Address	Data source	Phone	DOB
1	Enoch Y. Le	760 Mccallum Blvd	Addresses.com	123-4567	–
2	Enoch Y. Le	709 West Way	Whitepages.com	–	–
3	Enoch Y. Le	4060 Midrose Trl	Zabasearch.com	765-4321	–
4	Enoch Y. Le	7575 Frankford Rd	Zabasearch.com	123-4567	3/1968
5	Enoch Le	709 West Way	Addresses.com	–	–
6	Enoch Le	760 Mccallum Blvd	Zabasearch.com	716-2534	3/1968

15.3.2 Processing

Once web demographics are acquired from distributed data sources, it is likely that the returned data will consist of incomplete, irrelevant, fictitious, and duplicated records. The goal of the next procedure in processing the web demographics is to compile and consolidate the returned records into a centralized database. In order to generate reliable results for interpretation, it is crucial first to conduct the appropriate quality assurance/quality control (QA/QC) measures to screen out invalid records.

Querying any search engines, including people-finder sites, often yields irrelevant returns. For a valid record of web demographics, the mandatory demographic attributes include a first name or first initial, a valid surname, and a "geocodable" address (Word et al., n.d.; Chow et al. 2011). It is not uncommon to have incomplete data that may consist of records without a full name, a name without address, or the other way around. Thus, a preliminary check is needed to screen out invalid web demographics, including records with a company name (i.e., xx.Inc) or a "PO Box" address. In addition to the simple database functions to check null values, natural language processing (NLP) techniques in linguistic science can improve syntax parsing and semantics validation of essential demographic attributes (Métais 2002).

An important component in processing is duplicate detection and removal. Duplicate records are multiple records with varying degrees of similarity. There has been active research to remove duplicates in databases (Low et al. 2001), image processing (Cheng et al. 2011), and information redundancy (Li et al. 2005). In most applications, duplicate removal strikes a delicate balance between precision (the level of exactness necessary to be unique demographic records) and recall (completeness in terms of the total number of demographic records). In web demographics, a unique record can be defined as "a person having a unique combination of first name, last name, and any optional demographic attributes, including street address, age, and name(s) of other household members" (Chow et al. 2011). Table 15.2 illustrates the difficulty of distinguishing people through inclusion of optional demographic attributes, such as middle initial, date of birth (DOB), and/or phone number. Should the second and fifth records be the same person due to identical address? Should the fourth and sixth records be the same person due to identical DOB, or the first and fourth records because of identical phone number? The decisions regarding how to identify duplicates and how the demographic attributes would be merged imply different assumptions about migration, change of name/contact, and cultural practices. For example, Nagata (1999) reported a strong correlation between name change and succession to family headship and inheritance among Japanese families. Therefore, it is important to understand the cultural variation among targeted ethnic groups and realize the appropriate assumptions to be used in processing web demographics. Duplicate removal can also benefit from more research in the quantitative assessment of syntax and semantic similarity among the web demographics (Oliva et al. 2011).

A complicating factor in processing web demographics is the deliberate insertion of dummy records as a security measure by the data vendors. However, by matching the administration records, the detection of such artificial data is possible (e.g., the NUMIDENT file produced by the US Social Security Administration office or the Hundred Percent Unedited File of US Census 2000). Depending on the currency of the data, there may be temporal mismatch between the reference and web demographics. Unfortunately, an unmatched record in the web demographics can be a dummy record or a valid record indicating an illegal immigrant. There is simply no way to confirm unless the record is field validated (or verified by the data vendors). Moreover, these personal files from the government are regulated due to privacy and have limited access. Along this line of inquiry, (geographic) data-masking techniques, such as substituting, shuffling, encryption, etc. are common to protect the confidentiality of individual record without compromising the underlying pattern (Armstrong et al. 1999). Gujjary and Saxena (2011) proposed a neural network approach for data masking that would honor the semantics of original data. More research is needed to understand the impact of such database techniques and their appropriateness for population monitoring.

15.3.3 Validation

To demonstrate the potential of web demographics as an alternative data source to monitor population dynamics, it is important to validate web demographics in terms of coverage and accuracy.

Despite the vast amount of data available from diverse data sources, web demographics may represent a large pool of the population while omitting some people, a problem of any enumeration record. The exact coverage rate of web demographics among the general population, as compared to the actual number or the reference documented in census statistics, is not known. To illustrate the level of uncertainty, the Vietnamese-American (VA) population will be used as an example. Chow et al. (2010) employed 91 popular Vietnamese surnames to search for VA in Texas during 2009. The pilot study returned about 40.3% and 10.4% of the total VA count of 202,003 from WhitePages and Intelius (www.intelius.com) based on the American Community Survey (ACS) 2009 (Table 15.3). A follow-up search in Addresses (www.addresses.com) and Zabasearch returned 81,440 (40.3%) and 101,061 (50%)

Table 15.3 The coverage of web demographics of Vietnamese-Americans in Texas

	Intelius	WhitePages	Addresses	Zabasearch	Total[a]
Raw data	20,957 (10.4%)	81,354 (40.3%)	81,440 (40.3%)	101,061 (50.0%)	263,855 (130.6%)
Valid records	18,737 (9.3%)	78,460 (38.8%)	79,673 (39.4%)	94,632 (46.8%)	252,765 (125.1%)
Unique records	–	53,681 (26.6%)	58,232 (28.8%)	76,712 (38.0%)	188,625 (93.4%)
Geocoded records	–	53,425 (26.4%)	58,245 (28.8%)	74,865 (37.1%)	186,535 (92.3%)

[a]The total column represents the count of unique records from WhitePages, Addresses, and Zabasearch

records, respectively. The post-processing procedure mentioned in the previous section removed invalid and duplicate records (by consolidating with records from multiple people-finder sites) and yielded 53,425 (26.4%), 58,245 (28.8%), and 74,865 (37.1%) records from WhitePages, Addresses, and Zabasearch, respectively. As Intelius only returned city reference but not street-level addresses (Table 15.1), the web demographics of Intelius were not used in compiling the centralized web demographic database. Thus, the VA count from WhitePage, Addresses, and Zabasearch totals 186,535 (92.3%) unique records with addresses that can be address-matched in Texas.

The VA surnames have unique spellings distinguishable from surnames of other races/ethnicities (Lauderdale and Kestenbaum 2000), and hence a targeted search of web demographics among VA seemed to be suitable. While a high coverage rate of VA web demographics does not guarantee a similar coverage for other races/ethnicities, the literature suggests that surname analysis is effective in identifying ethnic minorities, including Hispanics (Perkins 1993), South Asians (Shah et al. 2010), Chinese (Quan et al. 2006), Middle Easterners (Nasseri 2007), and other Asian ethnic minorities (e.g., Koreans, Japanese, Indians, and Filipinos) (Lauderdale and Kestenbaum 2000). It is possible to utilize the popular surnames among other races/ethnicities to examine the web demographics returned from the people-finder sites. Nevertheless, intermarriage, name change, adoption, and self-identity can introduce uncertainties (Perkins 1993). For example, many studies used Hispanic surnames to identify Hispanic men and women with a high degree of overall accuracy (Perez-Stable et al. 1995; Morgan et al. 2004), but noticeable errors are observed among the Hispanic married women, Filipinos, and Native Americans (Barreto et al. 2008). More research is needed to examine the coverage rate of web demographics among other ethnic groups with a particular focus on married women (Wei et al. 2006).

It must be noted, however, that in some cases surname analysis is limited in identifying race/ethnicity. For example, the surname "Lee" is common among non-Hispanic Whites, African Americans, Koreans, and Cantonese-speaking Chinese. Thus, a targeted search using surnames shared by multiethnic races may return a mixed population. Nevertheless, such information may still be useful to monitor the dynamics of the general population, regardless of races/ethnicities. To estimate the representativeness of web demographics, the list of most frequently occurring surnames (US Census Bureau 2010) can be used to evaluate the coverage of web demographics for the population at large. Whether the goal is to search for the web demographics of a specific group or of the general population, it is important to "design a well-constructed list of representative surname samples that strike a fine balance between the anticipated type I or II errors in statistics" (Chow et al. 2011).

Aside from the coverage rate, the accuracy of web demographics is of paramount concern. An ideal accuracy assessment of web demographics should consist of a well-planned field survey that samples individual households over a region and within a similar time frame. However, the willingness of invited candidates to complete the survey and the level of trust necessary for disclosure of personal information may introduce uncertainties into the accuracy assessment. To evaluate the quality of web demographics, the assessment may be partitioned into demographic attribute correctness, positional accuracy, and temporal currency. No accuracy assessment has been reported to date in these regards.

As web demographics originate from public and private data, clerical and typographic mistakes may present errors. Each demographic attribute, such as name (first, middle, and last), age, household members, etc., can be evaluated by a simple accuracy measurement. For a qualitative attribute (e.g., name), any typographical mismatch can be considered as an incorrect entry. Thus, the accuracy of qualitative demographic attributes can be determined by the frequency of correct entries over total responses. Similarly, the quantitative attributes, such as age, can adopt descriptive statistics to explore the distribution of error between the surveyed and web demographics.

Positional accuracy of web demographics can be evaluated by the ground distance between the best-measured location of the residence and the position address, matched on the basis of the last address reported in the web demographics. Positional accuracy can be affected by both the source error and processing error (Goodchild 1989). For web demographics, any distortion in the data source implies an outdated or incorrect address being used, and the source error equates to the ground distance between the updated (if known) and the outdated addresses, geocoded by the same method. It goes without saying that the source error is intrinsically correlated to the temporal currency of web demographics. By connecting an address to a point in a predefined coordinate system, the process error indicates the performance of an address-matching process, which largely depends on the effectiveness of the matching algorithm and the quality of reference data.

Swift et al. (2008) reviewed eight geocoding systems, including both the desktop and web-mapping solutions (e.g., Google Maps Application Programming Interface), and reported that addresses with higher source error may benefit from a simpler address-matching algorithm, and vice versa. The positional accuracy of web demographics, in terms of both source and processing errors, can provide helpful insights into the reliability of spatial analysis that utilizes web demographics.

Temporal currency refers to the "degree to which the database is complete and up-to-date" (Goodchild 2008), and it may affect both demographic attribute correctness and positional accuracy. Temporal currency can be described by the dual aspects of the latest timestamp and relative timeliness in web demographics. The latest timestamp is the last known date/time when the demographics were verified or updated, and its assessment is relatively straightforward. Depending on when a life event occurs, such as relocation, an older timestamp may not necessarily equate to an incorrect or outdated entry. On the other hand, the difference between the exact moment of an actual event and the time when the web demographics reflected the corresponding update indicates the relative timeliness of a record. In reality, the exact time of a life event may not be captured by the web demographics, and the uncertain origin of a primary data source may result in a fuzzy relative timeliness. Future work can explore the relationships among coverage, demographic attribute correctness, positional accuracy, and temporal currency (i.e., the latest timestamp and relative timeliness if available) of web demographics and examine the spatiotemporal landscape of a given population.

15.3.4 Applications

A major motivation of gathering personal-level web demographics is to better understand the customer profile for marketing purposes (Linberger and White 1998). Some people-finder sites such as Intelius and PeopleFinders (www.PeopleFinders.com) sell value-added web demographics by offering various paid subscription plans for further personal details, including marriage status, financial history, etc. Corporate employers and legal services often utilize such information to acquire criminal checks and credit reports on prospective candidates. Conventional geodemographics have been used to derive indicators for social networking (Singleton and Longley 2009).

While the solicitation of web demographics is typically tailored for an individual, automating such a process makes it easier to acquire the web demographics of a population sample (Chow et al. 2011). A recent study compared the web demographics of VA in Texas during 2009 with the Census 2000 data to explore the potential of web demographics for population monitoring (Chow et al. 2010). By examining the data distribution of areas in which there are high discrepancies between web samples and Census 2000 references, various counties and census tracts that experienced significant population gain or deficit were identified. The spatial pattern of significant population change was found to be consistent with the urban-to-suburban migration typically observed around the city edge of major metropolitan areas in Texas, as well as the rural–urban migration in coastal fishery settlements (Chow et al. 2010). Thus, the results from this pilot study unveiled the potential of web demographics to map the spatial distribution of a population group and monitor the group's movement over time.

With a carefully designed surname list, it is possible to extend the systematic acquisition of web demographics to multiethnic races with distinguishable surnames. While the coverage and accuracy of web demographics have yet to be validated by a large-scale field survey, this alternative data source may prove complementary to the conventional census enumeration. As ethnic minorities are often subjected to undercount in census (Anderson and Fienberg 1999), web demographics of these population subgroups can be used to estimate the net under-/overcount by matching the enumeration record between the web and census demographics. The dual system estimation (DSE) adopts a capture-recapture method to adjust the estimation of population size (Wright 2000). It is not known; however, the degree to which web demographics may suffer from correlation bias, the problem of missing population in both demographic surveys, and how it may affect census adjustment.

In addition to a custom search of web demographics for ethnic minorities, future research can also examine its extension to the general population regardless of races/ethnicities by using a list of popular names, such as the 1,000 most frequently occurring surnames from Census 2000 (US Census Bureau 2010). Theoretically, an exhaustive web search of all surveyed surnames is similar in scope to Census 2.0 – an enumeration strategy that uses Web 2.0 technologies to survey the demographics of unique individuals from the Internet. In Census 2000, about 89.8% of the US population shared 151,671 surnames, which approximates 2.4% of all surnames surveyed (Word et al., n.d.). Because web demographics can be acquired at a much lower cost than a traditional census, with a flexible schedule and in a repeatable manner, it warrants continued research to examine the population displacement in response to a specific event, such as a disaster or an economic recession. Upon the validation of web demographics, a comparative study with other population distribution datasets, such as LandScan, will provide valuable insights for verification of the global database and foster possible integration of multi-scale study in microdemographics and macrodemographics (Bhaduri 2011).

Web demographics can be helpful to resolve community issues as well. Local communities can create limited demographic databases reflecting the values of their local neighborhood or communities. As indicated by Seeger (2008), one of the key challenges in VGI projects is to promote public awareness of and participation in such opportunities. The use of web demographics can help target relevant stakeholders and enhance the response rate for a VGI project through a cyberinfrastructure support system (e.g., www.citsci.org) (Newman et al. 2011). Web personal microdata enable spatial analyses at the neighborhood level in order to address problems within a community. This capability is particularly important in cultural communities (e.g., ethnic minorities) that need to allocate limited resources to the targeted group and require VGI for decision making, such as landscape planning (Seeger 2008).

The integration of other VGI, such as CommonCensus (http://www.commoncensus.org), can assist the investigation of the cultural landscape, including its factors, impacts, and implications to enhance community building and problem-solving. Such a community-based demographic database can contribute geographic and demographic dimensions to the formation of location-based social networks for applications such as geoportal development (De Longueville 2010) and epidemic monitoring (Firestone et al. 2011). In fact, WhitePages has recently launched a VGI platform (neighbors.whitepages.com) that allows individuals to claim their personal profiles and connect with their neighbors for community building. Such community profiles can improve the discourse of geodemographic analysis and enhance customized local deals, advertising, and marketing strategies based on place-based demographics and personal behavioral profile, such as usage history and search preference (Goss 1995). Thus, VGI-enabled local search will play an increasingly important role in mobile applications. A community-based demographic database can also enrich related research in "ecological momentary assessment" that tracked the activity space of volunteering individuals to unearth social interaction, mental health, and behavioral outcomes (Browning 2011).

Aside from linguistic or racial/ethnic connections, a surname may also reveal the physical bonding to a landscape, a hint to common ancestry, and a window to study population structure (Lasker 1985; Lasker and Mascie-Taylor 1990). Toponymic surnames (named after a place or a village) are common in some cultures (e.g., Bath in Britain) and are found to have higher frequency around areas of origin (Kaplan and Lasker 1983). Isonymy explores the pattern of common surnames over space; not surprisingly, the landscape of surnames is consistent with Tobler's first law of geography: near places have more similar names than those from afar (Tobler 1970). Longley et al. (2011) used surname frequency data from an electoral registry to study population dynamics of human migration. Their results from hierarchical clustering and multidimensional scaling identified regions with similar surname compositions and concluded that clusters of people who share common surnames tend to stay in proximate locations (Longley et al. 2011). Web demographics can serve as a contemporary source of surname data needed for isonymic analysis.

15.3.5 Privacy

The acquisition of personal data has been a long-standing issue with regard to information privacy, which is often perceived as a person's "ability to control the collection or dissemination of information about himself or herself" (Elwood and Leszczynski 2011). A question central to the issue of privacy over web demographics is the ownership of information. As personal data propagates from individuals to the primary data source(s) and to secondary users, what are the ownership "rights," as well as responsibilities, for each actor involved? Elwood and Leszczynski (2011) called for a reconsideration of privacy and protection practices in light of new multimedia (e.g., pictures from Google Street View) and the transformation of people's roles and relationships among civil, corporate, and state actors. In general, the civil actors of web demographics did not permit the disclosure of their personal data on the web, but they might have indirectly contributed to the assembly of web demographics by publishing their information through social networking sites and other online agreements (e.g., registering for an online sweepstake). Thus, the reconceptualization of privacy in light of web demographics may be different from the call for VGI and that of traditional geodemographics.

Privacy concerns can affect the quality of demographic surveying. Despite the constitutional mandate for a census in the USA, there have been reoccurring public controversies over the details of demographic data required, enumeration strategies, and channels of data dissemination, such as the now-defunct decennial census long form (Robbin 2001). Concerns about personal privacy would affect the response rate among a small but significant percentage of participating citizens (Singer et al. 1993). Web demographics are typically acquired from primary data sources without permission from the "volunteers." Hence, it is doubtful if the civil actors would trust or have confidence in the institutions responsible for the collection and dissemination of web demographics. Social networking sites have been identified as important data sources for a new model of geodemographic profiling (Singleton and Longley 2009). A recent review on social networking sites revealed that adult users are more concerned about potential threats to information privacy than younger users (Hugl 2011). It is not known how the differential privacy concerns among and within various population groups may affect the quality, in terms of both coverage and accuracy, of the web demographics.

Goss (1995) argued that not only are privacy issues at stake, but incomplete or inaccurate information presented in web demographics may result in socioeconomic mislabeling and a reduction of one's identity to "you are where you live." Moreover, the interconnection and intertwining of institutional and corporate databases lays down the infrastructure for "dataveillance" – the surveillance of individuals' behavior through an electronic data trail, including mobile and Internet usage. Technological advancements such as location-aware browsing and the culture of real-time sharing pave the way for more current updates for web demographics. In light of the ever-evolving web and mobile technologies, the societal and scientific implications of web demographics have yet to be unpacked.

Societal concerns and public discourse about information privacy drive the demand for appropriate practices of privacy protection. Some people-finder sites such as WhitePages adopt the opt-out approach for privacy protection. Any individual who wishes to remove his or her record from a published listing must request to be concealed from being "searchable." However, there is no guarantee that the removed information would be unavailable permanently because personal data may be reacquired and republished due to the updates of life events from primary data sources, such as moving or change of name (WhitePages 2011). Data shuffling techniques, such as geographic masking, can be used to protect the privacy and confidentiality of personal data on the web while balancing the need to analyze spatially aggregated demographics (Kwan et al. 2004). However, such practices can undermine the commercial objectives of people-finder sites and their potential applications for a targeted search of an individual.

15.3.6 Other Related Issues

In addition to the societal implications of web demographics, it is important to understand the role society has had on shaping the landscape of web demographics. A good example is the digital divide (the differential access to information and technology among population groups with varying levels of socioeconomic and demographic composition). As web demographics have originated from analog transactions and digital traces, it is possible that some population groups are less represented in the web demographics (e.g., young dependents, illegal immigrants, or otherwise disadvantaged groups). Theoretically, these population groups may still leave a trace of demographic transaction on public records (e.g., utility subscriptions, applications for credit cards, drivers' licenses). For broadband Internet usage, Prieger and Hu (2008) reported that race and ethnicity are important factors responsible for the digital divide while holding income, education, and demographic factors constant. They also found that competition among Internet providers may narrow the gap in the number of Internet subscribers and improve the quality of service, indicating that remote communities may experience physical barriers to digital access and representation. In light of these findings, racial/geographic disparity in web demographics may be partially attributed to the digital divide of Internet access. More research is needed to explore the common causes on coverage disparity of web demographics and how it may be empirically related to the digital divide.

Metadata is another area of interest for VGI research. Producers of VGI, unlike the conventional producers of geospatial data, may be motivated to document effective metadata. From a privacy protection perspective, however, metadata may be of particular concern in producing or updating web demographics. It is not known if a "user-centric approach" would improve metadata production for web demographics (Goodchild 2008). By adopting a Wikipedia approach of VGI (Sui 2008), an automatic timestamp can be produced in the metadata of web demographics whenever an individual record is verified or updated. An automatic timestamp is an example of object-level metadata (Chap. 4). Among the lessons to be learned from Wikipedia, the complete history of revision, the use of data mining, and the framework of flagging refined, unrefined, or problematic content are most valuable to the metadata of VGI (Hardy 2010). On the other hand, ancillary metadata, or metadata squared, of web demographics can be social media that introduces the swirling universe of other sociocultural characteristics about this person beyond the basic description of demographic attributes (Chap. 4). Thus, the VGI approach of metadata production is relevant to web demographics and can serve as a showcase for broader geospatial data.

Because the collection of demographic data relies on individual compliance, psychological and sociocultural factors may influence an individual's willingness to provide accurate and complete data. For example, the degree to which ethnic minority and immigrant individuals identify with the mainstream culture may affect the individual's level of compliance with the data-gathering process of the census (Berry et al. 2006). Cultural factors, as well as differences in education, citizenship, and age, are contributing factors to civil participation and engagement (Uhlaner et al. 1989). As migration is becoming more common in a global village, an individual's psychological acculturation strategy may provide useful insights to predict the individual's level of compliance with the gathering of demographic data (Berry et al. 2006). Thus, a better understanding of psychological acculturation strategies may be helpful to interpret the accuracy and coverage of web demographics among various racial and ethnic groups.

15.4 Summary and Concluding Remarks

This chapter began by examining web demographics as a kind of VGI. Using a preliminary VGI framework as an assessment, web demographics is similar to conventional VGI in terms of its chaotic structure of instant and cross-referenced information, produced by an ambiguous role of contributors within a compressed timescale (Goodchild 2007). However, it is perhaps most different in the motivation of "volunteers." Web demographics may serve as a case demanding a verdict to reevaluate the nature and the taxonomy of "volunteerism" in VGI.

In many social studies, the primary source, and very often the only source, of demographic data is the official census. However, census demographics are subject to a specific temporal interval (e.g., ranging from 1 to 10 years in general) and scale (i.e., aggregated into varying geographical areas). Demographic data "on demand" at the personal level were not available before the recent discovery of web demographics for population monitoring. Thus, although temporal currency has yet to be validated, the web provides gigantic "nearly" real-time personal level demographic data to anyone with Internet access. More research and development are needed to validate the coverage and accuracy of web demographics, enhance extraction rules of web demographics, identify and remove erroneous enumeration (i.e., duplicate, fictitious, and incomplete records), explore potential applications, and overcome probable shortcomings. The utilization of web demographics must be cautious with regard to their uncertainties, privacy issues, and societal impacts. Nevertheless, web demographics present an alternative data source for future research.

Acknowledgments The author is thankful to Yan Lin, who provided valuable assistance in preparing the statistics of web demographics acquired from several people-finder sites. Collaboration with colleagues and fellow students, including Niem Huynh, David Parr, John Davis, and Anne Ngu, on projects related to web demographics is instrumental to the ideas articulated in this manuscript. The author is in debt to Nancy Wilson, David Parr, and Niem Huynh for their editorial assistance and helpful reviews. The constructive comments from the reviewers greatly improved the quality of this manuscript. Any errors in the manuscript are solely the responsibility of the author.

References

Anderson, M. J., & Fienberg, S. E. (1999). *Who counts? The politics of census-taking in contemporary America*. New York: Russell Sage.
Anderson, M. J., & Fienberg, S. E. (2002). Why is there still a controversy about adjusting the census for undercount. *PSOnline*, March, 83–85.
Armstrong, M. P., Rushton, G., Zimmerman, D. L., et al. (1999). Geographically masking health data to preserve confidentiality. *Statistics in Medicine, 18*, 497–525.
Barreto, M., DeFrancesco-Soto, V., Merolla, J., & Ramirez, R. (2008). *Latino campaign ad experimental study*. Los Angeles, CA.
Berners-Lee, T., Hendler, J., Lassila, O., et al. (2001). The semantic web (Resource document). *Scientific American Magazine*. http://www.scientificamerican.com/article.cfm?id=the-semantic-web. Accessed 26 May 2009.
Berry, J. W., Phinney, J. S., Sam, D. L., Vedder, P., et al. (2006). Immigrant youth: Acculturation, identity, and adaptation. *Applied Psychology: An International Review, 55*(3), 303–332.
Bhaduri, D. (2011). Enhancing resolution of population distribution data in spatial, temporal, and sociocultural dimensions: Advances and challenges (Resource document). *Specialist meeting in future direction of spatial demography*. http://ncgia.ucsb.edu/projects/spatial-demography/docs/Bhaduri-position.pdf. Accessed 2 Jan 2012.
Browning, C. R. (2011). Future directions in spatial demography (Resource document). *Specialist meeting in future direction of spatial demography*. http://ncgia.ucsb.edu/projects/spatial-demography/docs/Browning-position.pdf. Accessed 2 Jan 2012.
Chang, C., Kayed, M., Girgis, M., Shaalan, K., et al. (2006). A survey of web information extraction systems. *IEEE Transactions on Knowledge and Data Engineering, 18*, 1411–1428.
Cheng, X., Hu, Y., Chia, L.-T., et al. (2011). Exploiting local dependencies with spatial-scale space (S-Cube) for near-duplicate retrieval. *Computer Vision and Image Understanding, 115*(6), 750–758.
Chow, T. E. (2008). The potential of maps APIs for internet GIS. *Transactions in GIS, 12*(2), 179–191.

Chow, T. E. (2011). Geography 2.0: A mashup perspective. In S. Li, S. Dragicevic, & B. Veenendaal (Eds.), *Advances in web-based GIS, mapping services and applications* (pp. 15–36). Boca Raton: CRC Press.

Chow, T. E., Lin, Y., Huynh, N. T., Davis, J., et al. (2010). Using web demographics to model population change of Vietnamese-Americans in Texas between 2000–2009. *GeoJournal*. doi:10.1007/s10708-010-9390-6.

Chow, T. E., Lin, Y., Chan, W. D., et al. (2011). The development of a web-based demographic data extraction tool for population monitoring. *Transactions in GIS, 15*(4), 479–494.

Coleman, D. J., Georgiadou, Y., Labonte, J., et al. (2009). Volunteered geographic information: The nature and motivation of producers. *International Journal of Spatial Data Infrastructures Research, 4*, 332–358.

De Longueville, B. (2010). Community-based geoportals: The next generation? Concepts and methods for the geospatial Web 2.0. *Computers, Environment and Urban Systems, 34*(4), 299–308.

Dobson, J. E., & Fisher, P. F. (2003). Geoslavery. *IEEE Technology and Society Magazine, 22*(1), 47–52.

Egghe, L. (2005). *Power laws in the information production process: Lotkaian informetrics*. Amsterdam: Academic.

Elwood, S., & Leszczynski, A. (2011). Privacy, reconsidered: New representations, data practices, and the geoweb. *Geoforum, 42*(1), 6–15.

Firestone, S. M., Ward, M. P., Christley, R. M., Dhand, N. K., et al. (2011). The importance of location in contact networks: Describing early epidemic spread using spatial social network analysis. *Preventive Veterinary Medicine, 102*(2), 185–195.

Goodchild, M. F. (1989). Modeling errors in objects and fields. In M. F. Goodchild & S. Gopal (Eds.), *The accuracy of spatial databases*. New York: Taylor & Francis.

Goodchild, M. F. (2007). Citizens as voluntary sensors: Spatial data infrastructure in the world of web 2.0. *International Journal of Spatial Data Infrastructure Research, 2*, 24–32.

Goodchild, M. F. (2008). Spatial accuracy 2.0. *Spatial Uncertainty: Proceedings of the Eighth International Symposium on Spatial Accuracy Assessment in Natural Resources and Environmental Sciences, 1*, 1–7.

Goss, J. (1995). We know who you are and we know where you live: The instrumental rationality of geodemographic systems. *Economic Geography, 71*, 171–198.

Gujjary, V. A., & Saxena, A. (2011). A neutral network approach for data masking. *Neurocomputing, 74*(9), 1497–1501.

Hardy, D. (2010). The wikification of geospatial metadata (Resource document). *Workshop on the "Role of Volunteered Geographic Information in Advancing Science"*. http://www.ornl.gov/sci/gist/workshops/papers/Hardy.pdf. Accessed 21 July 2011.

Hugl, U. (2011). Reviewing person's value of privacy of online social networking. *Internet Research, 21*(4), 1–17.

Kaplan, B., & Lasker, G. (1983). The present distribution of some English surnames derived from place names. *Human Biology, 55*(2), 243–250.

Kwan, M. P., Casas, I., Schmitz, B. C., et al. (2004). Protection of geoprivacy and accuracy of spatial information: How effective are geographical masks? *Cartographica, 39*, 15–28.

Landsbergen, D. (2004). Screen-level bureaucracy: Databases as public records. *Government Information Quarterly, 21*(1), 24–25.

Lasker, G. (1985). *Surnames and genetic structure*. Cambridge: Cambridge University Press.

Lasker, G., & Mascie-Taylor, C. (1990). *Atlas of British surnames*. Detroit: Wayne State University Press.

Lauderdale, D. S., & Kestenbaum, B. (2000). Asian-American ethnic identification by surname. *Population Research and Policy Review, 19*, 283–300.

Li, W., Liu, J., Wang, C., et al. (2005). Web document duplicate removal algorithm based on keyword sequences. In *Proceedings of 2005 IEEE International Conference on Natural Language Processing and Knowledge Engineering* (pp. 511–516). Piscataway: IEEE.

Linberger, P., & White, G. (1998). Geographic information on the web: Extracting demographic and market research information. *Proceedings of the Nineteenth Annual National Online Meeting, 19*, 235–242.

Longley, P. A., Cheshire, J. A., Mateos, P., et al. (2011). Creating a regional geography of Britain through the spatial analysis of surnames. *Geoforum, 42*(4), 506–516.

Low, W. L., Lee, M. L., Ling, T. W., et al. (2001). A knowledge-based approach for duplicate, elimination in data cleaning. *Information Systems, 26*(8), 585–606.

Métais, E. (2002). Enhancing information systems management with natural language processing techniques. *Data and Knowledge Engineering, 41*(2–3), 247–272.

Morgan, R. O., Wei, I. I., Virnig, B. A., et al. (2004). Improving identification of Hispanic males in Medicare: Use of surname matching. *Medical Care, 42*, 810–816.

Nagata, M. L. (1999). Why did you change your name? Name changing patterns and the life course in early modern Japan. *The History of the Family, 4*(3), 315–338.

Nasseri, K. (2007). Construction and validation of a list of common Middle Eastern surnames for epidemiological research. *Cancer Detection and Prevention, 31*, 424–429.

Newman, G., Graham, J., Crall, A., Laituri, M., et al. (2011). The art and science of multi-scale citizen science support. *Ecological Informatics, 6*(3–4), 217–227.

Oliva, J., Serrano, J. I., del Castillo, M. D., Iglesias, A., et al. (2011). SyMSS: A syntax-based measure for short-text semantic similarity. *Data and Knowledge Engineering, 70*(4), 390–405.

Perez-Stable, E. J., Hiatt, R. A., Sabogal, F., Otero-Sabogal, R., et al. (1995). Use of Spanish surnames to identify Latinos: Comparison to self-identification. *Journal of the National Cancer Institute Monographs, 18*, 11–15.

Perkins, R. C. (1993). *Evaluating the passel-word Spanish surname list: 1990 decennial census post enumeration survey results* (Resource document, U.S. Bureau of the Census, Population Division Working Paper No. 4). http://www.census.gov/population/www/documentation/twps0004.html. Accessed 15 July 2011.

Phillip, M. (2005). Why pay for value-added information? *World Patent Information, 27*(1), 7–11.

Prieger, J. E., & Hu, W. (2008). The broadband digital divide and the nexus of race, competition, and quality. *Information Economics and Policy, 20*(2), 150–167.

Quan, H., Wang, F., Schopflocher, D., Norris, C., Galbraith, P. D., Faris, P., Graham, M. M., Knudtson, M. L., Ghali, W. A., et al. (2006). Development and validation of a surname list to define Chinese ethnicity. *Medical Care, 44*, 328 333.

Robbin, A. (2001). The loss of personal privacy and its consequences for social research. *Journal of Government Information, 28*(5), 493–527.

Robinson, J. G., & Adlakha, A. (2002). *Comparison of A.C.E. revision II results with demographic analysis* (Resource document, U.S. Bureau of the Census, DSSD A.C.E. Revision II Estimates Memorandum Series #PP-41). http://www.census.gov/dmd/www/pdf/pp-41r.pdf. Accessed 12 July 2011.

Seeger, C. J. (2008). The role of facilitated volunteered geographic information in the landscape planning and site design process. *GeoJournal, 72*(3–4), 199–213.

Shah, B. R., Chiu, M., Amin, S., Ramani, M., Sadry, S., Tu, J. V., et al. (2010). Surname lists to identify South Asian and Chinese ethnicity from secondary data in Ontario, Canada: A validation study. *BMC Medical Research Methodology, 10*, 42. doi:101186/1471-2288-10-42.

Singer, E., Mathiowetz, N. A., & Couper, M. P. (1993). The impact of privacy and confidentiality concerns on survey participation: The case of the 1990 U.S. census. *Public Opinion Quarterly, 57*(4), 465–482.

Singleton, A. D., & Longley, P. A. (2009). Geodemographics, visualization, and social networks in applied geography. *Applied Geography, 29*(3), 289–298.

Sui, D. Z. (2008). The wikification of GIS and its consequences: Or Angelina Jolie's new tattoo and the future of GIS. *Computers Environment and Urban Systems, 32*(1), 1–5.

Swift, J. N., Goldberg, D. W., & Wilson, J. P. (2008). *Geocoding best practices: Review of eight commonly used geocoding systems* (Resource document, University of Southern California GIS Research Laboratory Technical Report No 10). http://spatial.usc.edu/Users/dan/gislabtr10_Eight-Commonly-Used-Geocoding-Systems.pdf. Accessed 2 Jan 2012.

Tobler, W. R. (1970). A computer movie simulating urban growth in the Detroit region. *Economic Geography, 46*, 34–240.

Tulloch, D. L. (2008). Is VGI participation? From vernal pools to video games. *GeoJournal, 72*(3–4), 161–171.

Uhlaner, C. J., Cain, B. E., & Kiewiet, D. R. (1989). Political participation of ethnic minorities in the 1980s. *Political Behavior, 11*(3), 195–231.

U.S. Census Bureau. (2010). *Genealogy data: Frequently occurring surnames from Census 2000* (Resource document). http://www.census.gov/genealogy/www/data/2000surnames/index.html. Accessed 15 July 2011.

U.S. Government Accountability Office. (2001). *Significant increase in cost per housing unit compared to 1990* (Resource document. GAO-02-31). http://www.gao.gov/new.items/d0231.pdf. Accessed 12 July 2011.

U.S. Government Accountability Office. (2008). *Census Bureau should take action to improve the credibility and accuracy of its cost estimate for the decennial census* (Resource document. GAO-08-554). http://www.gao.gov/new.items/d08554.pdf. Accessed 23 July 2011.

Wei, I. I., Virnig, B. A., John, D. A., & Morgan, R. O. (2006). Using a Spanish surname match to improve identification of Hispanic women in Medicare administrative data. *Health Research and Educational Trust, 41*(4), 1469–1481.

WhitePages. (2011). *WhitePages privacy central* (Resource document). http://www.whitepage.com/help/privacy_central. Accessed 20 July 2011.

Word, D. L., Coleman, C. D., Nunbziata, R., Kominski, R., et al. (n.d.). *Demographic aspects of surname from Census 2000, genealogy data: Frequent occurring surnames from Census 2000* (Resource document. US Census Bureau). http://www.census.gov/genealogy/www/data/2000surnames/surnames.pdf. Accessed 14 July 2011.

Wright, T. (2000). Census 2000: Who says counting is easy as 1–2–3? *Government Information Quarterly, 17*(2), 121–136.

Chapter 16
Volunteered Geographic Information, Actor-Network Theory, and Severe-Storm Reports

Mark H. Palmer and Scott Kraushaar

Abstract In this chapter, we will use actor-network theory to describe a decentralized, heterogeneous storm-spotting and storm-chasing network in the United States, which connects human-sensor observations in the field with operational meteorologists in the center of calculation and with television media as the recipients and public distributors of official severe-weather watches and warnings. In the first section, we present actor-network theory as a conceptual framework for describing the co-construction of society and technology and the centering processes associated with scientific laboratories and government agencies. In the second section, we demonstrate the use of actor-network theory (ANT) through a descriptive case study of the storm-spotter and storm-chaser network, which encompasses both decentralized and centralized processes. Finally, we analyze the centering processes that the National Weather Service (NWS) uses to mobilize, stabilize, and combine VGI storm reports with their existing technologies.

16.1 Introduction

Emerging research on volunteered geographic information (VGI) is beginning to conceive of humans as sensors and as enabling geospatial technologies. Goodchild argued that there are potentially six billion humans as sensors on this planet (Goodchild 2007a, b). Now, let us assume that Goodchild is speaking of sensors

M.H. Palmer (✉)
Department of Geography, University of Missouri, Columbia, MO, USA
e-mail: palmermh@missouri.edu

S. Kraushaar
Department of Geography, University of Missouri, MO, Columbia

as a combination of human senses and digital technologies. If this is the case, developed countries like the United States, the United Kingdom, and Western Europe have far greater access to digital technologies than people in the developing world; there is still a significant digital divide (Crampton 2010). But let us think abstractly for a moment and consider the staggering image of six billion humans (seven billion as of October 2011) who are somehow interconnected through an array of digital gadgets. Now consider the diversity of the world's population. Heterogeneity abounds within the diverse knowledge systems making up the world's population. Online maps in all languages and featuring a multitude of representations are emerging into the world – mapping is now at a pinnacle of human participation (Crampton 2010). But, citizens would have difficulty achieving their geographic or mapping goals and objectives without access to enabling technologies like the Internet, Web 2.0, geotagging, georeferencing, GPS, and graphics (Goodchild 2007a, b). Yet, if we think about this for a moment, the idea of humans as sensors takes on an organic, hybrid, and ecological quality. It is often assumed that humans and their senses are somehow separate from the technologies they use to Tweet, blog, or geo-tag online. Segregation of humans and technology can be intellectually limiting. Haraway's cyborg (Haraway 1987) and Goodchild's VGI humans as sensors lose some of their richness and power when viewed through the lenses of dichotomies like society/technology, human/sensor, or GIS/society. We believe this is so because it is very difficult to determine what is social about VGI and what is technological. Our thesis is that VGI is not an object but rather an interconnected network of heterogeneous humans, technologies, geographic information, and organizations. This means that VGI is not solely determined by social factors or by technological factors. Rather, VGI is equally shaped by society and technology.

In this chapter, we use actor-network theory (ANT) to describe a decentralized, heterogeneous storm-spotting and storm-chasing network in the United States that connects human-sensor observations in the field with operational meteorologists in the center of calculation and with television media as the recipients and public distributors of official severe-weather watches and warnings. In the first section, we present actor-network theory as a conceptual framework for describing the co-construction of society and technology and the centering processes associated with scientific laboratories and government agencies. ANT concepts used in this chapter include actor, intermediary, network, translation, and center of calculation. In the second section, we demonstrate the use of ANT through a descriptive case study of the storm-spotter and storm-chaser network encompassing both decentralized and centralized processes. We will describe the most significant actors and materials used to sense the environment and relay the information out into the world. What we construct is a partial network of human sensors. Finally, we analyze the centering processes that the National Weather Service (NWS) uses to mobilize, stabilize, and combine VGI storm reports with their existing technologies. On the surface, text bulletins appear to be objects. Center processes transform local information into a nameless and faceless network that extends back to the media and general public.

16.2 Actor-Network Theory

At the core of the science and technology studies (STS) approach to the co-construction of society and technology is actor-network theory (ANT). Initially developed by Bruno Latour, Michel Callon, and John Law, actor-network theory is a conceptual framework for "exploring collective sociotechnical processes," describing and explaining the "relationships between people, institutions, and artifacts connected by agreements and exchanges" (Harvey 2001, 30). ANT is also a method for tracing relationships, agreements, and exchanges. Through relationships and agreements, action is distributed to other spaces, at a distance. In GIS research, Martin (2000) used ANT as a framework to understand the interactions between GIS and society. The research showed the interactions between texts, people, money, technology, and control in the implementation of GIS in Ecuador. The paper suggests that "the best use of ANT for investigating GIS may be to continue exposing the social interactions behind GIS operations so practitioners, managers, theorists and researchers will be more sensitive to building stable GIS actor-networks" (Martin 2000, 735). Harvey and Chrisman (2004) have also drawn inspiration and conceptual frameworks from Deleuze's and Guattari's (1987) work on rhizomes. Both authors noted the difficulty of interpreting Deleuze and Guattari but also insisted that ecological concepts such as rhizome and strata are useful metaphors for research on GIS and society. Bruno Latour proposed that actant-rhizome is a more organic and better name for his project than ANT (Thrift 2000). ANT can inform non-ANT VGI research that focuses on the multiple combinations of humans, computer hardware, software, and data structures, creating new methods for analyzing data from the newly opened pool of sources and on understanding the effects of VGI on GIScience (Williams 2007; Mummidi and Krumm 2008; Bishr and Mantelas 2008; Goodchild 2007b; Gartner et al. 2007).

Nigel Thrift (2000) wrote that "geographers have become very interested in actor-network theory" (p. 5). ANT, as a component of the larger body of research conducted under the banner of STS, has caught the attention of geographers because the research framework can be used to understand the construction of technology and nature and is another means of conceptualizing space and place (Thrift 1996, 2000). As a method of analysis, ANT attempts to break up dualities and allows for theories of heterogeneous associations (Murdoch 1997a, b). Three review essays covering literature pertaining to heterogeneous associations (Murdoch 1997b), world city actor-networks (Smith 2003), and hybrid, indigenous ANT (Panelli 2010) recently emerged. The journals *Environment and Planning A* and *Environment and Planning D* are at the forefront of ANT-geography research. Two studies focused specifically on ANT, technology, and the geographies of relations (Bingham 1996; Hinchliffe 1996). Other geographic articles have proposed ANT as a framework for understanding economic shifts, arguing that networks are "a dominant organizational form in the post-Fordist era" (Murdoch 1995, 731), and have addressed the economic ecology of wetlands (Burgess et al. 2000), institutional geographies and ANT (Davies 2000), and plant-human interactions (Hitchings 2003).

ANT is not without controversy in geography. Scott Kirsch and Don Mitchell (2004) are concerned that strong versions of ANT present humans and nonhumans only as autonomous objects connected within networks. Humans have very little agency and are not subjected to either oppressive or liberating societal structures. However, the authors do agree that weak versions of ANT can contribute to Marxist insights and social theories that have informed materialist social theory and geography for the past 150 years. They also argue that actor-networks are often presented in terms of cause-and-effect relationships. However, Kirsch and Mitchell (2004) argue that the agency issue should not be a fatal flaw. Their own viewpoint is that power is situated in centers, individuals, and social relations, and "in some ways, actor-network studies have indeed been quite effective at showing how such centering processes occur" (Kirsch and Mitchell 2004). In the sections to follow, we will lay out some of the ANT concepts and methods associated with tracing actors, intermediaries, networks, translations, and centers of calculation.

16.2.1 Actors, Intermediaries, and Networks

Actors are authors. This is what differentiates them from the materials that flow between them. Actors combine, mix, degrade, compute, and predict materials to create the next inscription, simulation, model, or map (Callon 1991; Latour 1987). Actors are not preordained with characteristics. One must trace and read the materials that flow between them and trace the associated networks to gain more knowledge about actors. Callon stated that "It is precisely because human action is not only human but also unfolds, is delegated and is formatted in networks with multiple configurations, that the diversity of the action and of the actors is possible" (Callon 1999, 194). Actors, as mediators, transform and put materials into motion. For Latour, "an actor is what is made to act by many others" (Latour 2005, 46). The origin of an actor-network begins as a flow of intermediaries between actors.

Intermediaries are materials, like texts, that carry meaning and give networks their form (Callon 1991; Latour 2005). For example, Michel Callon wrote that "an intermediary is anything passing between actors which defines the relationship between them…examples of intermediaries include scientific articles, computer software, disciplined human bodies, technical artifacts, instruments, contracts and money" (Callon 1991, 134). Trying to identify and study intermediaries can be an overwhelming task, which can be overcome by limiting intermediaries to text inscriptions (social networking dialogues, reports, books, articles, notes), technical artifacts (machines, hardware, software), human beings (skills), and all forms of money and funding (Martin 2000; Callon 1991). Texts and technical artifacts, such as those associated with VGI, help to define the roles of human and nonhuman actors in the network (Murdoch 1995). VGI reports connect with other objects, texts, people, and places. In fact, "Words, ideas, concepts and the phrases that organize them thus describe the whole population of human and non-human entities… intermediaries describe their networks…and they compose them by giving them form" (Callon 1991, 135). Conceptualizing networks requires decoding the wording within intermediaries. Actors author such meanings within intermediaries. Meaning is inscribed within intermediaries such as maps, GIS layers, weather reports, natural resource assessments, and digital databases. Words within documents can be traced to other documents.

The combination of actors and intermediaries creates heterogeneous networks (Harvey 2001; Martin 2000; Callon 1991). From an ANT perspective, human sensors are heterogeneous materials that constitute a network. By following the networks, it is possible to locate and describe additional groups, identify other significant actors, and trace interconnected intermediaries (Callon 1991, 142). A network works as "a coordinated set of heterogeneous actors which interact more or less successfully to develop, produce, distribute and diffuse methods for generating goods and services" (Callon 1991, 133) as well as distributing VGI. The social and technological must be explained together, not as a dichotomy (Burgess et al. 2000). This is important here because texts initiate relationships between actors and allow them to define one another through interactions (Murdoch 1995). Actors can actually define one another through their interaction, including the authoring and circulation of intermediaries from one institution to another (Callon 1991, 135), and through free association between actors in networks that are extremely dynamic and only momentarily stable (Burgess et al. 2000; Davies 2000). Because they are dynamic, much work is required to build and maintain strong networks. Actors need to be enrolled or otherwise align closely with other actors and intermediaries to maintain stability. Throughout the process, scientists become powerful by enrolling participants to help them build durable networks (Murdoch 2006). Translations align actors, stabilizing or destabilizing networks.

16.2.2 Translation

Translations are the goals, objectives, and interests that pass between actors and those they enroll (Latour 1987). When extending networks, actors must be convinced that their goals and interests are aligned with those of scientists and their laboratories (Murdoch 1997a). To succeed in making networks strong and durable, actors will select only the people, places, and materials that help them reach their goals. For example, a scientist attempting to perform a translation might say, "You have a problem that I can solve, but you have to follow my instructions and guidelines precisely." Technical specialists, scientists, engineers, and others "speak in the name of new allies that they have shaped and enrolled; representatives among other representatives, they add these unexpected resources to tip the balance of the force in their favor" (Latour 1987, 259). Translations are inscribed into texts, criteria, technical objects, project guidelines, roundtable discussions, conference proceedings, newsletters, embodied skills, and countless other materials (Callon 1991).

Translation is the process of making two different actors equivalent (Law and Hassard 1999). This requires an alignment of intermediaries like storm reports, forest assessments, GIS data standards, or funding. Translation is "an idea that suggests that if scientific networks are to be extended through space and time, then actors of differing (natural and social) types must be 'interested' into the network – that is, their goals must somehow be aligned with those of the scientists" (Murdoch 2006, 62). Alignment requires some level of normalization (Callon 1991; Murdoch 2006), ordering (Davies 2000), or a set of criteria that everyone can agree to, like standardized data or criteria for classifying severe-weather phenomena. Actors will select only those translations that will help them maintain their networks and achieve goals (Latour 1987). Normalization and selectivity allow the networks to achieve their goals, increasing their ability to do work, to be powerful (Latour 1987).

Strong network alignment requires convergence of actors, intermediaries, and translations. Convergence measures the extent of the translation process through the circulation of intermediaries leading to agreement among actors (Callon 1991). Intermediaries are inscribed with rules, laws, standards, and meaning embedded by actors. Such entities are said to display strong coordination, which contributes to network stabilization. Successful translation processes lead to the stabilization of networks, and it is possible to trace these networks unproblematically. However, this alignment is not always the result of translation. Not all actors conform to network builders' translations; some actors resist (Burgess et al. 2000). Often there is controversy and conflict resulting from disagreement among actors about the translation process. This in turn can lead to betrayal when the actor refusing to accept the role assigned to him or her aligns with another competing network. These converging networks have boundaries that can be mapped to determine the reversibility or irreversibility of the network. While vital environmental conditions may open up, like children being able to create data, others stress that certain groups may be included or excluded from online spatial data (Zook and Graham 2007a, b; Harvey 2007; Goodchild 2007a). Convergence of actor-networks is attempted within the center of calculation.

16.2.3 *Center of Calculation*

The center of calculation concept describes the reach of scientific institutions that go out to the periphery and collect information. Information is brought back to the center and turned into scientific knowledge; one goal of science is to bring knowledge of the periphery back to the center (Latour 1987). Once back at the laboratory, the collected items can be studied, and more precise inscriptions made in the form of maps, manuscripts, models, and simulations. Contemporary scientists construct facts within their laboratories, allowing officials and managers to exert a degree of control on the physical environment. The development and implementation of models and simulations are very important components of the control process. By collecting and processing real-world data, scientists and technicians create virtual maps, models, and simulations, allowing them to experience the physical environment from within the controlled confines of laboratories.

Engagement with models and simulations allows scientists and engineers to perfect techniques like forecasting severe weather, controlling the flow of a river, or the management of natural resources before experiencing the real thing. Scientists create hundreds of models and run thousands of simulations in an effort to get the desired results in real-world situations (Latour 1987). Contemporary scientists construct facts within their laboratories, allowing officials and managers to exert a degree of control on the physical environment. The development and implementation of models and simulations are very important components of the control process. Through laboratory experiences, knowledge is created. The construction of scientific knowledge, tied to economic and political systems, gives those in the center an advantage over people, places, and things on the periphery. These conditions can convert a seemingly insignificant place into a center that can dominate the periphery from a distance (Latour 1987).

Centers of calculation are places where intermediaries holding information, VGI, data, maps, numbers, and inscriptions are brought together and organized. This is a historical condition. Empires send actors such as naturalists, cartographers, geographers, anthropologists, and technicians out into the world to collect and make initial inscriptions of people, places, and things located in faraway places. The process of going out, collecting, and returning to the center is known as a cycle of accumulation. There are often multiple cycles of accumulation. Each time an expedition goes out and returns, it brings back more information that can be used to return and claim additional natural resources or other materials. The accumulation of information about distant places gives the centers of calculation advantages over the periphery, allowing the centers to perform action from afar (Latour 1987). The collection of climatic data over the past 200 years is an extreme example of accumulation cycles (hourly, daily, weekly, monthly, yearly).

The center of calculation acts at a distance by mobilizing, stabilizing, and combining data with maps, GIS, and other materials. One geographer writes, "In the geography of actor-networks, a crucial issue is how actors are able to mobilize networks to act at a distance" (Murdoch 1995, 749). Ideas associated with centers of calculation have been incorporated into the works of some geographers (Palmer 2009, 2012; Kirsch 2002; Harris 2004; Martin 2000). It is important for the centers of calculation, like the NWS or the United States Geological Survey (USGS), to maintain the stability of the materials collected and inscriptions created "so that they can be moved back and forth [between the center and periphery] without additional distortion, corruption or decay" (Latour 1987, 223). For museums, specimens like plants, animals, bones, North American Indian medicine bundles, shields, weapons, housing, clothing, and maps need to be preserved so that the centers of calculation can produce more inscriptions and more knowledge about peoples and ecosystems on the periphery. Stabilization also involves taking local knowledge out of its context and placing it into a scientific classification scheme and incorporating standards so that new materials and inscriptions can be made combinable. Centers combine past and present materials and construct maps, database layers, tables, and charts (Latour 1987). Once the materials are stable, standardized, and combinable, the information and knowledge produced by anthropologists or geographers or geologists or meteorologists can be combined with seemingly distant entities such as economic institutions, missionary organizations, academic institutions, corporations, or broadcast media outlets to perform action at a distance (Latour 1987).

To briefly recap, GIS researchers use the ANT framework to describe the multiple combinations of human/nonhuman actors and their intermediaries that make up an actor-network. The glue that makes actor-network connections possible is called translations. Some translations are successful and actor-network convergence results. Other times, translations fail and actor-networks weaken. Powerful actors that reside within centers of calculation manipulate actor-networks to their advantage. Centers of calculation mobilize, stabilize, and combine information into standardized models, maps, or simulations so they can act at a distance. In the sections to follow, ANT will be used to describe an actor-network of human sensors consisting of actors including storm chasers, NWS meteorologists, television media, and multiple intermediaries, creating a general symmetry between society and technology.

16.3 Storm-Spotting and Storm-Chasing Actor-Network

The storm-spotting and storm-chasing network is an amalgamation of humans, skills, technologies, and texts that are dynamic and constantly changing. The primary actors are storm spotters, a heterogeneous group of storm chasers, operational meteorologists, and television media. All of the actors are defined by intermediaries that included skills, technologies, and texts. Below is a brief historical overview of storm-spotting and storm-chasing origins, followed by a generalized description of a decentralized, heterogeneous actor-network in the United States that connects storm spotters and chasers in the field, operational meteorologists in the center of calculation, and television media as the recipients of official severe-weather watches and warnings.

Technological innovations, public awareness, training, and scientific research influenced the development of storm spotting and chasing in the United States during the twentieth century. The origins of storm spotting can be traced back to World War II and the military's concern that ammunition depots and other installations might be at risk of being damaged or destroyed by severe weather (Bates 1962; Galaway 1985). In addition, technological innovations like television, radio, telephones, and weather radar-shaped storm spotting. Radio and telephone technologies "proliferated during [the early to mid-twentieth century], providing innovations that permitted rapid dissemination of warnings based on ongoing tornado events…notably telephone and radio, … an important component of the spotter network" (Doswell et al. 1999, 545, 547). Emphasis on warning the public about severe weather evolved in the 1950s. By the 1960s, emphasis was placed on public watches. More specifically, following the Palm Sunday tornado outbreak in 1965, "tornado forecasts officially became tornado watches," leading to the creation of the SKYWARN tornado-preparedness and storm-spotter program

(Galaway 1985, quoted by Doswell et al. 1999, 547), including storm-spotter training materials like manuals, booklets, and films.

Related yet different, storm chasing is "the art and science of meeting with a thunderstorm, for any reason" (Vasquez 2008, 1). It is the tracking and first-hand observation of severe weather, oftentimes the tornado being regarded as the primary target of a chaser. While storm chasing may be a global activity, in the United States, most of it plays out over the spring and summer months in "Tornado Alley" (Robertson 1999; Cantillon and Bristow 2001). Brooks et al. (2003) define Tornado Alley using the criteria of tornado frequency and reliability of the season to outline a region from west Texas that extends northeastward through central Minnesota. The core probably lies somewhere between north Texas, western Oklahoma, and western Kansas. This area contains roughly 40% of all US tornado events.

The idea of storm chasing emerged through scientific observation and ground truthing in the 1970s, 1980s, and 1990s. Field research, and the idea that the atmosphere is a laboratory, set the stage for multiple humans, using technology, to go into the field and literally act as human sensors for observing severe storms and tornadoes. By 1972, the National Severe Storms Laboratory (NSSL) launched the Tornado Intercept Project (TIP) and other projects throughout the decade (Golden and Morgan 1972). Veterans of TIP were involved in the creation of the first issue of *Stormtrack*, a publication about storm chasing. By 1980, NSSL had developed TOTO (TOtable Tornado Observatory), an instrument package designed for deployment in the direct path of a tornado. This project garnered the focus of a Public Broadcasting Service (PBS) special on *Nova* in 1985. The publicity helped to generate some additional interest in chasing (Vasquez 2008). By 1994 and 1995, the NSSL's Verification of the Origins of Rotation in Tornadoes Experiment (VORTEX) had taken off, followed by VORTEX2 in 2009 and 2010, which were some of the largest storm-intercept research projects to date (National Severe Storms Laboratory, NSSL 2011).

Historically, the NWS has relied on volunteer observers to report everyday weather and also to spot and report severe-weather events throughout the United States. Multiple technologies have played a major role in the sensing of local meteorological conditions and the communication of volunteered information across space. However, this is not solely a story of technology. Technologies like telephones, television, radar, or barometers do not solely determine the success of storm-spotting networks. Likewise, human society does not solely determine the success of networks either. In fact, it is extremely difficult to separate the technological from the social. ANT is used to describe a storm-reporting human-sensor network.

16.3.1 Actors-Intermediaries-Networks

Storm spotters and chasers are a diverse group of individuals. Storm spotters often work as civil servants or are concerned citizens who observe severe weather because it is their civic duty. They report to law enforcement, emergency managers, and the NWS. Next are four types of storm chasers. The first group can be classified as research scientists, graduate or undergraduate students. Some graduate students in meteorology, from around the world, gain experience and collect data for their dissertations in the field through programs like VORTEX and other mesoscale meteorology research, centered at the Oklahoma Weather Center in Norman, Oklahoma (Bluestein 1999). Undergraduates from around the United States are also connected with the VORTEX program (Palmer et al. 2000). A second group of storm chasers includes weather entrepreneurs and television reporters who view severe weather and tornadoes as commodities to be filmed or chased for a profit. Storm entrepreneurs and reporters post media in online venues for feedback, discussion, and to spread reputation. A few chasers sell video footage or operate tornado-chasing tour companies (Kraushaar 2011). Third, some storm chasers are recognized as nonprofessional scientists who volunteer in scientific activities such as the collection of data, analysis, or dissemination of a scientific project (Haklay 2010). And finally, a fourth group of storm chasers falls under the category of thrill seeker. For example, one storm chaser admitted that his goal is "getting up close and personal with a tornado. The sheer power mother nature can demonstrate is humbling to say the least" (Kraushaar 2011, 57). Because of thrill seeking, some government organizations and university research units try to distance themselves from being associated with storm chasing, even though chasing is very much associated with the legacy of scientific mesoscale meteorology. VGI storm reports may be dismissed due to mistrust based on the view that "science is best left to scientists, and it requires rigor, knowledge and skills that only professional scientists develop over time" (Haklay 2010, 4).

Meteorologists at the NWS maintain, standardize, and extend the storm-report network for the media and general public. The NWS is part of a vast bureaucracy within the National Oceanic and Atmospheric Administration (NOAA) of the US Government. Actors within the NWS are relevant and interested because much of the weather-forecast and weather-warning responsibilities are shouldered by them. The media, the general public, and industry within the massive severe-weather warning and emergency-management network rely upon the agency's issuance of weather watches and warnings. Operational meteorologists create both short- and long-term forecasts. Communicator or storm-report specialists work alongside meteorologists and monitor any storm-report communications coming into their forecast office.

Television media are an actor who warns the general public about severe weather but also sensationalizes weather as infotainment. Infotainment both informs and is a form of entertainment. Many television stations employ meteorologists or weathercasters. Reporters often act as storm chasers, guided from the television center by the station's meteorologist. Storms are also on the big screen. Scientific storm chasing was the inspiration for the infamous 1996 Hollywood film *Twister*. This blockbuster served as a launching pad for "temporary chaos" (Vasquez 2008, 5) as the popularity of storm chasing skyrocketed from the action-packed but distorted view of storm chasing (Robertson 1999). Subscriptions to *Stormtrack* magazine increased from 350 to more than 900 seemingly overnight, probably influencing the rise of the storm-chasing tourism industry. The 1990s also saw home videos of extreme weather explode in popularity with the media misrepresenting chasing and romanticizing the activity (Robertson 1999). While many *Twister* enthusiasts gave up on the activity by 2000, today there is a new saturation from many sources. The *Discovery Channel*, for instance, has a reality TV show, *Storm Chasers*, which premiered in 2007 and which documents chase teams as they attempt to drive modified vehicles into tornadic circulations. With all of this mainstream attention, storm chasing as a hobby continues to gain in popularity. In the words of a storm chaser, "Perhaps too many people are looking to make a name for themselves? I think there are a lot of people chasing now who have not spent years reading text books and studying basic severe weather meteorology but have been inspired by *Discovery Channel*" (Kraushaar 2011, 68).

Skills are intermediaries that help define storm chasers. Some storm spotters and chasers are educated, holding a math or science degree; others are self-taught citizen weather observers and scientists. Storm spotters use intermediaries like the NWS and SKYWARN storm-spotting manuals and materials. Spotters participate in yearly training at NWS forecast offices or at county emergency-management facilities. SKYWARN is an informal educational opportunity that is free of charge to anyone interested in becoming a spotter (NOAA 2011b). Storm chasers, on the other hand, often have access to higher-education resources and higher-level NOAA facilities like the Storm Prediction Center (SPC) and the National Severe Storms Laboratory (NSSL). However, much meteorological learning, by both storm spotters and chasers, occurs in the field. Training and field experiences enhance the spotter's and chaser's credibility with NWS forecasters and atmospheric scientists. Spotters and chasers are judged by the accuracy of their severe-weather forecasts and the ability to pinpoint the geographic region(s) most likely to contain a severe-weather or tornadic event. As a result, chase-day forecasting is a method of determining the credibility of a storm chaser, by others or through self-awareness. One chaser revealed that there "is still a desire to say 'hey, look what I did…' And because it's a big deal to me to be able to go back and find a report that matches what I might have seen (so to connect the dots on events). But mainly for less than brilliant reasons like image/self-gratification" (Kraushaar 2011, 52).

All of the actors depend upon technological intermediaries for traveling, remotely sensing, and reporting storms. Chasers drive their vehicles, equipped with laptop computers, mobile Internet, mapping software, GPS for accurate road and location data, digital cameras, and high-definition (HD) video cameras, to the base of severe thunderstorms. Positioned, spotters communicate with the media, emergency managers, law enforcement, or the NWS using cellular phones, handheld communication radios, or smartphones. Combined with digital technologies, human sensors emerge in the field. For example, some storm chasers, especially the entrepreneurs and reporters, use webcams to stream chases live over the Internet (Kraushaar 2011). While some citizen scientists carry intermediaries like "…handheld anemometers for measuring wind speeds, or have larger units mounted on their vehicles. The more research-oriented chasers may travel with mobile mesonets, more elaborate equipment with instruments to collect in-situ weather data, mounted on their vehicles. Mobile Doppler radar may also be used. Deployable sensor probes or camera probes may also be employed for either collecting scientific data or photos and video inside tornadoes. Having access to the Internet, live radar, real-time observations, and forecasts makes it very easy for new chasers to head out into the field and intercept a severe storm and tornado" (Kraushaar 2011, 40–41). Storm chasers use their instincts, senses, and technologies; they are human sensors. Chasers are "armed" with laptops and cell phones, but rely heavily upon their "eyes" and "instincts" to determine conditions and get in good position relative to the storm (Kraushaar 2011, 39–40)

Operational meteorologists, television meteorologists, and weathercasters monitor the atmosphere with the aid of intermediaries like terrestrial weather instruments, remote-sensing satellites in space, Doppler weather radars, and numerical computer models. Before calculating any kind of plan of action like the issuance of a severe-weather watch or warning, the NWS relies upon a network of sensors to make real-time observations and report back to forecast offices. Some of the sensors are static and mechanical, placed in specific geographic locations, so the origins of each observation including wind speed, direction, barometric pressure, temperature, and rainfall are known; the sensor's geographic location is predetermined and stabilized. All meteorologists use photos, videos, websites, and social media to verify their forecasts and storm warnings. These technologies as intermediaries are not disposable but rather are part of a sociotechnical network – a way of being in the world.

Texts (VGI) are important intermediaries that define the actors. Storm spotters and chasers author reports and diffuse them through various channels. As stated earlier, the NWS is interested in receiving reports from people in the field regarding severe weather to aid in the warning decision process. Specifically, they are interested in georeferenced information on strong winds, large hail, wall clouds, funnel clouds, tornadoes, flash flooding, damage, and dangerous winter weather for radar verification, warnings, and decision-making. Spotters, research chasers, entrepreneurs, citizen scientists, and thrill seekers author reports of tornadoes, wind, hail, storm damage, and flooding observations (NOAA 2011b). Meteorologists take visual observations and translate them into text messages that included severe-weather watches and warnings. These warnings flow to the general public through NOAA weather radio alerts, the Internet, radio, and television stations. Radio and television media outlets pass on a vernacular translation of severe-weather alerts through verbal communication, eye contact with the camera, graphic maps, and messages streaming at the bottom of television screens, to the general public.

The amalgamation of actors and intermediaries constructs what might be described as a human-sensor network. Within such a network there are reciprocal and symmetrical relationships between society and technology; a symmetry that blurs the lines between human societies and technologies. Human sensors are organic in that observations flow from the field to centers, while important scientific information, meteorological instrument measurements, and Doppler-radar reflectivity and velocity information flows from the NWS and television stations to the field. All actors are connected by the intermediaries mentioned above, and all are dependent upon one another with regard to early warnings and safety in the field. The goals and objects of the storm-report network are translated by the NWS, who has responsibility for forecasting and the issuance of severe-weather watches and warnings.

16.3.2 Translations

Meteorologists at the NWS translate goals, objectives, and interests in the storm-report network. The goal is to get storm spotters, chasers, and concerned citizens to report significant severe-weather events like tornadoes, flooding, damaging winds, and hail. The preferred method is to convince spotters and chasers to call in reports directly to the NWS. This is generally the quickest and easiest way to submit information and it facilitates good, two-way communication between the spotters, chasers, and the NWS communicator or storm-report specialist. Some storm chasers will call the NWS when they see a tornado or other critical event.

Next, the NWS would like chasers to submit reports via the Internet on eSpotter, to the Spotter Network, via Twitter or Facebook, all monitored by the NWS. This is a more decentralized approach to reporting. Spotters and chasers not only submit reports to the government. Many in the field have relationships with different media agencies to which they will relay their field accounts, photos, and video for rebroadcast. Others have working relationships with local emergency-management agencies. Some keep their followers on Twitter and Facebook updated with their chasing status in the field. The most prevalent method is use of the Spotter Network. The Spotter Network "brings storm spotters, storm chasers, coordinators and public servants together in a network of information. It provides accurate position data of spotters and chasers for coordination/reporting and provides ground truth to public servants engaged in the protection of life and property" (Spotter Network 2011). Launched by AllisonHouse, LLC in 2006, this free service has exploded to an estimated member base of 15,000 users today. Using a graphical interface on a computer, people can submit real-time reports based on their specific location using GPS coordinates. The Spotter Network also requires chasers to take an exam on the fundamentals of severe weather before they can join the group. Spotter Network is a decentralized network and alternative to reporting to the NWS.

Interest in the storm-reporting network is based on the ideas of civic duty, heroics, and participation as scientific observers in the face of severe-weather threats. This translation has a militaristic flare to it. The NWS encourages spotters and chasers to become citizen scientists and act as "the Nation's first line of defense against severe weather" (NOAA 2011c). Participating in the network is heroic in that chasers will "help meteorologists make lifesaving warning decisions." Spotters and chasers as citizen scientific observers are indispensable in that the NWS cannot do their job without them in an effort to "keep local communities safe" as the "first line of defense against severe storms" by "giving communities the precious gift of time-seconds and minutes [that] can help save lives" (NOAA 2011a, b, c, d).

The translations put forth by the NWS do not completely converge with apprehensive storm chasers. Some chasers believe it is important to assist the NWS in verifying and ground truthing severe-weather warnings and Doppler-radar signatures. Another reason to report storms revolves around the potential of material damage or human casualties. One storm chaser remarked, "I would feel terrible if I didn't call something in and someone lost a life in the storm, whether my report would have changed anything or not, at least I tried" (Kraushaar 2011, 48). On the other hand, there is some apprehension among storm chasers about reporting their observations and locations. Storm reporting among a cadre of trained and untrained storm chasers is political, and many chasers are wary of having their reports called into question. Other times, false reports lead to the "cry wolf" syndrome, causing the NWS to question the credibility of storm reports (Kraushaar 2011). A reverse argument about the NWS and verification is also made by another storm chaser whose perception is that "The national severe weather database is riddled with imprecise estimates, reports filtered for verification instead of scientific purposes, and reports of questionable integrity that are uncorroborated independently. This is a longstanding problem, one well-documented in several formal papers...Non-tornadic gustnadoes, dubious 'brief touchdown/no damage' reports at night in a forest, and so-called 'sheriffnadoes' (scud and other low-hanging features mistaken for tornadoes, often by poorly-trained law enforcement or other spotters) also may be kept in the system if they conveniently verify a local NWS warning, whether they really happened or not, for the sake of better-looking verification scores" (Kraushaar 2011, 69).

The translations presented by the NWS are only partially convergent. Furthermore, the NWS has to filter and act as a gatekeeper of information coming into their center of calculation.

16.3.3 The NWS Center of Calculation

In order for the NWS to act at a distance, meteorologists must mobilize, stabilize, and make storm reports combinable with their technologies and texts. This process is accomplished at the NWS center of calculation as a means of maintaining networks by centralizing and standardizing VGI report information that can be transformed into GIS polygons, radio alerts, and weather bulletin texts that are consumed by television media and the general public. NWS center of calculation has been accumulating severe-storm reports for over 60 years (Doswell et al. 1999). But recently, through the advent of enabling geospatial technologies like GPS, geotagging, georeferencing, and Web 2.0, the spatial resolution of VGI has also increased.

The NWS cannot bring the atmosphere into their forecast centers, so they rely on human sensors to submit bits and pieces of VGI reports to the center. Individual citizens are in great positions to offer local knowledge and real-time observations that are not easily remotely sensed. Local knowledge and expertise can replace centralization in cases where rapid, up-to-date information on local conditions is needed

(Goodchild 2008). However, the NWS is the official severe-warning organization in the United States, so for the time being there must be a reciprocal relationship between decentralized and centralized report and early-warning processes. NWS forecast offices are command and control centers equipped with multiple flat-screen televisions that are tuned to local broadcasts. Monitoring the broadcasts not only verifies that NWS information is being mobilized to the general public, but stations often use their own reporters as human-sensor storm chasers who stream real-time reports and most importantly videotape of the atmosphere at a given location. Reporters travel in vehicles or even helicopters. Live streaming reports that reveal geographic location (at the intersection of US 66 and Old School Rd.), estimated wind speed (strong inflow or outflow) and direction (southeasterly vs. northwesterly), how the air feels to the spotter/reporter (warm, moist, sticky as opposed to cool, dry air), and the visual verification of rotation are extremely important pieces of information to aid in severe-weather warnings and Doppler-radar ground truthing. NWS meteorologists are not in the field, but their auditory and observation skills are put to the test when monitoring television live feeds.

As mentioned earlier, phoning in storm reports directly to the NWS is the preferred method of communication. However, digital information is mobilized through the use of the NWS eSpotter program, NWSChat, Spotter Network, Twitter, and Facebook (Kraushaar 2011). NWS eSpotter is simply an online reporting system open the general public. However, individual citizens, companies, or others who are involved in chasing severe weather do not meet the qualifications for using NWSChat (https://nwschat.weather.gov). It is a tightly controlled government communication system where official talk flows between centers of calculation. Two promising but very heterogeneous digital landscapes are the social networking sites like Twitter and Facebook. Twitter and Facebook assist in mobilizing VGI reports and are monitored by communication technicians and meteorological technicians. The Twitterfall search engine is popular among meteorologists at the center. However, the challenge facing the technicians is determining what information is useful and what is not – what to let into the center and what to filter out. Keyword searches and hashtag searches (#wxreport) for wind, hail, or tornadoes can result in hundreds of tweet returns that do not always contain relevant information. Messages are often in narrative form, emotion is expressed, jokes are told, and egos are present. It would serve communication technicians to become expert text-discourse analysts and deconstructionists. Barring this, searching content is time consuming and unpredictable. For example, "Not only would you get relevant tweets about ongoing flooding or recent hail storms, but you would also get weather reports from all around the globe and tweets that [have] nothing to do with weather, but included the word hail or flood" (Brice and Pieper 2009, 2). However, the Twitter sources of VGI are potentially too valuable to ignore and can further help the NWS strengthen its position as a weather authority in an age of increasing privatization. Regardless of how the NWS center of calculation mobilizes information, they cannot use all of it.

NWS attempts to stabilize VGI reports through the standardization of reporting criteria and georeferences. Good reporting must provide detailed observations and georeferencing at a fine spatial resolution to be useful within the center of calculation.

Table 16.1 VGI report translation convergence/divergence

Convergence	Report
Convergent	lon = "−96.3691" lat = "34.3462" I was in a tornado 5:30 pm parked alongside lake thunderbird tree limbs covered my car
Convergent	lon = "−96.1724" lat = "34.2321" Significant damage to grocery store in little axe ok. Cars flipped. Cell tower down
Neutral	lon = "0" lat = "0" Chasing today in Southwestern Missouri currently in Branson MO
Neutral	lon = "0" lat = "0" Man I wish I could chase the storm with a classic hook N of Springfield MO
Divergent	lon = "0" lat = "0" Sitting in central OK streaming live video of an approaching severe cell so purchase videos on my homepage

Table 16.2 Spatial resolution of VGI storm reports

Spatial resolution	VGI	Usefulness
Lat. and long. coordinates	lon = −96.3691 lat = 34.3462	Excellent
Road intersections	Forum Rd. and Chapel Hill Rd., Columbia, MO	Excellent/good
Place-names	Columbia, MO	Good/fair
Regions	Boone County, MO	Fair
No georeference	N/A	Poor

For example, the NWS would like Twitter users to report "Damage from winds – briefly describing the damage; hail – including size of the hail; tornadoes and funnel clouds; and flooding" (NOAA 2011d). Reports containing the most convergent VGI translations in descending order include reports of occurrences (hail, tornado, etc.), reports of damage, relay of NWS information by spotters, personal opinions, and spotter, chaser, and media advertisements (see Table 16.1). Georeferences with fine spatial resolution are equally important to the center. Information about hail, extreme winds, flooding, and tornadoes must be located as accurately as possible in space. In descending order, the NWS center of calculation prefers latitude and longitude coordinates and road intersections over place-names, regions (counties or physiographic regions like "the Ozarks"), or no georeferences (see Table 16.2). When storm spotters, chasers, and citizens conform to the reporting criteria and lat/long coordinates through GPS or geotagging, the convergence between the decentralized network and the center of calculation is perfectly translated. Center communication specialists filter out less convergent reports.

The most accurate, fine-resolution georeferencing reports can be combined with Doppler-radar data to verify severe-weather warnings and with text products to be diffused publicly. NWS meteorologists translate their ability to precisely forecast storms, generate mesoscale numerical-model representations, simulate storm movements, and calculate the trajectories of severe storms and mesocyclones. Storm reports aid in ground truthing and verifying the accuracy of these products. Further connections are made between the field and the center as the NWS integrates Doppler-radar data into GIS-based warnings known as storm-based warnings.

Precise VGI reports are preferred when verifying the NWS polygon warning system called WARNGEN, a state-of-the-art Doppler-radar interface combined with geographic information systems (GIS). Individual storm reports are geotagged onto maps showing Doppler reflectivity, velocities, and the polygon warning area. The WARNGEN technology interface, implemented nationally, allows meteorologists to digitize localized polygons as a representation of a severe-thunderstorm- or tornado-warned area. Warnings are usually more accurate for areas where the spatial resolution is finer than county-wide warnings. The combination of VGI reports, Doppler-radar data, and geospatial technologies transforms the way that meteorologists monitor, represent, and issue severe-weather warnings. Local and particular information will become more important to the center of calculation as the severe-weather warnings become more precise.

Ultimately, the NWS center of calculation acts at a distance (Latour 1987) and extends its networks further by relaying transformed reports to the media. All relevant storm reports and the simulated worlds are at the fingertips of the meteorologists at the center. Through the integration of reports, criteria, and GIS, the center of calculation can generate text bulletins that exit the center and diffuse externally to the media and general public. As human sensors in the field provide severe-weather observations to the NWS, the information is transformed through the above-mentioned centering process, and the spotters/chasers/informants become invisible. The NWS center of calculation has that power because it is responsible for issuing many comprehensive forecast products to a wide audience, including the general public, at only a modest cost to taxpayers. Aviation, fire, weather, hydrologic, marine, tropical, climate, and public forecasts are created. The NWS produces event-driven, short-range products for alerting the public to all types of hazardous weather. These products include outlooks, watches, warnings, advisories, and special weather statements. The products allow the NWS to do work and have impacts on other dispersed spaces – as long as the networks remain stable and in place (Murdoch 2006).

16.4 Summary and Conclusions

In this chapter, we have demonstrated the utility of an ANT framework in the description of sociotechnical networks in the context of a volunteered geographic information system. Key concepts used here included actors, intermediaries, networks, translations, and centers of calculation. The dichotomy between humans and technology is a false one. Rather, humans and technologies are interconnected through actors and intermediaries that form networks. We have introduced readers to a storm-spotting and storm-chasing actor-network. All of the actors and intermediaries resemble an amalgamation of human sensors, not just isolated humans carrying external sensors. Mobile phones, concerned dialogues, observant people, digital video cameras, television speakers, and the Internet are the voice, eyes, and ears of the human sensors. All of the human sensors are decentralized and in a state of play with one another. The NWS, a centralizing force, monitors the airwaves, chatrooms, social networking sites, and televisions for translatable VGI severe-storm reports. NWS meteorologists do not have the luxury of free play. They are responsible, as officials, for warning the public about hazardous weather conditions. Meteorologists within the center of calculation receive, mine, and harvest raw VGI and transform it into a standardized severe-weather bulletin that can be confidently consumed by the media and the general public.

ANT is a conceptual framework, set of methods, and an ecological approach for understanding VGI. Through the lens of ANT, a general symmetry or reciprocity between humans and technologies stands at the forefront of an ecological theory. General symmetry between human and nonhuman actors can contribute ideas for the development of the human-sensor concept. As such, the human sensor is conceptualized as a "whole" network of interconnections and potential disconnections, a very dynamic organism. At the same time, ANT's method of tracing text, technologies, humans, and money can provide us with insights into the human sensor's motivations for volunteering geographic information, and issues pertaining to the quality of geographic data generated. Some storm spotters and chasers are preferred informants for the NWS. What conditions lead to this preferential treatment? What are the successful translations that lead to such a network? Are the network relations reciprocal? Finally, we believe that actor-networks could be categorized as environments. The storm spotting/chasing VGI environment is prescriptive, informal, and decentralized. One measure of success for the storm-report network is significant weather-event content and precise georeferencing. But how might we describe a Flickr VGI environment? In Flickr, the georeferencing information may be very precise (lat/long coordinates). Yet, what can we learn from the photographs and their content? Environments containing heterogeneous content may require us to think about "interpretive attributes" through the tracing and decoding of intermediaries.

References

Bates, F. C. (1962). Severe local storm forecasts and warnings and the general public. *Bulletin of the American Meteorological Society, 43*, 288–291.
Bingham, N. (1996). Object-ions: From technological determinisms towards geographies of relations. *Environment and Planning D: Society and Space, 14*, 635–658.
Bishr, M., & Mantelas, L. (2008). A trust and reputation model for filtering and classification of knowledge about urban growth. *GeoJournal, 72*, 229–237.
Bluestein, H. B. (1999). A history of severe-storm-intercept field programs. *Weather and Forecasting, 14*, 558–577.
Brice, T., & Pieper, C. (2009). Using Twitter to receive storm reports. Available online at: ams.confex.com/ams/pdfpapers/163543.pdf. Accessed 18 Feb2012.
Brooks, H., Doswell, C., III, & Kay, M. (2003). Climatological estimates of local daily tornado probability for the United States. *Weather and Forecasting, 18*(4), 626–640.
Burgess, J., Clark, J., & Harrison, C. M. (2000). Knowledges in action: An actor network analysis of a wetland agri-environment scheme. *Ecological Economics, 35*(1), 119–132.
Callon, M. (1991). Techno-economic networks and irreversibility. In J. Law (Ed.), *A sociology of monsters* (pp. 132–161). New York: Routledge.

Callon, M. (1999). Actor-network theory: The market test. In J. Law & J. Hassard (Eds.), *Actor network and after* (pp. 181–195). Oxford: Oxford University Press.

Cantillon, H., & Bristow, R. (2001). Tornado chasing: An introduction to risk tourism opportunities. In *Proceedings of the 2000 Northeastern Recreation Research Symposium* (pp. 234–239). Newtown Square: U.S. Department of Agriculture, Forest Service, Northeastern Research Station.

Crampton, J. (2010). *Mapping: A critical introduction to cartography and GIS*. New York: Blackwell.

Davies, G. (2000). Narrating the natural history unit: Institutional ordering and spatial strategies. *Geoforum, 31*(4), 539–551.

Deleuze, G., & Guattari, F. (1987). *A thousand plateaus: Capitalism and schizophrenia*. Minneapolis: University of Minnesota Press.

Doswell, C. A., III, Moller, A. R., & Brooks, H. E. (1999). Storm spotting and public awareness since the first tornado forecasts of 1948. *Weather and Forecasting, 14*, 544–557.

Galaway, J. G. (1985). J. P. Finley: The first severe storms forecaster (Part 1). *Bulletin of the American Meteorological Society, 66*, 1389–1395.

Gartner, G., Bennett, D., & Morita, T. (2007). Toward ubiquitous cartography. *Cartography and Geographic Information Science, 34*, 247–257.

Golden, J. H., & Morgan, B. J. (1972). The NSSL/Notre Dame tornado intercept program, spring 1972. *Bulletin of the American Meteorological Society, 53*, 1178–1180.

Goodchild, M. F. (2007a). Citizens as sensors: The world of volunteered geography. *GeoJournal, 69*, 211–221.

Goodchild, M. F. (2007b). Citizens as voluntary sensors: Spatial data infrastructures in the world of Web 2.0. *International Journal of Spatial Data Infrastructure Research, 2*, 24–32.

Goodchild, M. F. (2008). Commentary: Whither VGI? *GeoJournal, 72*, 239–244.

Haklay, M. (2010, Sept). Geographical citizen science – Clash of cultures and new opportunities. In *Proceedings of the workshop on the role of volunteered geographic information in advancing science*. GIScience 2010, Zurich, Switzerland.

Haraway, D. (1987). A manifesto for cyborgs: Science, technology, and socialist feminism in the 1980s. *Australian Feminist Studies, 2*(4), 1–42.

Harris, C. (2004). How did colonialism dispossess? Comments from an edge of empire. *Annals of the Association of American Geographers, 94*(1), 165–182.

Harvey, F. (2001). Constructing GIS: Actor networks of collaboration. *URISA Journal, 13*(1), 29–37.

Harvey, F. (2007, April). *Nowhere is everywhere? Towards postmodernist ubiquitous computing based geographic communication*. Paper presented at the annual meeting of the Association of American Geographers, San Francisco, CA.

Harvey, F., & Chrisman, N. (2004). The imbrication of geography and technology: The social construction of geographic information systems. In S. D. Brunn, S. L. Cutter, & J. W. Harrington (Eds.), *Geography and technology* (pp. 65–80). Dordrecht: Kluwer Academic.

Hinchliffe, S. (1996). Technology, power, and space-the means and ends of geographies of technology. *Environment and Planning D: Society and Space, 14*, 659–682.

Hitchings, R. (2003). People, plants and performance: On actor network theory and the material pleasures of the private garden. *Social and Cultural Geography, 4*(1), 99–113.

Kirsch, S. (2002). John Wesley Powell and the mapping of the Colorado Plateau, 1869–1879: Survey science, geographic solutions, and the economy of environmental values. *Annals of the Association of American Geographers, 93*(4), 645–661.

Kirsch, S., & Mitchell, D. (2004). The nature of things: Dead labor, nonhuman actors, and the persistence of Marxism. *Antipode, 36*(4), 687–705.

Kraushaar, S. (2011). *Ground truth: Volunteered geographic information and storm chasing*. Unpublished master's thesis, University of Missouri, Columbia.

Latour, B. (1987). *Science in action: How to follow scientists and engineers through society*. Cambridge: Harvard University Press.

Latour, B. (2005). *Reassembling the social: An introduction to actor-network-theory*. Oxford: Oxford University Press.

Law, J., & Hassard, J. (1999). *Actor network theory and after*. New York: Blackwell.

Martin, E. (2000). Actor-networks and implementation: Examples from conservation GIS in Ecuador. *International Journal of Geographical Information Science, 14*(8), 715–738.

Mummidi, L., & Krumm, J. (2008). Discovering points of interest from users' map annotations. *GeoJournal, 72*, 215–227.

Murdoch, J. (1995). Actor-networks and the evolution of economic forms: Combining description and explanation in theories of regulation, flexible specialization, and networks. *Environment and Planning A, 27*, 731–757.

Murdoch, J. (1997a). Inhuman/non-human: Actor-network theory and the prospects for a non-dualistic and symmetrical perspective on nature and society. *Environment and Planning D: Society and Space, 15*, 731–756.

Murdoch, J. (1997b). Towards a geography of heterogeneous associations. *Progress in Human Geography, 21*, 321–337.

Murdoch, J. (2006). *Poststructuralist geography: A guide to relational space*. London: Sage.

National Severe Storms Laboratory (NSSL). (2011). *VORTEX2 background*. http://www.nssl.noaa.gov/projects/vortex2/background.php. Accessed 12 Apr 2011.

NOAA. (2011a). What is skywarn? http://www.nws.noaa.gov/skywarn/. Accessed 20 Nov 2011.

NOAA. (2011b). Weather spotter's field guide. http://www.nws.noaa.gov/om/brochures/SGJune6-11.pdf. Accessed 20 Nov 2011.

NOAA. (2011c). America's weather enterprise: Protecting lives, livelihoods, and your way of life. http://www.weather.gov/om/brochures/Citizen_Scientist.pdf. Accessed 11 June 2011.

NOAA. (2011d). Skywarn Storm Spotters. http://www.arrl.org/files/file/Media%20&%20PR/EmergencyRadio_org/Skywarn.pdf. Accessed 15 June 2011.

Palmer, M. (2009). Engaging with indigital geographic information networks. *Futures, 41*, 33–40.

Palmer, M. (2012). Cartographic encounters at the BIA GIS center of calculation. *American Indian Culture and Research Journal*, (forthcoming).

Palmer, M. H., Stevenson, S., & Zaras, D. S. (2000, Jan). *Student evaluations of the Oklahoma Weather Center REU Program: 1995, 1998, and 1999*. Ninth symposium on education, Long Beach, CA. American Meteorological Society, pp. 24–27.

Panelli, R. (2010). More-than-human social geographies: Posthuman and other possibilities. *Progress in Human Geography, 34*(1), 79–87.

Robertson, D. (1999). Beyond twister: A geography of recreational storm chasing on the southern plains. *Geographical Review, 89*(4), 533–553.

Smith, R. G. (2003). World city actor-networks. *Progress in Human Geography, 27*(1), 25–44.

Spotter Network. (2011). http://www.spotternetwork.org. Accessed 15 Apr 2011.

Thrift, N. J. (1996). *Spatial formations*. London: Sage.

Thrift, N. J. (2000). Actor-network theory. In R. J. Johnson, D. Gregory, G. Pratt, & M. Watts (Eds.), *Dictionary of human geography* (pp. 4–6). Oxford: Blackwell.

Vasquez, T. (2008). *Storm chasing handbook* (2nd ed.). Garland: Weather Graphics Technologies.

Williams, S. (2007). Application for GIS specialist meeting. http://www.ncgia.ucsb.edu/projects/vgi/participants.html. Accessed 6 Sept 2010.

Zook, M., & Graham, M. (2007a). The creative reconstruction of the internet: Google and the privatization of cyberspace and DigiPlace. *Geoforum, 38*, 1322–1343.

Zook, M., & Graham, M. (2007b). Mapping DigiPlace: Geocoded internet data and the representation of place. *Environment and Planning B: Planning and Design, 34*, 466–482.

Chapter 17
VGI as a Compilation Tool for Navigation Map Databases

Michael W. Dobson

Abstract Volunteered geographic information, a crowdsourced approach to gathering geographic information, is being used in numerous map database compilation systems. Active and passive contribution systems exist, but it is the active systems, through which contributors can provide their personal local knowledge, that hold the greatest promise for improving the quality of spatial databases used for navigation and location-based services. Both open and hybrid map compilation systems have been developed in an attempt to benefit from VGI. We discuss the nature, limitations, and advantages of a range of crowdsourced compilation systems in an attempt to evaluate the influence of VGI in helping to improve various aspects of data quality.

17.1 Introduction

The quest for improved data quality in spatial databases used for mapping and navigation remains a challenging objective. Map, point of interest (POI), and business listings databases often fail to meet the quality standards mandated by their users. Data quality problems in these databases fall into categories that can be identified as completeness, logical consistency, positional accuracy, temporal accuracy, and thematic accuracy (International Organization for Standardization, ISO 2002).

It is clear that most data-quality problems in spatial databases result from compilation processes used to collect and maintain their content. Although the tools, techniques, and theory of map compilation have advanced in recent years, we have replaced older method-induced errors with other errors related to new methods, such as database synchronization, conflation, and overreliance on sophisticated

M.W. Dobson (✉)
TeleMapics LLC, Laguna Hills, CA, USA
e-mail: mwdobson@telemapics.com

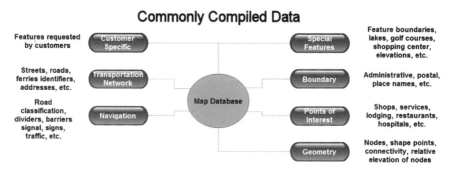

Fig. 17.1 The amount of data that may need to be collected to produce modern, navigable map databases is staggering. Current coverage demands are stressing our ability to collect these data in a manner that enhances data quality (Image courtesy of TeleMapics LLC)

software systems whose use often confounds issues of logical consistency. However, one critical, recurring weakness in compilation systems used to collect, categorize, and prepare spatial data for use lies in our inability to leverage local spatial knowledge to the advantage of data quality.

Compilation efforts related to spatial data used for mapping, navigation, and location services have always suffered from the lack of resources that can be allocated to collecting, curating, and organizing these data. Numerous categories of data are collected, and each of these requires detailed collection strategies to produce an accurate, comprehensive, and up-to-date database (Fig. 17.1). See Chap. 14 by Coleman (2012) in this book for a comprehensive evaluation of VGI and its use in conventional topographic base-mapping programs. While most feature categories have been standardized (see, e.g., geographic data file (GDF) now ISO14825: ISO 2004), unique categories of data may be requested by licensed customers or, in some cases, by influential user groups.

As major online companies such as Google and Microsoft, as well as the providers of navigation map databases such as NAVTEQ and TomTom (formerly Tele Atlas), attempt to extend their mapping products to include world coverage, the problem of allocating resources to collect the required data will become untenable. It is possible, however, that this problem can benefit from the tactical use of volunteered geographic information that involves the gathering of spatial information through a form of mass collaboration commonly known as crowdsourcing and sometimes referred to as Wikinomics (Tapscott and Williams 2008) or VGI (Goodchild 2007).

17.2 VGI in Respect to Map Database Compilation

Crowdsourcing is an architecture of participation that is based on volunteering, sharing (social networking), cooperation, and collective action. Some of its benefits are that it does not require coordinated management, an operational budget, an overarching business structure, or that its data be prepared in respect to specific use cases. Crowdsourcing attempts to leverage the wisdom of crowds through the use of software applications that help to self-synchronize otherwise latent groups to accomplish previously unattainable objectives.[1] Note that many people are interested in maps and consider themselves casual geographers or cartographers and have the desire to contribute data that could be used to improve the data quality of map databases. Before the Internet and the associated development of software to self-synchronize communities with similar interests, these potential map editors did not have access to systems that allowed them to contribute to map databases in a direct and satisfactory manner.

VGI is about finding ways to discover, incentivize, elicit, and use information that online communities interested in maps and mapping possess about local geographical space. Our interest here is in crowdsourcing, focused on the collection of spatial data to augment, update, or extend map databases. It is a process that allows those interested in improving the quality of maps to contribute data or edit maps based on their knowledge of local circumstances. What is important here is that the applications provide contributors with access to a database and accept community edits to that database but do not require significant training in map or data compilation to become a contributor to these systems (Heipke 2010).

17.2.1 How Is Crowdsourcing Applied to Map Compilation?

Current practice involves two types of crowdsourcing for map compilation purposes. These processes are categorized as passive community input and active community input.

17.2.1.1 Passive Community Input

Passive community input involves the use of probe data automatically gathered from user devices, such as GPS receivers or similar location technology, used to record the path of users (or their vehicle) during their daily journeys. This "probe" or "floating point data" records the geometry of the user's path, elevation, and speed of movement. These recordings can be a very practical method of measuring traffic, as well as vehicle-based congestion. Other data, such as speed limit, signal and stop sign location, and turn restrictions, can be inferred from interrogating the path and behavior of the probe as it moves between origin and destination.

These passive data are normally collected on an anonymous basis, and the companies collecting these data often, but not always, require the users to agree to

---

[1] See Surowiecki (2005) for an impressive essay on the "Wisdom of Crowds" and Shirky (2008) for a fascinating discussion of the crowd and its self-synchronization.

have their probe data collected. In addition, some companies take additional steps to assure confidentiality, such as dropping the first and last 2 min of the path over concerns related to identifying residences, work location, or other destinations that might reveal details related to an individual's identity.

Overall, the passive contributions have provided data in volumes previously unavailable in map compilation efforts. For example, users agreeing to participate in TomTom's Map Share tracking program contributed over two trillion GPS points within the first 2 years the program was in existence. Currently, TomTom collects over three billion new measurements each day (TomTom 2011). According to TomTom representatives, aggregating this massive number of data points helped them average out data collection errors and enhance their navigation database with an extremely high level of positional accuracy (Dobson 2010a). A further advantage of passive community data is that it rarely contains "data spam" that is purposefully contributed to crowdsourced systems as a method of degrading the usefulness of the database. What little data spam enters the system through passive contributions can usually be discovered through algorithms tuned to discover, compare, and evaluate logical inconsistencies in the contributed data.

Passive community input, however, is effectively blind to many of the attributes whose collection is a required part of the map compilation process. Attributes such as street names, route numbers, addresses, dividers, water features, points of interest, and various types of boundaries cannot be discovered through the collection of passive probe data. Instead, the collection of these types of data attributes benefits from the use of active community input.

17.2.1.2 Active Community Input

Those who actively contribute are the primary source of detailed local knowledge in crowdsourced map compilation systems, and these inputs potentially provide a significant source of improvement to data quality in these types of systems (Heipke 2010). In theory, the process of evoking active community input relies on the goodwill of a user who decides to directly contribute compilation information by using an online system to submit edits or map data that might resolve a mapping problem experienced by them while using these data in a local area. The chapter by Harvey (2012) in this book raises important questions about potential biases in these data related to the form of the agreement under which these data are contributed.

In some active systems, the contributors may lack the relevant local knowledge to resolve the issue and instead digitize the geometry of streets and roads in these areas referencing satellite imagery. The assumption is that local users will later add the attributes to these digitized strings that are necessary to flesh out the database. Contributors who digitize streets provide labor, but do not contribute the attribute data that converts a maze of lines into a functional map database. Unfortunately, crowdsourced compilation systems depend on the success with which they can attract contributors who are able and willing to provide data or edits based on their personal local knowledge of an area, rather than those contributors providing labor in the form of digitizing services.

There are at least two types of active contributed observations of interest for map compilation: "direct" and "indirect." For the purpose of crowdsourced systems used for map compilation, direct observation means that the person contributing local knowledge about a place is either at the location at the time of the contribution; has visited the location in order to gather notes, photos, and data for the contribution; or has past familiarity with that location (i.e., in the past, has worked there, lived there, or frequented the area for some purpose and created a contribution based on this memory). Indirect observation relies on the analysis of imagery and other sensed location data to extract road and related attribute data for purposes of creating or updating map compilations. This is a "shallow" task that requires little or no local knowledge to complete and may be error prone due to lack of on-the-ground familiarity with local geography.

In general, observations made by persons presently or recently at the location for which they are contributing map data are more likely to be accurate than observations contributed by those who have familiarity with an area but are relying on memory. Similarly, it is generally true that a person who is at the location for which he or she is contributing data is more likely to provide higher-accuracy input than someone who is relying on memory supplemented by satellite or aerial imagery of the location. A person who is not at the location but viewing "street-side" imagery (such as that provided by Google, NAVTEQ, or Earthmine) might be able to use this type of imagery and attribute data derived from the imagery to provide higher-accuracy updates than a person actually at the location. The accuracy of imagery-based inputs would depend on the skill of the analyst and the quality, comprehensiveness, and capture date of the imagery and associated sensor output for the location being interrogated. Given the capabilities of current compilation systems, capturing the details of local information through contributions from users who have direct and recent personal local knowledge of these areas may be a preferred method that is accurate in respect to position, currency, and theme.

17.3 Types of Crowdsourced Systems with Active Contributions

Active crowdsourcing is most often used in open and hybrid compilation systems. Open systems used for map compilation are nonproprietary and trust their users to contribute the types of spatial data they desire, as well as to edit their works or the works of others without restriction. In addition, open systems make their data freely available for use, with certain restrictions on derived works that the community feels are required to preserve the integrity and value of the parent database. On the other hand, hybrid systems, except for those employed by some government agencies, are generally proprietary, allow their crowdsourcing system to funnel edits and contributions only for limited types of data, claim rights of ownership in and to the data, and contractually restrict most uses of their databases. In addition and most importantly for this discussion, hybrid compilation systems often ingest crowdsourced edits but include further rigorous analysis before these inputs are committed to the database. Hybrid compilation systems are so named because they mix crowdsourcing with traditional compilation techniques that may include other data sources (both licensed and unlicensed), as well as the use of field survey vehicles equipped with leading-edge technology capable of capturing road and street data and its attributes with extreme levels of accuracy.

An open map database compilation system, such as OpenStreetMap (OSM), accepts all contributions by its users, pushing the edits in an unchanged and unevaluated state to the live instance of their database. Those who contribute data edits to open systems are regarded as trusted, and their input is readily accepted for publishing, although it is subject to future revision by other members of the crowd. This behavioral approach is based on the assumption that more local eyes help reveal map edits or augmentations that are considered spam, erroneous, incomplete, or unsatisfactory to other users for some reason. Open crowdsourced databases are considered to be self-healing over time. The speed with which the changes can be published and evaluated by other participants is critical for potentially increasing the data quality resulting from crowdsourced compilation. New edits can be reviewed and edited by others, or a previous instance of the data for the location in question can be reinstated if the posted corrections are rejected during communal evaluation. For example, MapQuest, in its use of OSM data for some of its online navigation websites, publishes specific types of updates, originally made to the master OSM database, to its local instance of the database, in a matter of minutes. More complex types of data may require additional processing by MapQuest (e.g., those related to routing attributes) but are usually posted in less than a day, allowing for immediate review by others in the community (Dobson 2010b).

Hybrid systems mix crowdsourcing with traditional map database compilation techniques. The operational difference is that the contributions from crowdsourcing may be used only as indicators of change and signal the system management module that further research needs to be undertaken. Public contributors to these systems may not be regarded as trusted but merely considered as another valuable source of data. These data are then researched further before being accepted and pushed to a live database which has passed the quality-assurance procedures required for distribution.

For example, the Swiss Federal Office of Topography (swisstopo) has integrated the concept of crowdsourcing into their system of national databases known as the topographic landscape model and digital cartographic model (Guélat 2009). Swisstopo appears to have taken the view that crowdsourcing provides a new source of content for upgrading their national databases, as well as the possibility of opening a new form of relationships with their customers. Their system allows the uploading of GPS tracks, attachments, notifications of errors, surfaces, and lines. This crowdsourced input is channeled through an intranet connected with a central revision layer whose dispatcher distributes the comments to the appropriate databases. The crowdsourced data is considered to be a source of change detection that may or may not require field investigation before acceptance into the national databases.

In the commercial arena, Google, through its Google Maps platform, is the leading user of a hybrid compilation that features active crowdsourcing. Google's initial online mapping effort, an attempt to support its local advertising business, was based on navigable map databases the company licensed from NAVTEQ and Tele Atlas, as well as a number of minor suppliers. In part, the transformation of Google to a map compiler rather than licensor was related to its concern that the commercial databases it licensed did not, in all cases, meet the company's need for accuracy or coverage, nor were they up-to-date[2].

Google was an early adopter of crowdsourcing. Google Map Maker was developed to help the company provide map data in countries where its map suppliers were unable to provide coverage. The Map Maker application allowed registered users to create and update maps of areas of personal interest to them. Google's agreement with its users gives the company a license in and to the data that its users create. The company then upgraded the maps originally created in Map Maker with traditional map compilation technology to create its own, proprietary map base. Google quality-engineered the data contributions of their users (by updating them with fine-resolution imagery), conflated the data with that from authoritative geographic data sources, tested the transportation networks on the map for connectivity, and then published the results in Google Maps, its corporate map franchise for consumer use.

In 2009 Google replaced the map data for the United States and Canada from NAVTEQ and Tele Atlas with coverage compiled and created by Google. The majority of these data have been created by Google in a traditional compilation process and not through crowdsourcing. Google has strategically added a "report a problem" button on the bottom of Google Maps that allows users to suggest changes for Google Maps and also to use Google Map Maker as a tool for contributing edits on a limited number of data types for which the company desires crowdsourced input.

TomTom (formerly Tele Atlas) has achieved considerable success in crowdsourcing through its use of the Map Share program that has become an integral part of the TomTom personal navigation device (PND) experience. However, TomTom has chosen to focus on its collection of passive probe data at the expense of actively contributed data, possibly due to financial issues (Dobson 2010a).

Before its 2008 acquisition by TomTom, Tele Atlas compiled their database using traditional tools. Their compilation efforts surveyed information from over 50,000 data resources including aerial photos, tax data, government partners, utility, fleet and postal drivers, strategic partners, proprietary Web crawlers, their fleet of mobile survey vehicles, and end-users (Dobson 2007). Tele Atlas used a two-pronged approach to creating their database. They compiled what they could from other sources and drove streets and roads (mainly using sophisticated, mobile sensing platforms) to verify information that cannot be provided through source compilation. Their belief at the time was that they needed to mix fieldwork with other standard compilation techniques to create the data needed for a navigation-quality database. At that time, the company was receiving only 600 customer error correction forms a week (using an online input gathering tool), but the time it took to research, confirm, and edit the reports was 45 days (Dobson 2007). The post-acquisition use

---

[2] During his keynote address at Interactive Local Media (ILM) 2007, John Hanke, director, Google Earth and Maps for Google, while discussing the quality of the data provided by Google's map data suppliers, looked up at the audience and said, "You have no idea how bad these data are."

of Map Share dramatically increased the number of active community inputs yet helped to decrease the time required to research active edits significantly since each was transmitted in a controlled format, tagged with a GPS coordinate, and displayed on a map image of the area recorded at the time the user touched the screen of the TomTom GPS to initiate the error correction process.

NAVTEQ, similar to TomTom, has been very successful in collecting sources for its inventory of probe data from its many partners in the transportation and navigation markets. It will likely benefit from additional probe data through the various map programs residing on many Nokia phones. Conversely, NAVTEQ has been less successful in the area of active crowdsourcing, as it collects insignificant numbers of contributions through its online Map Reporter interface and currently lacks an alternative, popular, customer-facing interface that could be used to attract active crowdsourced map data.

It is important to note that Google, NAVTEQ, and TomTom limit the types of edits that their users are allowed to contribute. As opposed to Google, NAVTEQ and TomTom tend to use active community data as measures of change detection. In essence, receipt of an edit sparks research into the accuracy of the edit, and a positive evaluation, at this stage, initiates a search for an authoritative or trusted source that could validate the change information[3]. In the rare case that the situation cannot be confirmed by data mining, the location is queued for examination by a field representative or a field vehicle during other research in the area. The distinction in method is important, as NAVTEQ's and TomTom's customers regard the data these companies license as authoritative and able to meet the specifications required by their applications. While TomTom and NAVTEQ have embraced crowdsourcing, they have not embraced the integrity of the crowd, preferring instead to evaluate the data on their terms, although this practice does not impugn the potential reliability of active crowdsourced data.

Crowdsourcing, in its various guises, has been embraced by major and minor players in the world of mapping, who desire to compile quality map databases. It is unlikely that these implementations of crowdsourcing have been embraced without considerable evaluation, but is the application of crowdsourcing/VGI a panacea that will finally bring us improved map quality?

17.4 Does Crowdsourced Map Compilation Work as Advertised?

Suroweicki (2005) indicated that "wise crowds" possess diversity of opinion, independence, decentralization, and aggregation (i.e., they have found a method that allows them to collaborate). When these characteristics exist, the crowd's judgment, according to Suroweicki, is likely to be accurate. For example, although each person's

---

[3] See David Stage (2009) for an interesting distinction between "authoritative" and "trusted" data sources. Stage considers those who collect and distribute authoritative data to be trusted.

evaluation of specific map data is likely to contain both information and error, group evaluation of contributed map edits (aggregation) in crowdsourced systems is thought to reduce error and to create a map database that over time should become more accurate. Attaining this goal depends on the diversity of opinion, independence, and decentralization in the crowdsourcing population, as well as the influence of the method used for soliciting the contributions. In essence, if the crowd cannot satisfy these conditions, its judgments are unlikely to be accurate. These general rules for the efficacy of crowdsourcing are useful for evaluating best practices related to whether crowdsourcing systems are unbiased and likely to be successful in meeting the goals set for them by the contributors. However, directional measures such as these do not really assess how accurate or erroneous crowdsourced systems might be in respect to specific uses of these data.

Some have suggested detailed methods for analyzing these data. For example, the spatial data quality indicators recommended by Van Exel et al. (2010) would focus on user quality, feature quality, and their interdependence as a method of evaluating crowd quality in crowdsourced map compilation activities. Their initial recommendations for using local knowledge, experience, and recognition to evaluate user quality will be very difficult to quantify in any meaningful sense. In addition, their approach to "feature quality" involves measures of lineage, positional accuracy, and semantic accuracy, although the authors admit these elements are "… not considered consistent for crowdsourced data[4]."

Conversely, efforts at approaching the accuracy of crowdsourced systems through measures of fitness of use are ill advised. Defining use for real-world applications is a complex and error-prone process that is often invalidated by the lack of understanding of the legal and financial ramifications of qualifying data as capable of supporting a specific task.

Detailed research into the accuracy of crowdsourced map compilations remains in its infancy due in part to the novelty of VGI and also due to problems with approaching the measurement of the resultant data. Van Oort (2005) presented a comprehensive description of data quality standards, but a comprehensive analysis of the various quality elements has not yet been conducted on a major crowdsourced compilation system. The works of Haklay (2008, 2010), Girres and Touya (2010), and Zielstra and Zipf (2010), focused on OpenStreetMap and its positional accuracy and completeness, have moved us toward more comprehensive analyses of crowdsourced compilation systems.

17.4.1 How Does Crowdsourcing of Map Data in Open and Hybrid Map Compilation Systems Measure Up?

We focus first on OSM. Its crowdsourcing experience has been a successful demonstration of the ability to compile a map database that is popular, well regarded, and

---

[4]See Van Exel et al. (2010), p. 3.

freely available under an open license. However, the notion that an open crowdsourced map database measures up in terms of data quality may be misleading. Many contributors to pure crowdsourced systems for map compilation, such as OSM, place only an informal emphasis on data specifications. Specifications may be lacking, fuzzy, considered too restrictive, regarded as too complex, and ignored by many of the contributors to the system. Girres and Touya (2010) discuss the problematic aspects of the heterogeneity of OSM data for France and conclude that the logical inconsistencies in the data impede its use, other than for simple mapping (as opposed to use in a GIS).

Some might argue that all spatial databases are heterogeneous with respect to data quality. For some aspects of data quality, this is undoubtedly true. Conversely, the traditional compilation procedures underpinning commercial map database compilation efforts have a record of producing databases suitable for navigation, location-based services, GIS, and other applications that require a significantly higher bar than that met by OSM. Requirement to update commercial databases is, in many cases, contractual. This distributes attention to correcting errors in areas on a systematic basis, as opposed to error correction based on personal interest, as is the practice in open crowdsourced compilation systems. Perhaps more interesting is the question of whether or not the crowdsourced input used in open VGI map compilation systems consists of contributions that are independent of prejudice and not influenced by groupthink while truly reflecting decentralized inputs of local knowledge.

From the perspective of theory, the success of crowdsourced map compilation systems for country or multi-country areas depends on the contributions of a relatively large number of participants distributed across the extent of the domain. Further, the basic notion that "With more eyes, map errors become shallow" (apologies to Linus Torvalds) requires some degree of spatial redundancy in terms of the distribution of volunteers. One person contributing map data alone for a local area does not provide the iterative approach to data evaluation or the editing that is crucial to producing acceptable data quality in open crowdsourced map compilation systems. The need to update map data over time reflects the unending nature of spatial change and adds to the continual need for new contributors. Finally, the notion that these open map databases are self-healing over time (errors will only be corrected when someone notices) requires a constant flow of new contributors to update the database coverage, as well as to replace those contributors who have lost interest or become inactive.

By January of 2010, only 6 years from its founding, OSM had attracted over 200,000 contributors (Zielstra and Zipf 2010), but it appears that volunteering for crowdsourcing and contributing on a frequent basis may be the exception rather than the rule. The research of Budhathoki et al. (2010), based on data collected in 2009, indicated that the vast majority of contributors to OSM did so only once (44%), with the number of one-time contributors lowest in Europe (approximately 42%) and hovering near or above 60% in Africa, Asia, North America, and South America. Perhaps more troubling from a local knowledge perspective, Haklay (2008) provided a graph indicating that over 70% of the data then in the OSM coverage of England was the work of approximately 50 contributors, a trend that appears in the continental areas presented in the research of Budhathoki et al. (2010), where a small number of people (0.6% globally) appear to have dominated the process with each contributing over 100,000 nodes.

In addition, Budhathoki et al. (2010) provide an interesting portrait of the OSM contributor. At the time of the study (2009 data), males comprised 96% of the mappers, of whom 64% were under the age of 40. Most contributors were highly educated (49% had college degrees), while 21% had college and postgraduate degrees and another 8% had doctoral degrees. Although the sample size was modest ($N=426$), the results may indicate that the OSM crowd is significantly unlike the general population in terms of sex ratio, education, or age distribution. While it is natural that the generation that has grown up with the Internet would be attracted to OSM and crowdsourcing, it is equally clear that this group may not reflect the independence and diversity of opinion suggested by Suroweicki (2005) that constitutes an "intelligent" crowd.

Yogi Berra, the famous American baseball player known for his folk wisdom, is reputed to have said "You can see a lot just by looking." A visual examination of the OpenStreetMap database indicates that while the database has an amazing degree of coverage, the coverage is neither complete nor is it comprehensive. The analysis of Zielstra and Zipf (2010), in their important study of OSM in Germany, indicated that map coverage varied significantly between urban and rural areas, as well as between large cities and smaller cities. Haklay (2010) noted the same types of discrepancies in the OSM database of England, and similar discrepancies were noted in the OSM database for France by Girres and Touya (2010). These types of coverage limitations raise questions of the widespread applicability of crowdsourced data when it is used for map compilation in open systems.

It is unclear whether those who do contribute data to OSM focus their contributions on local knowledge. Neis et al. (2012) note that the relatively high number of streets in the OSM database of Germany lacking a name or a route designation may result from these transportation links being digitized from satellite images by users who were lacking local knowledge. Other research by Haklay et al. (2010) shows that numerous contributors evaluate and edit the same instances of spatial information but do so due to differences of opinion on the positional accuracy of previously contributed work.

Might the division of labor in OSM's database result in variations in data quality and completeness that inhibit the use of these data other than across limited urban areas, where there are numerous contributors willing to volunteer local data? If the majority of OSM contributors are spending their time digitizing imagery for the database, as opposed to contributing GPS traces and attributes from paths along which they have traveled or know something about, how likely is it that the OSM effort benefits from local knowledge to the same extent that it benefits from free digitizing?

Haklay (2010), in a study focused on OSM coverage in London, indicated that there is a social bias in the contributions to the OSM database that is based on affluence and reflects a participation gap. Obviously one needs a computer, an Internet connection, and plenty of spare time to digitize road networks, suggesting an economic privilege that is not available to all members of society. The finding suggests that the distribution of those able to contribute local knowledge to OSM may not mirror the area of coverage targeted for the database. If this limiting scenario is a possibility, then who will attribute data in the areas not populated by those who fit the profile of OSM contributors? Other research by Girres and Touya (2010) indicated that data posted to the OSM database of France may be focused on captured objects of interest to the contributors at the expense of the comprehensiveness of the OSM database.

It may be that the self-synchronization of OSM contributors masks what is an intended, but fatal, lack of direction impeding the comprehensive edit and update of the OSM database. For instance, there is no reliable way to redirect the spatial focus of the existing group of contributors, since the system, by definition, is self-organizing. Further, it is unknown how long a period of time it will take an open database to reach a level of completion for large coverage areas that is comparable in quality to the databases produced by the leading commercial firms or by government agencies. Nor is it possible to know how long it will take to correct an error in the database, since this depends on whether a contributor is interested enough in the location to edit it. Finally, the lack of data standardization and effective quality control adds another layer of complexity to the problem of analyzing the completeness of open crowdsourced databases.

Supporters of the open compilation model would argue that because there are no specifications for the database, there is also no such thing as a completed work. While all spatial databases are incomplete, or inaccurate to some degree at any given point in time, it is difficult to appreciate the purpose of building a specification-less database that defies any attempt to manage its contributors to produce a database with some level of uniformity and completeness.

OSM was developed, in part, because of the dissatisfaction with the high cost and punitive licensing restrictions associated with commercial spatial databases and those of some governments. If OSM was founded to remedy this dissatisfaction, one might reasonably presume that the goal in building the OSM database should to be to endow it with functionality and data quality similar to, if not better than, that which it was designed to replace.

Two of the supposed advantages of crowdsourced data compilation systems are that they have no formal management structure (or the associated expenses) and that they lack an overarching business structure. The lack of both may, in fact, be a limitation when applied to map compilation.

One of the advantages possessed by commercial mapping firms is that they have customers who are willing to use their data based on their belief that it meets the customers' need for a specific level of data quality that supports a particular application. Meeting that expectation requires the setting of standards for data and the implementation of a data compilation system capable of satisfying these proposed uses of the database. Ensuring that the data meets customers' needs is the responsibility of marketing and editorial teams tasked with ensuring that customers' needs are discovered, established as compilation goals, and engineered into the database standards.

Government agencies that compile spatial data do so using procedures designed to meet the data requirement described in legislation that defines the mission of the agency and requires the agency to collect data in compliance with the specified directives. Open crowdsourced systems appear to lack the processes, checks, and balances common in other map database compilation efforts. This lack of structure may contribute to the relative absence of logical consistency in open databases. As a consequence, it appears that the lack of structure in the compilation process is a major factor impeding the success of all attempts at open map compilation using crowdsourcing.

The issues described above lead to the conclusion that OSM may fail in its attempt to create a passably uniform global database. It is likely that VGI-based compilation efforts will continue to struggle to complete sustainable databases in locations other than major metropolitan areas, except perhaps in Germany, and possibly England[5]. Perhaps the open model for crowdsourced map compilation will be effective only in certain places. If so, then there appears to be a potential role for a hybrid compilation system that includes crowdsourcing in the areas neglected by open systems. It seems logical that the use of VGI and crowdsourcing for map compilation should be an attractive notion for commercial map database providers.

17.4.2 What Is the Status Quo of Commercial Efforts to Use Crowdsourcing?

Commercial mapping firms have several advantages that do not accrue to open crowdsourced compilation efforts. Foremost among these is the dependence of these companies on successful marketing and licensing of a product that meets the acceptance criteria of their customers. NAVTEQ, for example, spent a significant sum on database creation and distribution in 2007 but licensed their database for the sum of \$853,287,000 during the same period (NAVTEQ 2008)[6]. In some sense, the willingness of customers to pay for the ability of the NAVTEQ database to meet their product needs is a surrogate measure of the authority and trust with which the NAVTEQ database is regarded.

Most commercial map database firms command some degree of authority and trust based on their willingness to set data quality standards, collect data reflecting these standards, institute discipline in correcting errors (verification and editing), establish targeted cyclical update programs, and demonstrate that errors discovered by customers will be researched, corrected, and pushed into their distribution network in a reasonable amount of time (Fig. 17.2).

In compilation systems used by commercial firms, data quality is the result of data collection according to specifications that are adhered to by teams trained in compilation and data collection techniques, and verified by professional data-quality editors who inspect and evaluate these data. Data that do not meet the standard are flagged during the compilation process and either immediately re-collected or placed in the formal update queue. Most companies institute an ongoing data

---

[5]The progress of OSM in developing a crowdsourced database in the United States, a country with a rich history of publicly available, low-cost, nonproprietary spatial databases, appears to be less impressive than in other areas of the world.

[6]2007 was the last year for which publicly available financial data was available for NAVTEQ. Due to its acquisition by Nokia in 2008, NAVTEQ ceased to have public responsibility to report on its financial health.

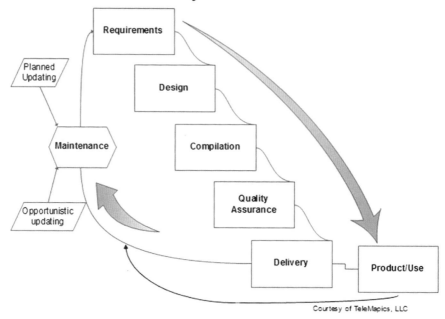

Fig. 17.2 The update cycle used by commercial firms and many government agencies often experiences a beneficial effect related to the use of the database product by customers or constituents. These external factors often help sharpen the research focus and agendas related to the targeting of map updating (Image courtesy of TeleMapics LLC)

sensitivity process where areas suspected of significant spatial changes are flagged for verification and update, if required. Other changes in data quality are found by systematically working research teams through all coverage areas over time, based either on an established production cycle or on estimates of the rates of data change in particular areas. The overall assignment of compilation activities is designed to maximize the time value of money while increasing the integrity of the database. It is this attempt to actively harmonize the information in the database that distinguishes commercial database compilation efforts from the relatively unstructured efforts that surround the update of open systems dependent on crowdsourcing for an iterative data evaluation, collection, and posting cycle.

While commercial databases might be more reliable and have more uniformity in terms of data compiled using traditional compilation techniques, even with the benefits of significant spending, these databases often are out of date, inaccurate, noncomprehensive, of variable quality, and possibly too expensive to maintain. NAVTEQ (2008), for example, in their 2007 10-K Annual Report to the Security and Exchange Commission of the United States, revealed that their database creation and distribution cost for the year 2007 was $395,778,000, an increase of over $120 million from the prior year (p. 36). Even with this massive expenditure, it appears that the NAVTEQ database is not without its weaknesses (Dobson 2010c, d).

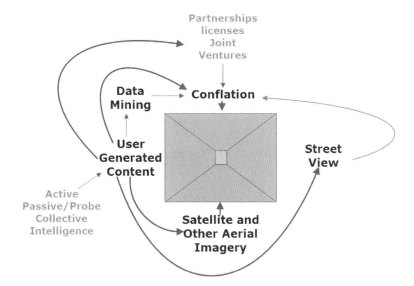

Fig. 17.3 Google's compilation is a classic example of the hybrid process. It fuses traditional compilation techniques with crowdsourcing through the use of active and passive community inputs (Image courtesy TeleMapics, LLC)

However, the advantages of the commercial approach to compilation combined with active crowdsourcing in a hybrid compilation system could be deployed to improve data quality in commercial navigation databases. The focus on this hybridization in commercial environments should be on methods for incentivizing users to contribute information on errors and omissions based on their personal, local knowledge. It is likely that commercial firms would be interested in using crowdsourcing as a source of local knowledge related to map attributes that might be incorrect or incomplete, as well as information on recent developments postdating their most recent data sweep of the area. While open crowdsourced systems could adapt the commercial compilation techniques, I believe that there currently are too many cultural/processual management barriers for this to happen.

Google, through Google Maps, is currently the leading example of the hybrid approach to map compilation, using active crowdsourcing for both the original build of its map database, as well as the revision of its database. As shown in Fig. 17.3, Google collects input based on partnerships with authoritative sources of spatial data, as well as ventures with other map data suppliers. It combines these data with crowdsourcing (both active and passive contributions), imagery (satellite, aerial, and Street View), augmented by data mining and conflation to produce an effective fusion of sources to create the database for Google Maps.

It is likely that Google's original database build process (Fig. 17.4) was imagery driven and leveraged crowdsourcing coupled with data mining and the conflation of authoritative data sources to create its original database. In the revision process (Fig. 17.5), however, it appears that crowdsourcing is the key input, queuing up the areas in which updated imagery and authoritative data are required to improve the quality of the database.

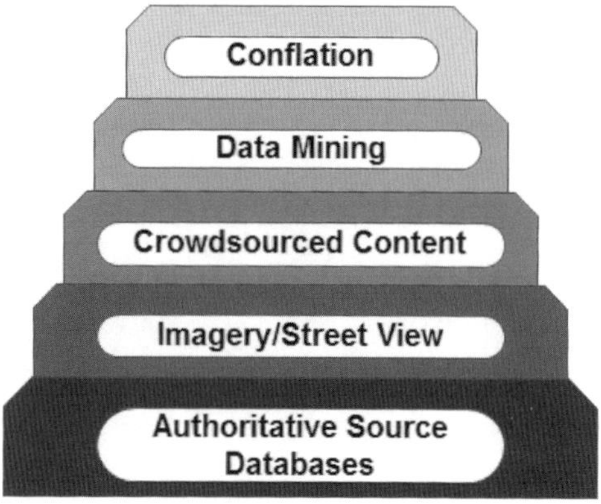

Fig. 17.4 The build process used by Google to create their map database of the United States relied on traditional compilation techniques more than on crowdsourcing (Image courtesy of TeleMapics LLC)

Fig. 17.5 Google relies on VGI/crowdsourcing as the prime revision and change-detection tool for their US database (Image courtesy of TeleMapics LLC)

Crowdsourcing was a relatively unimportant consideration during the build of the Google Maps proprietary database of the United States (Fig. 17.4). Database sources that Google considered authoritative on national and local scales (federal agencies for national coverage augmented by authoritative or trusted state, regional, and city sources) were acquired and conflated to provide a base that could be updated by reference to current imagery (satellite, aerial, and Street View). Since the Google model for the United States portion of its database was to control the geometry, connectivity, and data attribution for the initial database build, crowdsourcing played a limited, indirect role in the initial development effort. It is likely that the main use of crowdsourcing was to target areas identified in customer complaints as being of low quality that related the company's previous use of commercial map databases. Data mining was used to search the Internet for additional information on the geography of interest. This included the mining of Google's own search click-stream to unearth potentially relevant information for augmenting the database. Conflation processes were used to relate and integrate the various spatial data collected.

It is Google's revision process for the US database where crowdsourcing is used as the main update agent, principally through the use of Google Map Maker, although this may not have been the company's original intent (Fig. 17.5).

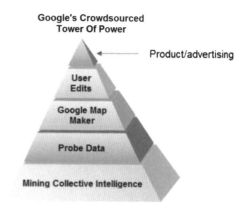

Fig. 17.6 Google's use of both active and passive community updates to revise its own maps is a critical aspect of its success in the local advertising market and a fundamental aspect of its strategy for keeping its maps up-to-date in local communities (Image courtesy of TeleMapics LLC)

In terms of crowdsourcing, Google benefits from its enormous reach and its popularity as a search engine, which serve to attract users to Google Maps and Map Maker. It is the classic case of more eyes helping to reduce error, and Google can attract more of these than any other commercial map database provider. This is quite fortunate for Google, since the initial quality of its US database was quite low (Dobson 2009).

Originally Google attempted to resolve this problem through the development of new algorithms, but they soon changed their approach (Dobson 2010e). Google realized that improving the quality of their US database would require analysis by those with the beneficial local knowledge required to correct errors that were too elusive or confounded to be discovered and remedied by software methods. As a consequence, in 2011 the company turned to crowdsourcing/VGI through the use of Google Map Maker as a method of soliciting users of Google Maps to provide critical local detail to Google's database (Fig 17.6; Dobson 2011c).

Access to the Map Maker functionality is provided to those willing to correct the Google Maps' US map. However, instead of the peer review process which is used by Google in many countries, the contributions of users in the United States who edit the map are evaluated by a team of "trusted reviewers" who are Google employees. Google's "trusted reviewers" appear to have limited knowledge on which to base their evaluation of the edits for local geographies. For instance, these reviewers are not located in the United States. In addition, each of Google's reviewers not only evaluates user data contributions across the entire United States but also edits user data contributions for numerous countries around the world (Dobson 2011b). At some time in the future, Map Maker contributions for the US map may be evaluated by the peer community of contributors located in and familiar with local detail across the United States. For now, Google appears unwilling to relinquish its editorial purview and in doing so may impede the degree to which local knowledge can be used to improve their Google Maps database in the United States.

Unlike OpenStreetMap where contributors can edit any aspect of the map database, contributors using Google's Map Maker are allowed to edit only select features and, in some cases, only certain aspects of select features (Dobson 2011a). For example,

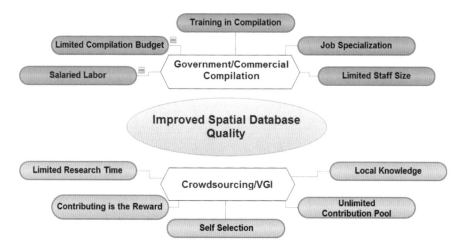

Fig. 17.7 Both VGI and traditional compilation systems are influenced by human factors. Blending these differences is required to produce an effective hybrid compilation system (Image courtesy of TeleMapics LLC)

incorrect boundary information describing the jurisdiction within which the feature is located is not editable by contributors, nor is some information related to street geometry. In effect, Google has decided the categories of edits that it will allow and reserves other data categories for editing based on data compiled from sources Google considers more authoritative than local users. Google has likely taken these steps in an attempt to ensure logical consistency within their database. Regardless of the limitations in its approach, Google appears to be doing a better job of soliciting VGI than its competitors and has the strongest chance of adapting crowdsourced data with its algorithmic approach in an attempt to produce a superior database.

At present, no commercial firm is able to produce the kinds of benefits that, at least theoretically, should result from the use of VGI in their compilation process. The reason for this shortcoming is the inability to find a management platform that blends the limitations and advantages of traditional compilation techniques with the limitations and advantages of crowdsourcing in a manner that produces a superior compilation system (Fig. 17.7). It is clear that the VGI can provide significant benefits, but only if government and commercial entities can find the appropriate technique for integrating the benefits of VGI in a method that accommodates the limitations and benefits of the parent system.

17.5 Summary and Conclusions

This chapter has examined the role of volunteered geographic information as a sustainable compilation tool for the purposes of creating map databases used for navigation. While professional and commercial mapping organizations have many advantages in creating detailed navigation databases, it is clearly the case the their efforts fall short in terms of data accuracy, specifically in respect to coverage and the currency of the data mapped. Crowdsourcing in the form of active and passive community input provide a potential, renewable source of local data that may provide more consistent map databases than other alternatives. It is our conclusion, however, that a hybridization of map compilation techniques combining standard approaches used by government and commercial entities with the benefits of volunteered geographic information will likely produce map databases with qualities exceeding the basic capabilities of either method used separately.

As a final note, the use of crowdsourcing in today's hybrid compilation systems is too restrictive. Governments and commercial entities producing spatial databases should consider the following recommendations to reap increased benefits from the use of VGI:

- Add improved structure/taxonomy to features and objects available to crowd-sourced editing.
- Increase the types of features and objects that can be edited through crowdsourcing to benefit from local knowledge.
- Provide multiple paths/solutions/software tools as an alternative to current methods that limit the flexibility of users who are attempting to contribute local information.
- Remove the constraints of the current edit systems for more diverse crowdsourced data. Provide a form or otherwise solicit the contributors' suggestions on the need for corrections in objects, particularly complex objects (e.g., overpasses, tunnels), for which edits and modifications are not allowed.
- Provide metadata for aerial, satellite, and street-side imagery to allow contributors to understand the potential temporal limitations of the imagery accessible during an editing session.
- Track the IP addresses of users on an anonymous basis to determine a coarse location for the user, and use this as a factor in establishing an initial categorization of whether contributors, and particularly editors, might have relevant local knowledge related to the location being edited.
- Establish user history profiles that cross-compare edit location, edit quality, and user location to help evaluate contributions.
- Tune data spam filters to focus on logical consistency issues.
- Make the editorial review for evaluation of contributed data more transparent and helpful. Follow best practices and do not allow edits to be ignored, rejected without reason, or responded to with potentially derogatory or irrelevant comments ("You don't have the experience to make this type of correction.")
- Establish a hierarchy of authoritative sources for database features using measures of data quality and temporal considerations against which all crowdsourced contributions can be evaluated to help select those data and attributes that would most benefit from crowdsourcing.

Jenkins et al. (2009) describe the challenges of participatory cultures. VGI is a prime example of a participatory culture that provides a unique and challenging environment both for the contributors, as well as for those who hope to benefit from assets developed by these groups. While it is unclear whether online participatory cultures are long-term phenomena or will be time-bound to the generation that founded them, these systems currently have a significant influence on map compilation systems. As envisioned by Jenkins, participatory cultures provide low barriers to contribution, support for creating and sharing, informal mentorship, belief that contributions matter, and some degree of connection through a social network[7]. Clearly crowdsourced map compilation systems are an example of a participatory culture and one that may forever change the nature of spatial database compilation.

References

Budhathoki, N., Haklay, M., & Nedovic-Budic, Z. (2010). Who map in OpenStreetMap and why? Presentation at State of the Map, Open Street Map. http://www.slideshare.net/nbudhat2/sotm-us-2010-nama-r-budhathoki. Accessed 23 Oct 2011.

Coleman, D. (2012). Potential contributions and challenges of VGI for conventional topographic base-mapping programs. In D. Sui, S. Elwood, & M. Goodchild (Eds.), *Volunteered geographic information: New development and applications*. Berlin: Springer.

Dobson, M. (2007). TomTom, Tele Atlas and map updating, the road to UGC (Exploring Local Blog. TeleMapics LLC). http://blog.telemapics.com/?p=40. Accessed 25 Oct 2011.

Dobson, M. (2009). *Field checking the Google base* (Exploring Local Blog. TeleMapics LLC). http://blog.telemapics.com/?p=211. Accessed 25 Oct 2011.

Dobson, M. (2010a). TomTom, Tele Atlas and map share. *Exploring Local Blog*. TeleMapics LLC. http://blog.telemapics.com/?p=323. Accessed 27 Oct 2011.

Dobson, M. (2010b). MapQuest on Botox, Tele Atlas on Detox. *Exploring Local Blog*. TeleMapics LLC. http://blog.telemapics.com/?p=334#comments. Accessed 28 Oct 2011.

Dobson, M. (2010c). Map accuracy, Google, NAVTEQ and free. *Exploring Local Blog*. TeleMapics LLC. http://blog.telemapics.com/?p=241. Accessed 23 Oct 2011.

Dobson, M. (2010d). Nokia news and NAVTEQ does it again. *Exploring Local Blog*. TeleMapics LLC. http://blog.telemapics.com/?p=279. Accessed 26 Oct 2011.

Dobson, M. (2010e). Google plays whack-a-mole. *Exploring Local Blog*. TeleMapics LLC. http://blog.telemapics.com/?p=282. Accessed 28 Oct 2011.

Dobson, M. (2011a). Google map maker's edit and authority system part 1. *Exploring Local Blog*. TeleMapics LLC. http://blog.telemapics.com/?p=371. Accessed 28 Oct 2011.

Dobson, M. (2011b). Google map maker's review and authority system part 2. *Exploring Local Blog*. TeleMapics LLC. http://blog.telemapics.com/?p=374. Accessed 24 Oct 2011.

Dobson, M. (2011c). Google map maker goes crowdsourced in the United States on "Judgment Day". *Exploring Local Blog*. TeleMapics LLC. http://blog.telemapics.com/?p=368. Accessed 25 Oct 2011.

Girres, J., & Touya, G. (2010). Quality assessment of the French OpenStreetMap dataset. *Transactions in GIS, 12*(4), 435–459.

Goodchild, M. (2007). Citizens as sensors: The world of volunteered geography. *GeoJournal, 69*(4), 211–221.

Guélat, J. (2009, August). *Integration of user generated content into national databases – Revision workflow at swisstopo*. 1st EuroSDR Workshop on Crowdsourcing, Federal Office of

---

[7]Jenkins et al. 2009, pp. 5–6

Topography, Wabern, Switzerland. http://www.eurosdr.net/workshops/crowdsourcing_2009/presentations/c-4.pdf. Accessed 20 Oct 2011.

Haklay, M. (2008). *Understanding the quality of user generated mapping*. Department of Civil, Environmental and Geomatic Engineering, University College London. PowerPoint Presentation. http://www.slideshare.net/mukih/osm-quality-assessment-2008-presentation?from=ss_embed. Accessed 15 Sept 2011.

Haklay, M. (2010). How good is volunteered geographical information? A comparative study of OpenStreetMap and ordnance survey datasets. *Environment and Planning, B, 37*, 682–703.

Haklay, M., Basiouka, S., Antoniou, V., & Ather, A. (2010). How many volunteers does it take to map an area well? The validity of Linus' Law to volunteered geographic information. *The Cartographic Journal, 47*(4), 315–322.

Harvey, F. (2012). To volunteer or to contribute locational information: Truth-in-labeling for crowdsourced geographic information. In D. Sui, S. Elwood, & M. Goodchild (Eds.), *Volunteered geographic information: New development and applications*. Dordrecht: Springer.

Heipke, C. (2010). Crowdsourcing geospatial data. *ISPRS Journal of Photogrammetry and Remote Sensing, 65*, 550–557.

ISO (International Organization of Standardization), Technical Committee 211. (2002). *ISO 19113, Geographic information – Quality principles*. Geneva: ISO. http://www.isotc211.org/. Accessed 15 Oct 2011.

ISO (International Organization for Standardization). (2004). *ISO 14825, Intelligent transport systems – Geographic data files (GDF) – Overall data specification*. Geneva: International Organization of Standardization.

Jenkins, H., Purushotma, R., Weigel, M., Clinton, K., & Robison, A. (2009). *Confronting the challenges of participatory culture: Media education for the 21st century*. Cambridge, MA: MIT Press (Resource document).

NAVTEQ Corporation. (2008). *Form 10-K annual report filed with the securities and exchange commission of the United States, Washington D.C.* Financial Disclosure. NAVTEQ. http://www.sec.gov/Archives/edgar/data/834208/000110465908014358/a08-2455_110k.htm. Accessed 24 Sept 2011.

Neis, P., Zielstra, D., & Zipf, A. (2012). The street network evolution of crowdsourced maps: OpenStreetMap in Germany 2007–2011. *Future Internet, 4*(1), 1–21. doi:10.3390/fi4010001.

Shirky, C. (2008). *Here comes everybody*. New York: Penguin Press HC.

Stage, D. (2009, February). Authority and authoritative data: A clarification of terms and concepts. *Fair & Equitable*, 13–16. http://www.iaao.org/uploads/Stage.pdf. Accessed 20 Sept 2011.

Surowiecki, J. (2005). *The wisdom of crowds: Why the many are smarter than the few and how collective wisdom shapes business, economies, societies and nations*. New York: Anchor Books Edition.

Tapscott, D., & Williams, A. (2008). *Wikinomics: How mass collaboration changes everything*. London: Penguin.

TomTom. (2011). *TomTom makes the largest historic, traffic database in the world available for governments and enterprises via its online web portal*. Press Release. TomTom B.V. http://corporate.tomtom.com/releasedetail.cfm?ReleaseID=545204. Accessed 23 Jan 2012.

Van Exel, M., Dias, E., & Fruijtierm, S. (2010). *The impact of crowdsourcing on spatial data quality indicators*. http://www.giscience2010.org/pdfs/paper_213.pdf. Accessed 14 Oct 2011.

Van Oort, P. (2005). *Spatial data quality: From description to application*. PhD thesis. *Geodesy 60*, NGC (Netherlands Commission for Geodesy) Delft, the Netherlands. http://www.ncg.knaw.nl/Publicaties/Geodesy/pdf/60Oort.pdf. Accessed 27 Oct 2011.

Zielstra, D., & Zipf, A. (2010, May). *A comparative study of proprietary geodata and volunteered geographic information for Germany*. 13th AGILE International Conference on Geographic Information Science, Guimarães, Portugal. http://agile2010.dsi.uminho.pt/pen/ShortPapers_PDF%5C142_DOC.pdf. Accessed 10 Oct 2011.

Chapter 18
VGI and Public Health: Possibilities and Pitfalls

Christopher Goranson, Sayone Thihalolipavan, and Nicolás di Tada

Abstract Recent advances in technologies that allow for the collection of volunteered geographic information (VGI) are providing new opportunities for health research. These technologies provide for the collection of time-sensitive, fluid data from a broad pool of subjects using sophisticated yet easy-to-use data collection tools – principally the smartphone and other location-aware devices. Never before has it been so easy for health researchers to collect and analyze real-time location-based data. The result of which can provide continuously updated datasets that often capture a more than just a snapshot of events or environmental factors. These technologies also allow the researcher to create novel datasets that do not presently exist. However, the use of such technologies to collect potentially identifiable data poses risks to both the researcher and the subject. The tools introduce new challenges and ethical problems if used improperly for health research. This chapter investigates both the potential of VGI in public health research while discussing some challenges of using technology platforms that can leverage and provide collection tools for volunteered geographic information.

C. Goranson (✉)
Parsons Institute for Information Mapping, The New School, New York, NY, USA
e-mail: goransoc@newschool.edu

S. Thihalolipavan
Bureau of Chronic Disease Prevention & Tobacco Control, New York City Department of Health and Mental Hygiene, New York, NY, USA

N. di Tada
Innovative Support to Emergencies Diseases and Disasters (InSTEDD),
Palo Alto, CA, USA

18.1 Introduction

Recent technological advances combined with a more geographically aware population are introducing new opportunities and challenges for data collection for public health purposes. Volunteered geographic information (VGI) generally refers to data volunteered by individuals, including a geographic component, that can be later disseminated using various tools (Goodchild 2007). VGI data itself can also be thought of as any data provided by an individual that includes some geographic context and which allows for the aggregation and dissemination of data based on this information. Even throughout this book and chapter, the Goodchild's is the prevailing definition, the "volunteering" of geographic information did not suddenly start in the last few years – rather our ability to rapidly collect, conceptualize, and use this information has grown significantly. This is important because some of the traditional methods by which individuals share geographic information can transfer knowledge about the individual quickly and in ways that was not possible or was much more difficult only a few years ago.

Today, new tools are providing ways for individuals, communities, corporations, and governments to harness volunteered geographic information without necessarily needing a fully functional geographic information system (GIS). Terms like GPS, GIS, and location-based services are more commonplace in the media and used much more frequently by the general public. The popularization of tools like Google's My Maps, Google Earth, Microsoft's Virtual Earth (now Bing Maps), foursquare, OpenStreetMap, and others has made spatial processes and data more accessible. Google's My Maps provides free tools for the creation of vector data. Users can create features on an existing base map by adding attributed locations as points (an address), lines (a popular jogging path), or polygons (favorite neighborhood to hang out in; location of a popular street fair). Users can also overlay and georectify digital images in products like Google Earth. New cloud-based services like those available through ArcGIS.com are even providing free access to some of the functionality found in common GIS desktop software. Cloud-based GIS services are providing more opportunities for collaboration among GIS users, and relatively little instruction is necessary in order to leverage these tools. GPS units, once commonplace, are now being replaced by similar functionality in smartphones. As a result, many spatial processes, once requiring expensive GIS software, hardware, GIS datasets, and technical expertise, are now exposing traditional GIS tools to a broader audience. Some of these advances and increase in interest have partially been achieved through better usability and improved user interface design. These advances have all contributed to the removal of some traditional barriers to entry in GIS. Some believe this has even contributed towards the creation of two new disciplines – neogeography and neocartography. Whether or not these tools and techniques lend themselves to entirely new disciplines remains to be seen. It is clear that the methods and access through which geographic data is collected, analyzed, and disseminated will never be the same. The use of VGI promises to do what KML and subsequent adoption as an Open Geospatial Consortium standard (OGC 2011) did for GIS data.

Tools like InSTEDD's GeoChat allow for groups or individuals to share, comment on, and collect location-based information, later aggregating it to create datasets capable of display in a GIS. GeoChat is designed to work essentially as a group chat; the program was originally designed to allow groups of individuals to communicate during an emergency and report an information (InSTEDD 2011). The application provides an easy way to catalog location-based, temporal information using just a cell phone equipped with SMS texting capabilities. Products like OpenDataKit provide an open-source set of tools that allow users to create and disseminate surveys using smartphones (ODK 2011). Using such systems, users can easily create their own forms for data collection activities in the field. Because the cost of development of the form and systems is easily scalable, they can be adapted by small or large groups.

For the public health researcher, these tools may at first glance seem trivial, but they are providing new opportunities to gather information from individuals and communities using cheap and efficient methods. Anyone who can send a simple text message on their cellular phone can inform a study or provide geographically referenced data and thereby potentially create information that is not available in any standard dataset. The short message service (SMS) protocol allows for very short bursts of data to be shared with not only individuals but groups of individuals, parsed and aggregated to create bigger datasets and viewed on a map when geocodable information is included. This approach towards data collection and aggregation can provide much needed quantitative and qualitative data, capturing a variety of variables that would enable increased understanding of physical and/or built environments. For urban areas, these tools may aid in better understanding an individual's perception of safety, identify or document environmental health concerns, better understand what a neighborhood means to the individual, collect data during health events or emergencies, or better understand how the food environment is viewed by the individual across various socioeconomic status (SES) measures. These new, rich data sources would provide a tapestry of information that better reflects the realities of those inhibitors that presently block or discourage access to a variety of health services, identify health behaviors, provide for a better understanding of the built environment's impact on health, and will help to identify outreach strategies at a neighborhood level.

18.2 The Potential for VGI in Public Health Practice

In 2011, the New York City Department of Health and Mental Hygiene, with student volunteer support from CUNY's School of Public Health at Hunter College, undertook a study to evaluate the concentrations of alcohol advertisements throughout New York City (NYC DOHMH 2009). The study was partially based on earlier results gathered through a similar sampling performed that summer with support from local community groups by the department's Bureau of Alcohol and Drug Use, Prevention, Care and Treatment. In both studies, GeoChat was used to catalog alcohol advertisements into three broad categories: type of advertisement, type of alcohol, and brand. In the fall of 2011, a strategy for gathering location-based information on the advertisements was devised in which a random sample of 30 ZIP codes across three income categories (low, medium, high) were selected, and teams were sent out to capture the location of alcohol advertisements using nothing more than a cell phone and a copy record for backup and notes. The result was the creation of a point dataset that represented the alcohol advertisements throughout the 30 selected ZIP codes.

Surveys provide another logical opportunity for the attributed collection of geographic data – either administered by a survey subject directly through an application or administered to them by field staff. In survey design, VGI provides a way to collect results that leverage geography. In New York City, the Department of Health and Mental Hygiene conducts an annual survey of approximately 10,000 New Yorkers called the Community Health Survey (NYC DOHMH 2009). In this survey, respondents are asked a series of questions – information from questions such as "Have you ever been told by a doctor that you have diabetes?" provides neighborhood-level estimates for a variety of health indicators. In all, approximately 35 of these indicators in any given year can be mapped. In all cases, ZIP codes are the only geographic indicator other than the phone prefix (which is not generally used to identify place except in cases where the provided ZIP code does not exist or cannot be properly aggregated). Because the survey needs to maintain a high level of statistical validity, ZIP code level data is rolled up into groups of two or three ZIP codes which comprise what are called United Hospital Fund (UHF) neighborhoods. These neighborhoods are good approximations for place at a subborough level (boroughs in New York City are counties) but are often criticized for not being at a fine-enough scale to identify very small compositional differences between neighborhoods. UHF neighborhoods often group disparate areas together, which makes the term "neighborhood" a bit of a misnomer. Still, the Community Health Survey provides detail-rich data which guide many public health programs. VGI holds some unique promise for improving the geographic accuracy of surveys like the Community Health Survey. For one, using VGI allows the survey participant to provide location-based information that is not dependent on their ZIP code or home address. Location-based services can capture a point returned by the GPS in the phone or by triangulation using cell phone towers. Alternately, if it is determined that such information is too sensitive, participants in a survey can be aggregated to a predefined or flexible grid. In the case of a predefined grid, participants and their respective answers are grouped into grid boundaries or existing polygon neighborhood definitions, like the NYC Projection Areas created by the NYC Department of City Planning. In this way, the participant's exact location is only necessary to assign a neighborhood definition. Alternately, if a location is recorded, the exact information can be loaded into any administrative boundary, where the underlying population is sufficiently large to ensure statistical validity in the dataset.

In the case of epidemiological investigations, VGI may provide opportunities to crowdsource accurate spatial representations of travel patterns and behavior. For example, in a tuberculosis or food-borne illness investigation, if a contact is suspected, the user may turn on a tracking mechanism that reports all place history, times, and length of stay for a period of time. When compared to others that are identified as existing in that same cohort, relationships can be identified that were previously unnoticed. The information feed can also be passive. Location-aware individuals can behave as public health "lookouts" or sentinels. When one of these individuals becomes ill or otherwise affected by a public health concern, the entire place histories can help to identify spatial relationships that were not evident before.

Understanding the relationship between proximity to healthy food and health is something that VGI can help to address. By enlisting local community groups and volunteers, a robust dataset can be sourced that reflects locations of quality and type of food sources, as well as the opinions and insight that a community can bring to better understanding an area's local tapestry. Existing government data sources and initiatives can be enhanced and further populated by the VGI experience. In New York City, the recent Food Retail Expansion to Support Health (FRESH) initiative identified food desserts through a Supermarket Need Index (SNI) (Smith et al. 2011). The SNI is an index that reflects a number of variables in the calculation of need, including population, access to a car, poverty, number of fresh fruits and vegetables, obesity, and diabetes. The initiative and follow-up work between the NYC Department of City Planning, NYC Department of Health and Mental Hygiene, and NYC Economic Development Corporation provide insight into how many variables contribute to the creation of a high-need area. However, there are other indicators of place that are not widely accounted for. One example is the definition of a neighborhood itself; the NYC Department of City Planning's Projection Areas serve as better approximations of neighborhoods than ZIP codes, but it is clear that neighborhoods do not start or stop at distinct Project Area boundaries.

18.3 Privacy Concerns and Health Data

Clinical and public health practitioners are well aware of the importance of protecting the confidentiality of personally identifiable information. However, the collection and disclosure of personally identifiable information in this novel geographic context are less understood by many public health researchers. Privacy is always a top issue in technology circles – major privacy breaches and a lack of transparency into how users' data is used have recently reignited some public awareness of privacy concerns. As technology giants begin harnessing larger and larger datasets that include spatial information, new data sources provide yet further opportunities for the exploitation of personally identifiable information (Forbes 2011). Companies like Google are realizing that in order to compete in big data environments, simplifying the process by which data can be connected is important. In 2012, Google got "rid of over 60 different privacy policies" and replaced them with "one that's a lot shorter and easier to read." The stated goal was to create a more seamless experience for users (Google 2012).

It is not clear that these companies are fully aware of the precedent they are establishing when first releasing potentially personally identifiable information into the wild and later placing restrictions on data access or protections in place once problems have been identified. A recent CNET investigation found that Microsoft had collected spatial data on laptops, cell phones, and Wi-Fi devices and released that information on the web without taking precautions that other companies with similar datasets (e.g., Google) had (CNET 2011). In public health research, the notion that removing attribute data from a person's geographic footprint and thereby only linking things by geographic and temporal proximity should not be seen as a sufficient mode of protecting an individual. Reverse geocoding, a process of identifying a street address from a point on a map, provides plenty of opportunities for identifying an individual, as do time-stamped geographic data or travel paths (Brownstein et al. 2005). Therefore, it is up to the public health researcher to ensure that such requirements for privacy are met. John Snow's legendary Broad Street map of cholera cases in 1854 would present problems in a peer-reviewed publication today. However, spot maps are still a popular way to depict emerging cases during epidemiological investigations.

One clear way to address this problem is through simple education of the public health researcher. A street address, for example, can already be considered confidential information, but the context of the information, say a spreadsheet versus an online map, does not change the confidentiality of the data. While the risk associated with the publication of a spot map identifying the location of HIV-infected patients, for example, may be very obvious to health researchers today, ethical challenges in applying GIS in a public health setting remain. Furthermore, with the democratization of geographic information, more personally identifiable datasets are created, collected, and analyzed by individuals that have no formal health education, training, or work experience. Therefore, it is quite possible that health-related applications emerge which are created either by individuals without a health sciences background or an incentive to adequately protect privacy.

Ethics in geographic information science remains an underexplored topic. While GIS is heavily used in oil and gas exploration, environmental sciences, and military intelligence, the investigation of appropriate and ethical use in the literature is weak (Goodchild 2011). Recently, however, there have been initiatives, papers, and presentations on ethics in GIS, of which the most visible and rigorous exploration of these issues are a series of graduate seminars through a National Science Foundation Grant (Penn State 2010).

18.4 Ethical Norms in Public Health

For the public health researcher, collecting data using VGI is a novel method of data capture but can use many of the standard approaches the researcher is likely already familiar with. As discussed thus far, tools that provide for the collection of VGI allow for the opportunity to collect spatial information by a subject or group of subjects. In traditional public health data collection practice, it may be common to collect standard US Postal Service information including a person's country, state, ZIP code, residential address, and name. Using a survey tool like OpenDataKit, a user can be prompted to enter all the same information through their smartphone. Spatial information can also be collected directly from the device. If the collection of geographic information is continuous over a period of time, it is possible to realistically ascertain a residence, workplace, and commuting pattern of an individual. Google's Latitude does just that – by passively monitoring a user, with their permission, the application will build a personal history profile, even providing statistics on how often one stays at work, where the individual travels most frequently, and during what times of the week.

Because a person's geographic history can be collected and retained directly in the device, there is the risk that this location-based information can be collected without the person's awareness or consent. Even if consent is provided, the individual may not be aware of the extent of additional information that can be ascertained about him/her simply by providing a steady stream of location-based information. The recent uproar surrounding the disclosure that the "Carrier IQ software was being used by phone manufacturers and carriers to monitor performance without implicit knowledge of the individual" is one recent example and has led to the creation of draft legislation towards a Mobile Device Privacy Act (Ars Technica 2012).

By retaining detail-rich spatial and temporal data histories, the researcher can determine the place and time of many events surrounding the subject. While this information can be immensely useful for understanding patterns, links, and relationships between people and place, it can also compromise a person's identity. As mentioned previously, it would be wrong to assume that the removal of personally identifiable information except for the subject's latitude/longitude, travel paths or footprints, or travel history is itself sufficient for protecting patient confidentiality. It is easy to see how one could readily determine with reasonable accuracy an HIV patient's particular habits by simply reverse geocoding the estimated point of residence and cross-checking the other place visits with facilities in the area. This might lead one to determine a particular individual visiting an HIV clinic, or a domestic violence victim visiting a shelter or support group. If such information is disclosed (the location of an Alcoholics Anonymous meeting, domestic violence shelter, STD treatment facility), the exposure of just one subject could risk exposure of others who have not provided consent. In this manner, people essentially become sentinels for behavior, and while public health uses abound, threats to privacy and security do too. If consumers were made aware of the power of their own geographic footprint, they would in turn be more likely to protect it the same way they do their home address. Today, geographic data collected by location-aware devices is almost always coupled with temporal data, which can make it exponentially more powerful. The question of where someone lives or works becomes, "Where were you last Saturday evening?"

Consumers are however prone to quickly accept these caveats in the interest of using the latest application, and marketers of these products are happy to make that process easy. A legally binding document or agreement would simply be skipped over by a quick selection of the "I Accept" button, whereby consumers would be willing to give up highly personal information in exchange for the value of a service. Furthermore, certain services might be constructed whereby they were essential, leaving the customer little opportunity to consider an alternative with which they would not sacrifice such information.

The most recent draft of the Personal health record (PHR) by the Office of the National Coordinator for Health Information Technology states that "Health information stored in a personal health record is under the control of the patient" (ONCHIT 2010a, b). This draft goes on to highlight how different vendors should use a unified method, similar to that of a nutritional label on food packaging to alert customers to privacy and security. A similar approach or model for the use of VGI might also be appropriate, thereby putting the control of one's personal geographic history back in the hands of the individual. It is also important that individuals also retain some responsibility and understand what they are responsible for. One example is found in the HONcode of the Health on the Net Foundation. The HON Code of Conduct is referenced in Microsoft's HealthVault Account Privacy Statement. It provides guidelines for the ethical dissemination of health information and relies on user's "sense of responsibility" to report health-related websites that deviate from this standard (Microsoft Health Vault 2011a, b; HON 2011).

18.5 De-identification of Geographic Information Under the Privacy Rule and the Institutional Review Board

The Health Insurance Portability and Accountability Act (HIPAA) of 1996 established a Federal Privacy Rule facilitating more rigorous regulation of the use of protected health information (NIH 2007). Protected health information rules apply to "individually identifiable health information" but have not kept up with new data available with emerging technology (NIH 2007). The Privacy Rule presently identifies geographic subdivisions, including the street address of an individual, as data that must be removed from a record, but it does not provide explicit guidance around an individual's given latitude/longitude or personal travel history (NIH 2007). It is clear that this information could potentially be classified as a unique identifying number (NIH 2007). Since new technologies, including those now found in smartphones, provide ample ways of identifying a person's location, and numerous applications use this information to connect an individual with services, it can be very difficult to decouple the individual from their location. Traditionally, statistical measures have been employed on datasets, thereby aggregating individual-level data securely to a larger administrative boundary. However, data can also be aggregated to a much smaller area (a grid cell or series of grid cells, the size of which is determined by the GIS analyst) for analytical purposes. This may or may not provide ample protection for the individual, since the grid of cells still needs to be generalized enough to not give away sensitive location information but granular enough to provide better definition than other administrative boundaries.

There are however ongoing efforts to advocate for such privacy concerns. The Office of the National Coordinator for Health IT has established a Chief Privacy Officer. This position is responsible for providing advice on the implementation of technology within HITECH programs and advising the National Coordinator on privacy issues (ONCHIT 2011). The Health Information Technology for Clinical and Economic Health (HITECH) Act provides additional protections for health information already covered by HIPAA. The new protections are geared towards making information available to the patient and expanding patient rights to such information while also protecting disclosures of health information to insurers, business associates, and marketers without previous patient authorization (ONCHIT 2011).

Along another front, an institutional review board (IRB) can provide a systematic check that seeks to protect study subjects from unethical behavior and undue risks. Ethics standards as spelled out in the National Institutes of Health Clinical Research Training state that because human subjects are "a necessary means to the end of greater knowledge," that there is the potential for exploitation (NIH 2012). Geography tells us a lot about study subjects, and the collection of such data and the enrollment of subjects should periodically be checked, just as it would be for any other IRB-approved study. Guidelines effectively reduce the risk of this happening. Informed consent is used to help protect subjects and dictates that subjects clearly understand and agree to the study's goals and objectives. Subjects need to have a clear sense about how information collected about them is used later – something that is clearly lacking from many privacy statements. Existing protections, as defined by and provided for a traditional IRB review and approval process, must be able to take into account such privacy concerns from these emerging technologies. In particular, individuals sitting on IRB panels must understand the ramifications of practitioners collecting geographic data in public health studies and must provide direction for ensuring that such data collected using VGI cannot be used for something other than how it was originally intended. The expiration of VGI data or some other mechanism may be one way to help ensure that data collected for one purpose is not later repurposed for something else.

Researchers using VGI must question the risks of tools developed in an environment that does not require institutional review board (IRB) approval. Preferable researchers without access to an IRB or similar institution would seek out with a partner with one that is willing to review their proposed work since IRBs have the ability to review research projects conducted outside of their organization (FDA 2011). However, it is unlikely that many researchers unaffiliated with an IRB would voluntarily seek regulation, partially because IRB review and approval is often a long process that requires significant upfront documentation and – very clearly – the informed consent of study subjects. It seems unlikely that such stringent measures would be placed on geographically centered research being performed outside of a research institution with an IRB.

Given the history of privacy concerns around public health and the protections already in place, could a board of GIS professionals serve to provide some oversight and proactive guidance on the use of such technologies? If those leveraging the power of VGI will not seek out human subject protection on their own, it may fall on the GIS professional to ensure that human subject abuses never take place. If standards are not put into place, it is likely that real problems with unrestricted collection of spatial and temporal data will not be uncovered until explored further by the courts. Use of GPS devices by police is coming under increased scrutiny, but such inspection is often late (NYT 2012). VGI, when used in public health settings, must be approached in terms of existing protections and potential for future abuse. In particular, because VGI provides a mechanism for collecting increasingly accurate spatial and temporal data about an individual, privacy must be protected above all.

18.6 Summary and Conclusions

As geographers and public health practitioners, it is our duty to inform the general public as to the value of geography. We must educate the general public to the inherent risks and rewards of sharing geographic information with private companies, nonprofit organizations, government, and individuals. Location should be viewed as one would view their social security number – something over which they should not lose control nor share broadly with others. When it is shared, it should be with full disclosure of the risks and – as much as possible – the unintended consequences that might remain. As it can be common for tools and even data to be used for something other than originally intended, it is vital to develop an appropriate framework of recommendations for the collection, analysis, and use of the information. Such guidelines would provide some mechanism by which the creators of such systems can be made aware of such concerns while empowering those in the public health community to identify and address such applications.

Volunteered geographic information will thrive in an open community largely because it is, fundamentally, volunteered. However, location-based information collected on large subsets of the population is largely done without adequate disclosure to individuals. Unfortunately, it may not be until we see abuses of such technology that we see a need to further regulate the collection of such data.

Inherently, one's location belongs to the individual – not to the cell phone company, not to the government, nor to any other application provider. Until we treat identifiable information coming from the individual with the adequate level of care and ensure that individuals understand the ramifications of providing such location-based information, it is unlikely that future abuses will be curtailed. In an electronic age, it is only too easy to accidentally release personally identifiable information, and an individual's geographic footprint only provides an additional measure by which it can readily be disseminated.

Perhaps the easiest approach is to remove the passive monitoring of one's location, when the benefits to the individual clearly do not outweigh the risks. Location needs to be treated as privileged information owned first and foremost by the individual and no one else. Active participation in an application that requires geographic information in order to work correctly still requires receipt of consent, but to what extent should the individual be able to control and later remove one's own data from further analytical use?

Institutional review boards clearly show one alternative to vetting the use of VGI in research, but, as pointed out earlier in this chapter, the likelihood that all applications using location for research purposes would abide by such regulation seems remote. Clearly the power must sit with the individual, and the individual must be willing and able to exercise some control over their own geographic footprints. Federal, state, and local initiatives, along with adoption of electronic health records by large employers, may make possible successful merger of volunteered geographic information and health information technology in a way that does not sacrifice privacy. The rapidity of the use, acceptance, and development of VGI in developing countries is phenomenal. Need for efficient, low-cost, and user-friendly technology has outweighed privacy concerns and thus facilitated the adoption of VGI for a wide range of uses. Areas that are lacking the necessary infrastructure for more common public health informatics deployments can utilize their own light-weight, web-based systems across networks that are in place, the cell phone network being an example of this. In fact, some countries have better infrastructure for cell phones than sanitation (*The Telegraph* 2010). In these scenarios, have we sacrificed privacy and perhaps data quality for convenience?

There will surely be abuses of personal geographic information, and there will be cases where sharing one's location or location history will unwittingly implicate or otherwise harm an individual, group of individuals, or organizations. The more aware public health researchers, geographers, and to a greater extent the general public are to the challenges that remain in securing such geographic information, the more likely it is that we as practitioners are able to avoid and mitigate future damages from such exposure. Personal geographic information should remain under the individual's control, and mechanisms like the HIPPA privacy rule may assist individuals in understanding not only their rights but how important personal information is used (DHHS 2011).

Acknowledgments The authors would like to thank the New York City Department of Health and Mental Hygiene (NYC DOHMH) for contributions and feedback on this chapter. Christopher Goranson works for the Parsons Institute for Information Mapping, of The New School, and previously worked for the NYC DOHMH. Sayone Thihalolipavan works for the NYC DOHMH, and Nicolás di Tada works for Innovative Support to Emergencies Diseases and Disasters (InSTEDD).

References

Ars Technica. (2012). "Mobile device privacy act" would prevent secret smartphone monitoring. http://arstechnica.com/tech-policy/news/2012/01/mobile-device-privacy-act-would-prevent-secret-smartphone-monitoring.ars. Accessed 31 Jan 2012.

Brownstein, J., Cassa, C., Kohane, I., & Mandl, K. (2005). Reverse geocoding: Concerns about patient confidentiality in the display of geospatial health data. *AMIA Annual Symposium Proceedings, 2005*, 905.

CNET. (2011). Declan McCullagh. Microsoft's web map exposes phone, PC locations. http://news.cnet.com/8301-31921_3-20085028-281/microsofts-web-map-exposes-phone-pc-locations/. Accessed 2 Aug 2011.

DHHS (U.S. Department of Health & Human Services). (2011). Understanding health information privacy. http://www.hhs.gov/ocr/privacy/hipaa/understanding/index.html. Accessed 13 Aug 2011.

FDA (U.S. Food and Drug Administration). (2011). Institutional review boards frequently asked questions – Information sheet. http://www.fda.gov/RegulatoryInformation/Guidances/ucm126420.htm. Accessed 13 Aug 2011.

Forbes. (2011). Facebook's privacy issues are even deeper than we knew. http://www.forbes.com/sites/chunkamui/2011/08/08/facebooks-privacy-issues-are-even-deeper-than-we-knew/. Accessed 14 Aug 2011.

Goodchild, M. (2007). Citizens as sensors: The world of volunteered geography. *GeoJournal, 69*(4), 211–221.

Goodchild, M. (2011). Firenze: The Vespucci Institute. *9th Summer Institute on Geographic Information Science.*

Google. (2012). *One policy, one Google experience.* http://www.google.com/policies/. Accessed 30 Jan 2012.

HON (Health On the Net Foundation). (2011). The HON code of conduct for medical and health web sites (HONcode). http://www.hon.ch/HONcode/Conduct.html. Accessed 6 Aug 2011.

InSTEDD. (2011). *GeoChat.* http://instedd.org/technologies/geochat/. Accessed 14 Aug 2011.

Microsoft Health Vault. (2011a). Microsoft HealthVault account privacy statement. https://account.healthvault.com/help.aspx?topicid=PrivacyPolicy&culture=en-US. Accessed 6 Aug 2011.

Microsoft Health Vault. (2011b). Welcome, Google Health users. http://www.microsoft.com/en-us/healthvault/google-health.aspx. Accessed 6 Aug 2011.

NIH (National Institutes of Health). (2007) Health services research and the HIPAA privacy rule. http://privacyruleandresearch.nih.gov/healthservicesprivacy.asp. Accessed 13 Aug 2011.

NIH (National Institutes of Health). (2012). Clinical research training on-line – Based on a presentation by E. J. Emanuel, M.D, Ph.D. http://www.cc.nih.gov/training/training/crt.html. Accessed 13 Jan 2012.

NYC DOHMH (Department of Health and Mental Hygiene). (2009). Community health survey: Survey data on the health of all New Yorkers. http://www.nyc.gov/html/doh/html/survey/survey.shtml. Accessed 13 Aug 2011.

ODK (OpenDataKit). (2011). *About.* http://opendatakit.org/about/. Accessed 24 June 2011.

OGC (Open Geospatial Consortium). (2011). KML – OGC KML. http://www.opengeospatial.org/standards/kml. Accessed 1 Aug 2011.

ONCHIT (The Office of the National Coordinator for Health Information Technology). (2010a). http://healthit.hhs.gov/portal/server.pt/community/joy_pritts_-_chief_privacy_officer/1798/home/17792. Accessed 13 Aug 2011.

ONCHIT (The Office of the National Coordinator for Health Information Technology). (2010b). Building trust in health information exchange: Statement on privacy and security. http://healthit.hhs.gov/portal/server.pt?CommunityID=2994&spaceID=11&parentname=CommunityEditor&control=SetCommunity&parentid=9&in_hi_userid=11673&PageID=0&space=Community Page. Accessed 13 Aug 2011.

ONCHIT (The Office of the National Coordinator for Health Information Technology). (2011). Draft personal health record (PHR) model notice (2011). http://healthit.hhs.gov/portal/server.pt/community/healthit_hhs_gov__draft_phr_model_notice/1176. Accessed 5 July 2011.

Penn State (John A. Dutton e-Education Institute). (2010). Ethics education for geospatial professionals. https://www.e-education.psu.edu/research/projects/gisethics/. Accessed 24 June 2011.

Smith, L., Goranson, C., Bryon, B., Kerker, B., & Nonas, C. (2011). Developing a supermarket need index. In J. A. Mantaay & S. McClafferty (Eds.), *Geospatial analysis of environmental health* (Geotechnologies and the Environment, Vol. 4). Dordrecht/New York: Springer Science+Business Media B.V.

The Telegraph. (2010). India has more mobile phones than toilets: UN report. http://www.telegraph.co.uk/news/worldnews/asia/india/7593567/India-has-more-mobile-phones-than-toilets-UN-report.html. Accessed 18 June 2011.

NYT (New York Times) (2012). Justices say GPS tracker violated privacy rights. http://www.nytimes.com/2012/01/24/us/police-use-of-gps-is-ruled-unconstitutional.html. Accessed 30 Jan 2012.

Chapter 19
VGI in Education: From K-12 to Graduate Studies

Thomas Bartoschek and Carsten Keßler

Abstract Volunteered geographic information (VGI) is making its way into an increasing number of fields within geographic information science. This development has raised a need to cover VGI at various educational levels, which has led to a number of new classes on VGI ranging from elementary through secondary school to undergraduate and graduate university curricula. In this chapter, we give an overview of the state of the art of VGI in education at these different levels. We outline different ways of introducing VGI in class. Specifically, we have investigated the long-term effects of using VGI in education to find out whether students who have come across VGI in class remain interested in the topic and engage in the communities. For this purpose, we have created a survey that was circulated among students of past VGI classes at different levels. The evaluation of the 202 completed surveys gives an overview of motivations and impediments with respect to different VGI platforms. We conclude with recommendations for the future development of curricula covering volunteered geographic information.

19.1 Introduction

Geographic information systems (GIS) and applications such as location-based services have made their way to the general public. Education systems in many different countries have realized their relevance and potential and have therefore developed new strategies to implement the use of geographic information as an integral part of their curricula, mostly starting in secondary schools (Bartoschek et al. 2010). The general aim is to foster competences such as spatial orientation and

---

T. Bartoschek (✉) • C. Keßler
Institute for Geoinformatics, University of Münster, Münster, Germany
e-mail: bartoschek@uni-muenster.de; carsten.kessler@gmail.com

spatial learning and thinking while using GIS as a support system (National Research Council Committee on Support for Thinking Spatially 2006).

More recently, volunteered geographic information (VGI) (Goodchild 2007) has started playing an increasingly important role in the teaching curricula at different educational levels, from elementary and secondary schools (together referred to as *K-12 education* in this chapter) through undergraduate to graduate classes at universities. The idea of using VGI in education has already been around since 1995, long before the term *volunteered geographic information* was coined. At this time, Al Gore initiated the *Global Learning and Observations to Benefit the Environment*[1] program (GLOBE) that supports students and teachers in investigations of environmental issues. GLOBE is still operative and is currently supported by NASA and the US National Science Foundation, among others. Evidently, this innovative idea took close to 15 years to be adopted on a broader scale.

The authors of this chapter have gained experience in teaching at all the mentioned levels: Thomas Bartoschek has been coordinating the GI@School[2] program and worked with students in elementary and secondary schools, mostly through short projects in the context of geography or computer science classes and by training (future) geography teachers. Carsten Keßler has been teaching various classes covering VGI, including Introduction to GIS and project seminars around OpenStreetMap. In this chapter, reports on our personal experience teaching these classes serve as an entry point for detailed review of the current use of VGI in education. We give an overview of literature on the use of VGI in education, followed by a classification of the multiple ways in which VGI is being used in learning and teaching: (1) students working with VGI collected by others, (2) students producing VGI, and (3) students developing VGI-related applications. Based on this classification, we give an overview of classes covering VGI taught at the K-12 and university levels.

The main contribution of this chapter, however, is a survey with 202 students from all age groups that have participated in VGI-related classes at some point. In this survey, we have investigated the long-term effects of using VGI in education. We were particularly interested in which VGI projects the students were familiar with and how they learned about them, in which of them they have already actively participated and whether they are still doing so, and finally what their incentives for participation are. The analysis of the completed surveys has shown interesting correlations between the given answers and participants' age and gender, respectively. These insights can help to improve the curricula for future classes by matching the motivations for participation of broader student groups.

In the next section, we give an overview of the literature on VGI in education, followed by an overview of our personal experience and a classification of VGI in education in Sect. 19.3. We then outline our survey and provide a detailed breakdown of the results in Sect. 19.4, followed by conclusions in Sect. 19.5.

---

[1] See http://globe.gov/about.
[2] See http://www.gi-at-school.de.

19.2 Related Work

Education systems all over the world have realized the relevance and potential of GIScience. New strategies have been developed to include geographic information as an integral part in the curricula, mostly starting in secondary school (Bartoschek et al. 2010). Moreover, the use of computers and mobile devices with positioning technologies such as GPS receivers, smartphones, and tablets in classrooms is increasing. By connecting both facts, the way for volunteered geographic information in K-12 education should be an easy one. However, there are few publications on the use of VGI in the classroom, and most publications related to VGI in the area of education focus on OpenStreetMap. The same applies to VGI in undergraduate and graduate studies.

An analysis of publications from the Learning with Geoinformation conference series, held yearly since 2006, shows that VGI-related topics first came up in 2009 (4 out of 30 articles) but never reached a high level of articles (see Table 19.1) and were even decreasing again while other topics (GIS, remote sensing, mobile applications) have remained consistently present.

Tschirner (2009) speaks about the way from classical wall maps to OpenStreetMap in geography classrooms but does not give any examples. Wolff and Wolff (2009) also refer to OpenStreetMap and argue that the use of OSM in school or university fosters early experience with geodata and geoinformation and can be very motivating. Stark and Bähler (2009) present their project *Map your World*, where Swiss high school students use PDAs and GPS receivers to contribute to OpenStreetMap and OpenAddresses, a VGI platform for collecting geocoded address data for analytical use. Schubert and Bartoschek (2010) introduce a concept for a didactic seminar in geography education, where VGI (OpenStreetMap again) is a small part of the curriculum. Wolff and Wolff (2010) present VGI projects for undergraduates in geography and history where historical objects, such as monuments, are integrated into OSM. Andrae et al. (2011) show possibilities of collaborations between universities and high schools in the creation and application of a web-based portal for POI collection. Hennig and Vogler (2011) propose a system for participatory spatial

Table 19.1 Breakdown of papers presented at the *Learning with Geoinformation* conference series by topic area

	VGI related	GIS	Web-based GIS	Virtual globes	Remote sensing	Geoinformation	Mobile	Spatial thinking	Other	Sum
2006	0	6	1	2	0	2	1	1	3	16
2007	0	7	5	3	1	5	1	0	0	22
2008	0	8	2	3	4	2	1	2	1	23
2009	4	8	2	4	4	1	2	1	4	30
2010	3	8	1	2	4	2	3	1	3	27
2010	4	6	3	2	5	2	1	1	3	27

planning, where students in high school and university contribute data via Scribble Maps[3] to enhance city development processes in Salzburg, Austria.

An analysis of the titles and abstracts of talks at the annual international OpenStreetMap conference *State of the Map* since 2006[4] shows that there were hardly any education- or learning-related topics, except for the papers presented by Rieffel (2010, 2011) and a talk by Hale (2010) reporting work with US school children mapping their neighborhoods. The OpenStreetMap Wiki offers a page[5] dedicated to education where some projects are listed, mostly in high schools. The page defines the role of OpenStreetMap as follows:

> OpenStreetMap is being used within education, in schools, universities and colleges in a wide range of disciplines. Some projects involve only the use of existing OpenStreetMap data and others result in additional data within the OpenStreetMap dataset. The OpenStreetMap project has relevance to geography, mathematics, ecology, community planning and technology. Students are able to not only observe and record their mapping explorations, but can also contribute data to the project. Contributing to OpenStreetMap can be used when teaching computing skills, gaining valuable knowledge in the fields of GIS, planning and community development. (OpenStreetMap 2012)

Besides that, an OpenStreetMap curriculum[6] is being presented that focuses on introducing the topics that OSM addresses: mapping, open-source technologies, crowdsourcing, and community efforts. The curriculum consists of four units: State of Mapping, Crowd Sourcing, Introduction to OpenStreetMap, and Integrating OpenStreetMap. These can be applied in high school or at the university level. There is also a page on using OSM in home education.[7] Topics such as understanding maps, experiencing GPS technology, understanding the local environment, and elementary school lesson plans are briefly outlined.

OSM also plays a role in another recent education project addressing map design and complexity. While OSM has generally reached an acceptable level of detail, the map design addresses adults. The level of complexity and symbology is not suitable for learning to read and understand maps. The development of map literacy and spatial understanding of children will benefit from an OSM version designed specifically for children, following their perception and representation patterns of space (Rieffel and Bartoschek 2012).

Finding courses related to VGI at the university level has turned out to be difficult, mostly due to course titles and descriptions that do not give insight into all topics covered in the courses. It is hence impossible to claim completeness for the list of courses mentioned here; the classes mentioned in the following cover those about which we could find information online.

---

[3] See http://scribblemaps.com/.
[4] See *YEAR.stateofthemap.org*, e.g., http://2011.stateofthemap.org.
[5] See http://wiki.osm.org/wiki/Education.
[6] See http://wiki.osm.org/wiki/Education\ #The_OpenStreetMap_Curriculum.
[7] See http://wiki.osm.org/wiki/OpenStreetMap_and_Home_Education.

The City University of New York – School of Public Health offers a graduate course on *GIS in Public Health* since 2010, which is mainly dealing with VGI data collection (Chap. 18 by Goranson, this volume). Students spend some time learning the technology for data collection (GeoChat), select a research question, and collect data in the field using VGI. After preprocessing, the data will be further analyzed in ArcGIS. In a separate Public Health course, the students set up their own Ushahidi[8] deployments around a research question and solicited participation from the public.

Since 2010, the Université Paris-Est Marne-la-Vallée (Marne-la-Vallée, France) offers a weekly course on *Production de Données Géographiques Collaboratives* (*Production of Collaborative Geographic Data*) for master's students in information systems and web applications and in geographic information. It is an introductory course explaining this new approach to collecting geographic data. The students revise related work and summarize the main advantages and weakness of this type of data. The process of contribution to OSM is discussed and tested. In a final VGI development project, some of the students choose to use OSM data and the API while others choose sources like Google Maps or Géoportail to implement crowd-sourcing applications.

In summary, following a review of literature, it has been shown that the current role of VGI in education, websites, and courses is still a minor one. Very few references to VGI in K-12 and university education are available, even in GIScience education. Most of the courses, lectures, and projects on VGI are focused on OpenStreetMap, not taking into account the variety of available applications.

In the next section, we will outline our personal experience teaching classes related to VGI at the Institute for Geoinformatics, University of Münster.

19.3 VGI in the IFGI Curriculum

The Institute for Geoinformatics at University of Münster[9] (IFGI) is one of Europe's biggest research and education centers in the field of GIScience. The course program covers both BSc and MSc programs in geoinformatics. Additionally, there is an international MSc in Geospatial Technologies and a graduate school offering an international PhD program on semantic integration of geospatial information.[10] IFGI offers numerous GI-related courses for students in other programs, such as computer science, landscape ecology, geography, or didactics of geography.

Moreover, IFGI started a high school education and cooperation initiative called GI@School in 2006. GI@School has developed from a student-driven initiative into

---

[8] See http://www.ushahidi.com/.
[9] See http://ifgi.uni-muenster.de/.
[10] See http://irtg-sigi.uni-muenster.de/.

an integrated part of the curriculum and project work at IFGI with the purpose to transfer fundamental and hands-on knowledge in geoinformatics to K-12 education. It has established a functioning network of schools, teachers, and students, where the exchange with public authorities, industry partners, and partnerships with local schools plays an important role. In addition, the GI@School project develops teaching and learning modules on geoinformatics for schools. These modules can be integrated into class units of different lengths (90 min up to full project days or weeks) being held in schools or at IFGI. VGI is one of the key topics in GI@ School's work.

The experience with VGI in IFGI's educational work at the K-12, undergraduate, and graduate levels, as well as from works presented in the related work section, allows us to define a certain classification of the ways on how VGI can be used in learning and teaching. The simplest and most often first contact with VGI (also shown in the survey) is when students are working with VGI collected by others (1). This might be when browsing on the OpenStreetMap website, analyzing geotags from Flickr, or even importing vector data from OpenStreetMap into a GIS environment for further analysis. A different case of use is when students are producing VGI (2) as in data collection and submission for OpenStreetMap or geotagging of Flickr pictures. In the last case, students may be developing VGI-related applications (3), where available VGI is part of the application (as base map, as POIs) or where the application is made for VGI collection:

1. Working with VGI collected by others:
   (a) Using products based on this VGI (e.g., map from OpenStreetMap)
   (b) Using this VGI to create other products (e.g., in a GIS)

2. Producing VGI:
   (a) Collecting data in the field and submitting them to OSM
   (b) Enriching existing nonspatial information (e.g., geotagging)

3. Developing applications:
   (a) based on VGI (as in 1)
   (b) for VGI collection (as in 2)

We will refer to this classification in further sections, where we present examples of VGI-related activities at IFGI.

19.3.1 VGI in K-12 Education

VGI plays a major role in the approach GI@School takes for K-12 education. Some of the practical modules of GI@School, being held in high schools, cover the underlying principles and practical use of GPS technology. As first actions to implement VGI modules after introducing GPS, we started with OpenStreetMap mapping parties and mapped the school ground and surroundings with the students (2a). They were divided into groups that would each adopt one part of the area to be mapped. The groups annotated their mapped points and tracks with paper and pencil. Back in class, we used the Java OpenStreetMap Editor[11] to annotate and upload the data to OpenStreetMap. Seeing the mapped features appear on the OpenStreetMap website after some minutes was very motivating for the students, as they immediately received feedback about their successful contribution to the project. Over the last 3 years, we have organized 12 mapping parties in German high schools and 8 on an international level (India, Rwanda, Brazil) and have reached around 800 high school students.

Since 2009, we have organized some project weeks with the focus on web-based VGI projects. One outstanding result is the platform TiMiC[12] (*TiMiC is Mobility in Cities*). It was implemented in 2009 by a group of high school students working at IFGI once a week for 2–3 h over the course of 4 months. This VGI-based system allows users to submit traffic events, such as accidents or road works, via SMS to the system, which then appear on a Google Map interface (3b). The students participated in a national competition and won the first prize for IT-related projects.

In the GeospatialLearning@PrimarySchool project,[13] children of ages 6–12 work with the XO laptop, developed by the One Laptop Per Child (OLPC) initiative, and a connected GPS receiver to map geographic features such as trees, water, and vegetation. Due to XO laptop's capability to create ad hoc networks, the children can collaboratively work on the data collection. While tagging outdoors, they see each other on a digital map (from OpenStreetMap) and see their tagging activities, so they collaboratively create a map of the surrounding area (1a, 2a). The collected data can be exported to KML and integrated into other systems (1b).

19.3.2 VGI in Undergraduate Programs

In undergraduate classes, VGI has been a central topic over the last 2 years in the Introduction to GIS class taught every semester at IFGI. This is a compulsory class for all first-semester geoinformatics bachelor students, as well as third-semester geography and landscape ecology bachelor students. Due to the broad range of student backgrounds and the spectrum of topics around GIS covered in this class, the coverage of VGI is limited to OpenStreetMap. The students learn about the organization of the OpenStreetMap community and how the data is collected (1a, 2a). The more technical aspects cover the structure of the OpenStreetMap files and the versioning approach taken in OpenStreetMap. Our experience suggests that this part

---

[11] See http://josm.openstreetmap.de.
[12] See http://www.timic.de.
[13] See http://52north.org/GeospatialLearning.

is challenging for most students, as most of them do not have any experience with the eXtensible Markup Language (XML).

The Introduction to GIS class is a combined lecture and lab, where the students have to solve practical tasks every week and do hands-on work in groups of two. The tasks for the VGI lecture are to register to OpenStreetMap and add several features using one of the web-based editors. By completing these tasks, the students see how editing works and what different types of features can be created. Specifically, they learn how the tagging process works, including checking the OpenStreetMap Wiki for the current consensus on the tags to use for the features they add (2b). Recording and uploading GPS tracks are not part of the class; instead, the students learn how to import the OpenStreetMap data into ArcGIS and how to combine the data with other data (1b). This way, the students also learn about both the advantages and disadvantages of using VGI in their project, namely, that OpenStreetMap offers data that are hardly available from any other easily accessible source but that the accuracy and completeness may vary from one area to another (and even from one feature to another). From most students, we get positive feedback about the VGI lecture. Especially the fact that their own added data is almost immediately shown on the main OpenStreetMap website is very motivating for the students.

19.3.3 VGI in Graduate Studies

At the graduate level, VGI has become an important part of the curriculum for the MSc students in geoinformatics and the Erasmus Mundus students for the MSc in geospatial technologies at IFGI. While these classes are in principle open to students from other programs, students from other programs hardly ever participate in classes related to VGI, probably due to the technical nature of these course offerings. Most graduate classes related to VGI are organized as study projects, which have the goal to develop a software project over the course of the semester in small teams of three to five students. One of the recent study projects was also centered on OpenStreetMap. No specific task was given, though, so that the first task during the semester was to form groups of three to five and come up with an innovative project that is realistic to implement by the team within the given time frame.

The four projects implemented by the teams showed an impressive range of ideas. Team one developed a web application for smartphones that allows users to tag points of interest on the go (3b). Team two implemented a service that allows users to print maps out of OpenStreetMap for hiking or biking trips (3a), including the calculated route (Fritze et al. 2011). The service automatically generates a multipage PDF file from a given GPS track (i.e., the planned route) and optimizes the print layout around the track. Team three came up with the idea for an indoor counterpart for OpenStreetMap called OpenFloorMap (Lasnia et al. 2011). Their implementation consists of an Android app to measure rooms using smartphone sensors and a web application that allows users to integrate their room measurements into an existing building model (3b). This project is still developed further[14] and will also be the subject of an upcoming graduate class. Finally, team four adapted the idea developed for LinkedGeoData (Auer et al. 2011), which provides the OpenStreetMap data as a semantically annotated Linked Data set. One shortcoming of LinkedGeoData at the time of the study project[15] was that it was never up to date, as the dataset was rebuilt infrequently from an OpenStreetMap dump. The team therefore built a live wrapper that provides a semantically annotated view of the current state of the OpenStreetMap dataset (3b), including querying and annotation capabilities (Trame et al. 2011).

The approach to let the students develop their own project ideas, instead of giving them a prepared task, led to a set of innovative and diverse projects that the students really identified with. The general level of motivation was very high. As the references above show, three out of the four projects have led to a conference publication. The students therefore did not only engage with their software project but also learned how the submission and reviewing process of a conference works, which was a new experience for most of them. On the development side, the students acquired consummate skills in generating, handling, and processing VGI.

19.3.4 Application Development and Testing

A recent study project on geospatial learning has produced two VGI-related prototypes that focus on the use in schools. In the "Participatory App," school children can mark points of interest related to one topic (chosen collaboratively or by the teacher) on a map, based on OpenStreetMap. This can happen on a desktop PC in class, where they tag places they know about, or with a mobile device with GPS (3b). Scenarios under consideration are marking dangerous traffic situations on their way to school or places where they find trash in public or parks. These data, collected by several classes or schools, are a valuable input to the city council, pointing them to littered parks, for example. This can be an incentive for the children to participate, as their work can change their living conditions.

A second group presented a tool for educational geotagging (3b) in school. Here, the teacher has to prepare appropriate feature descriptions which will be automatically integrated into the system. The students developed an XML file for tree classification. Geotagging happens in the field with a mobile device. The children approach a tree and click a button in the system, which marks their position on a map. A wizard, generated from the tree XML file, leads the user through the classification process by asking questions ("Is it a broad-leafed tree or a conifer?") and showing pictures of leaf types. By answering a series of such questions, they get to the exact tree classification and also learn about the classification process.

---

[14] http://www.openfloormap.org

[15] LinkedGeoData (http://linkedgeodata.org) also provides a life wrapper now.

Usability tests were conducted with elementary and high school children for both applications, where user interface problems could be solved. The test participants revealed that they were highly motivated in the data collection process. An ongoing project on "Educational Map Apps" will present more VGI prototypes of this kind for smartphones or tablets.

In this chapter, we have outlined IFGI's education activities, referring to the incorporation of VGI across multiple education levels, from K-12 to graduate studies. We have introduced a classification for the integration of VGI in education and have classified IFGI's activities into courses and projects where (1) students work with VGI collected by others, (2) students produce VGI, and (3) students develop VGI-related applications. We have shown that a broad implementation of VGI activities is possible and can be integrated into curricula on all levels. We have presented synergies in VGI application development in graduate studies for use in undergraduate and K-12 education.

19.4 Survey and Evaluation

The goal of the web-based survey presented in this section was to find out more about the impacts of the use of VGI in education. The participants were high school and university students who have already worked with VGI in some form, either through the GI@School program (K-12) or in undergraduate and graduate classes. We were particularly interested in the influence of educational use of VGI on their knowledge about and current use of VGI.

19.4.1 Survey Design

The questionnaire consisted of 15 multiple choice questions and was divided into five parts. Questions 1–5 and 15 focused on personal information, such as age, gender, educational background (high school or university), the favorite/main subjects (for high school students) or the field of study (for university students), and the private use of social networks with the possibility to share location information (such as "check-ins"). Questions 6 and 7 asked if the participants are familiar with different VGI applications and how and in which context they got to know them. Taking into account that the term VGI is not very common, especially in high school and undergraduate classes, we asked these and further questions in the form of a matrix, presenting seven common VGI applications. We decided on photo geotagging platforms (giving the examples of Flickr, Picasa, and Panoramio), OpenStreetMap, Wikimapia, GeoCommons, Google My Maps, Google Building Maker/SketchUp, and CrisisMappers. Questions 8–10 and 12 focused on motivation, incentives, and impact in the use and contribution of VGI; question 11 asked about the usability of the mentioned VGI applications, and questions 13 and 14 asked about the students' opinion on the use and contribution of VGI in education.

19.4.2 Participants

At the K-12 level, the questionnaire was sent out to ten German schools that participated in IFGI's GI@School program and worked with VGI (mostly in the form of OSM mapping parties). Two additional German schools were contacted via the OSM education projects website. At the undergraduate and graduate levels, students attending VGI-related classes at IFGI were contacted. These were students in the Geoinformatics (BSc, MSc), Geospatial Technologies (MSc), Landscape Ecology (BSc, MSc), Geography (BSc, MSc, MEd), and Computer Science (BSc, MSc) programs. The survey link was also sent to international colleagues teaching VGI-related courses.

In total, 202 students participated in the study, of which 26.5% were female. Thirty-three percent of the participants were in high school; the remaining 67% of students were in university. Most participants from high school mentioned mathematics or geography as their main or favorite subjects, while languages and computer science were not very common (see Fig. 19.1). Over 50% of the participants at the university level are enrolled in courses closely related to GIScience (Geoinformatics and Geospatial Technologies) (see Fig. 19.2).

19.4.3 Evaluation

OpenStreetMap is the best known VGI application to the survey participants; only 4% of the participants did not know this project. Due to strong use of OpenStreetMap in IFGI's education activities, this seems to be obvious: 70% of the participants had their first contact to OSM in high school or in university. In comparison, Wikimapia, GeoCommons, and CrisisMappers are not very known to the participants: 65%, 68%, and 85%, respectively, did not know these platforms (see Fig. 19.3). These three seem to be particularly new for the high school students: 77% (Wikimapia), 96% (GeoCommons), and 100% (CrisisMappers) had not come across these platforms before. It is noticeable that photo geotagging platforms and Google My Maps were discovered in their spare time by close to 50% of the participants and by less than 20% in the education context.

As OpenStreetMap seems to be the most commonly used VGI application in education, we take a closer look into the educational use. In high school, OpenStreetMap is being used mainly in projects and regular lessons (2a) by the participants. At the university level, projects and introductory classes are also the most common cases of use, but thesis work and other contexts also seem to be relevant (see Fig. 19.4). This may be due to the fact that the use of OpenStreetMap data (1a) and software development based on OpenStreetMap (3a, 3b) are being fostered at IFGI. Nonetheless, a large part of the participants used (1a) or contributed (2a) to OpenStreetMap in a private context.

The main motivation for using and contributing to OpenStreetMap is education, as already found out in question 7. But other highly rated motivations are the general interest in the project (30%) and the belief in the social impact of OSM (17%).

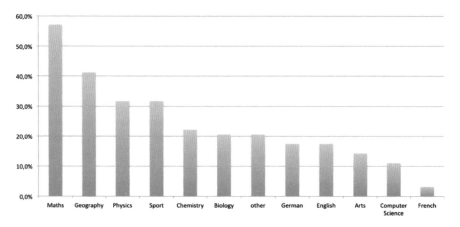

Fig. 19.1 Favorite and main subjects of K-12 students (N=63)

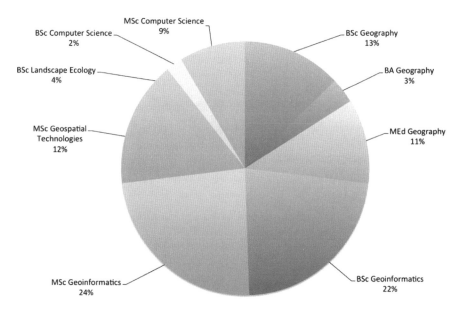

Fig. 19.2 Study programs of university students (N=106)

Wikimapia, GeoCommons, and CrisisMappers were not known to most of the participants. Most participants had "just tried out" geotagging photos and Google My Maps. Besides OpenStreetMap, Google SketchUp and the Building Maker were used by 33% of the participants in an educational context (see Table 19.2).

A closer look at the participants' behavior after working with VGI in school or university reveals that a high number of participants have never used the platforms again. In particular, CrisisMappers (85.7%), Wikimapia (71.4%), and GeoCommons (74.5%) were hardly ever used again.

19 VGI in Education: From K-12 to Graduate Studies

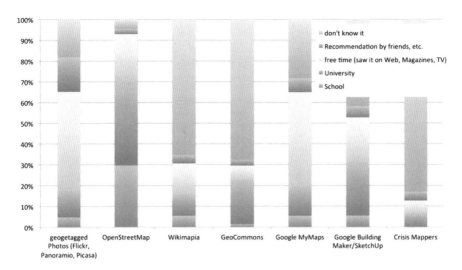

Fig. 19.3 "Which of these projects/applications do you know and how did you find out about them?" ($N=156$)

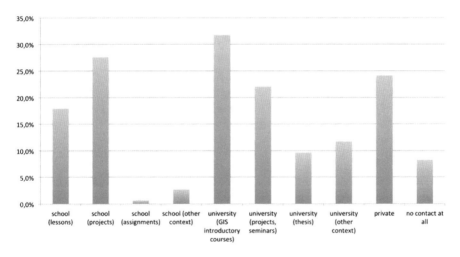

Fig. 19.4 Context of interaction with OpenStreetMap ($N=145$)

Table 19.2 Education as a motivation for VGI use and contribution

Answer options	Education (abs.)	Education (rel.) (%)	Responses
Geotagged photos	20	15.0	133
OpenStreetMap	76	38.8	196
Wikimapia	11	13.6	81
GeoCommons	15	20.5	73
Google My Maps	23	20.0	115
Google Building Maker/SketchUp	31	31.3	99
CrisisMappers	3	5.2	58

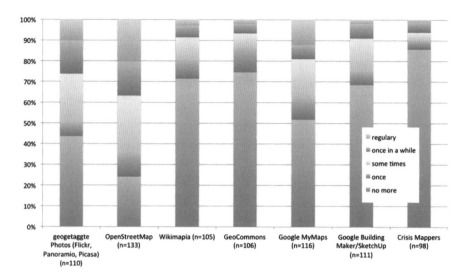

Fig. 19.5 Frequency of project activity after educational use

In comparison, OpenStreetMap reaches only low numbers. Of the participants, 24.1% have never used OSM again after their contact with the application in an educational context. OSM is to be seen as the most positive example, as 20.3% of the participants still use it regularly, 16.5% once in a while, and 27.8% use it sometimes. Photo geotagging platforms (50%) and Google My Maps (40%) are still used by nearly half of the participants more than once after educational contact (see Fig. 19.5).

Seeing the educational background is interesting in the case of OSM. There are big differences between high school and university students in terms of likelihood of further use. While 75.6% of the university students worked more than once with OSM after contact in education, only 46.3% of the high school students did so (see Fig. 19.6).

Figure 19.7 shows the reasons for ceasing or continuing to contribute to the VGI platforms. It only takes the participants into account that stopped after educational use. Participants who never used the platform or still use it were ignored in this analysis. OSM and Google SketchUp/Building Maker are being considered as too time consuming by around 20% of the participants. OpenStreetMap seems to be the most complicated platform to contribute to; this reason was given by 19.2% of the participants. Privacy issues are a relevant reason for photo geotagging platforms and Google My Maps, but the most striking reason is the missing revenue for the participants, chosen by 40–50% for all platforms.

This leads to the question of which incentives would get students at the high school or university level to participate in and contribute to VGI projects in general. Better usability is the highest rated incentive (62.8%), followed by friends being active in the same project (47.3%). More acceptance and acknowledgement in society

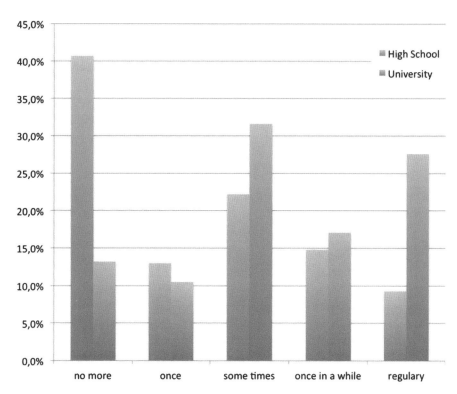

Fig. 19.6 Frequency of OpenStreetMap activity after educational use, compared to level of education (*N* = 130)

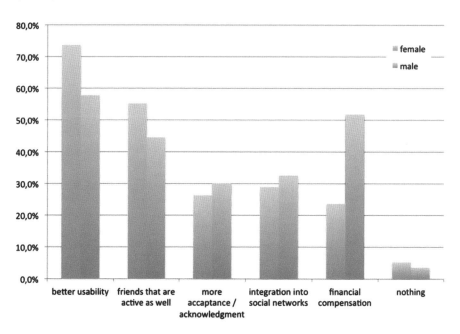

Fig. 19.7 "Which incentives would make you participate in the projects/applications mentioned before?" (*N* = 121)

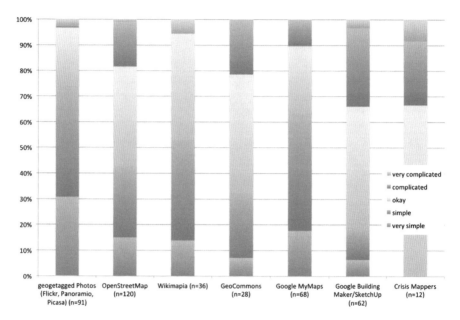

Fig. 19.8 Usability ratings for use and contribution

do not seem to be very important for the participants (28.9%). Financial compensation is interesting for the participants at this point as well (43%), but a closer look reveals some major differences in gender (see Fig. 19.7). Financial compensation seems to be particularly important for male participants (51.8%), but not for female (23.7%). Female participants stress the importance of usability (73.7% vs. 57.8%).

A closer look at the perceived complexity of the aforementioned VGI platforms shows that few participants find the platforms very complicated to use. CrisisMappers is only considered to be "okay" to "very complicated" (on a 5-step scale, ranging from "very simple" to "very complicated") but was only known by 12 participants. Google SketchUp/Building Maker has quite similar values. Only 17.7% of the participants considered it being simple or very simple. In contrast, the photo geotagging platforms and Google My Maps are considered simple or very simple by 73.6% and 63.2% of the participants, respectively. OpenStreetMap, known and used by most of the participants (120), was rated complicated or very complicated by only 18.3% (see Fig. 19.8).

The last questions referred to the use of VGI in education in general as a quantitative and qualitative measure. Fifty-two percent of the participants think that VGI data and applications are being used too little or far too little in education, while 41.5% think its coverage in education is just right. It has to be taken into account that the participants were recruited from groups that were exposed to VGI at a relative high level (GI@School students, IGFI students). This is not very common, especially in K-12 education, also in undergraduate and graduate studies, as shown in the related work section. Of the participants, 70.2% think that the use of and learning with VGI in education make sense or make a lot of sense, which confirms the importance of our efforts to include these projects in education. In both questions, the results were very similar for high school students and university students and for male and female students.

In summary, the results of the survey indicate that there is still significant room for improvement for the long-term effects of VGI in education. The majority of students does not engage in any of the bigger VGI communities after their contact with VGI in class. This picture contradicts our own impressions sketched in Sect. 19.3 – VGI is always a topic for which students are highly motivated. In the end, there seems to be a significant difference between activities that are fun in class and activities that are fun in one's spare time.

The questions on usability and motivations give an indication of how long-term engagement could be increased, namely, by (1) making the tools less complicated and hence more accessible for less tech-savvy users and (2) increasing the user base. While usability is a technology issue that is constantly improving, the user base will eventually get broader over time as VGI becomes more of a mainstream topic. Easier to use interfaces should also flatten the learning curve and should make the VGI communities more accessible for less tech-savvy users.

19.5 Conclusions

Looking at the emergent impact of VGI in society and science (Haklay 2010), it is surprising that the topics and methods concerned with VGI are integrated into curricula only slowly and only to a limited extent, as shown in the related work section. The Institute for Geoinformatics has put a focus on VGI-related education and partly developed a flow, where students in graduate programs develop VGI applications, undergraduate students are exposed to VGI data analysis, and K-12 students work on data collection (see Sect. 19.3). The K-12 students act as usability testers of newly developed applications by IFGI students and contribute to scientific projects as data collectors or as part of participative planning processes (see Sect. 19.3.4). An integrative approach to university-school cooperation as a means to bring VGI to educational use, as presented here, seems to be a good solution, as curriculum development is a rather slow and inflexible process. VGI applications are often not known to teachers or even students in GIScience-related programs, as the survey has shown. To formalize the multiple ways VGI can be used in learning and teaching, a classification of use cases was presented: (1) students working with VGI collected by others, (2) students producing VGI, and (3) students developing VGI-related applications. These three classes can be subdivided in different contexts as shown in Sect. 19.3.

The conducted survey (Sect. 19.4) revealed that OpenStreetMap is the most known VGI application among the participants. Since IFGI's VGI-related education efforts are often based on OSM, this seems to be obvious. However, OSM is also one of the applications that offers more than just data collection. Due to easy access

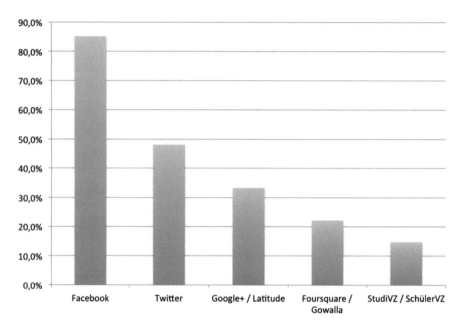

Fig. 19.9 Social network use among the participants ($N = 117$)

to the raw data (under a Creative Commons Attribution-ShareAlike 2.0 license), it can be used in various ways: as a geodata source, for data quality analysis, as a base map in other applications, or for studies in cartography (Rieffel and Bartoschek 2012). This makes OSM an extremely flexible tool in education, also shown in the variety of educational contexts in which it is already being used (see Fig. 19.4).

VGI has already found its way into the experience of a major part of the students. Photo geotagging applications and Google My Maps are being used in spare time by nearly 50% of the survey participants. This shows a general acceptance of participation in and contribution of geographic information. However, the further use of and contribution to VGI-related applications after the contact in school or university are still quite low. It seems to be difficult to keep the students contributing, especially in the K-12 context (Fig. 19.6). The missing revenue (Fig. 19.7) shows that teachers should emphasize the impact and importance of VGI and especially OpenStreetMap in society. Options to reach this goal could be focusing on applications in developing countries and for humanitarian activities.[16] Another aspect, especially in education, is the use of participatory VGI applications, where VGI can be used as an instrument in planning processes. This is a strong motivation in the educational context. The students know that all their work around VGI (from collection and submission to use) makes sense and will be used for a good purpose that they will benefit from. A closer look at the incentives (Figs. 19.8, 19.9)

---

[16] See http://hot.openstreetmap.org/

implies that VGI platforms still need to improve their usability. The activity of friends or integration into social networks is important and feasible as well, in contrast to financial compensation. Seeing the participants' strong activity in social networks (see Fig. 19.9), one could assume that some kind of a mixed form of VGI and social network platform could keep the students on track with contribution.

From the educator's perspective, we can say that working with and contributing to VGI projects are a very encouraging part of class. One can find several modern learning paradigms in this activity.

In the constructionism theory, based on Piaget's constructivism theory of childhood learning (Piaget 1926), children learn by doing and making in a public, guided, collaborative process including feedback from peers, not just from teachers. Students explore and discover instead of only consuming prepared knowledge (Papert and Harel 1991). Papert states that this happens "especially felicitously in a context where the learner is consciously engaged in constructing a public entity." It is meant to see learning as "building knowledge structures." In our case, the meaningful product is VGI; it is created through a collaborative process, and there is feedback of a whole community.

As constructionism may be a rather extreme approach to learning, situated learning may fit better the educational reality. Here learning is a product of an activity, context, and culture, wherein realistic tasks shall be carried out collaboratively (Brown et al. 1989). Situated learning with VGI also brings the educational and the computational worlds together and leads to modern computational approaches, such as situated computing. Situated computing can be understood as a paradigm for mobile computer users based on their physical context and activities carried out as a part of their daily life (Hirakawa and Hewagamage 2001). The growing use of mobile devices in education, even at the K-12 level, is also a foundation for future VGI applications in education, where the whole process of data collection, data processing, and use of VGI can be carried out in the field.

References

Andrae, S., Erlacher, C., Paulus, G., Gruber, G., Gschliesser, H., Moser, P., Sabitzer, K., & Kiechle, G. (2011). OpenPOI – Developing a web-based portal with high school students to collaboratively collect and share points-of-interest data. In T. Jekel, A. Koller, K. Donert, & R. Vogler (Eds.), *Learning with geoinformation 2011* (pp. 66–69). Berlin: Wichmann Verlag.

Auer, S., Lehmann, J., & Hellmann, S. (2011). LinkedGeoData: Adding a spatial dimension to the web of data. In *ISWC'09: Proceedings of the 8th International Semantic Web Conference* (pp. 731–746). Berlin/Heidelberg (2009): Springer-Verlag.

Bartoschek, T., Bredel, H., & Forster, M. (2010). GeospatialLearning@PrimarySchool: A minimal GIS approach. In R. Purves, & R. Weibel (Eds.), *GIScience 2010 extended abstracts, sixth international conference on Geographic Information Science*. Zurich, Switzerland. http://www.giscience2010.org/pdfs/paper_166.pdf.

Brown, J., Collins, A., & Duguid, P. (1989). Situated cognition and the culture of learning. *Educational researcher, 18*(1), 32ff.

Fritze, H., Demuth, D., Knoppe, K., & Drerup, K. (2011). Track based OSM print maps. In A. Schwering, E. Pebesma, & K. Behncke (Eds.), *Geoinformatik 2011: Geochange* (Vol. 41 of *ifgiprints*. Münster, Germany). Heidelberg: AKA.

Goodchild, K. (2007). Citizens as sensors: The world of volunteered geography. *GeoJournal, 69*(4), 211–221.

Haklay, M. (2010, September). *Geographical citizen science – Clash of cultures and new opportunities.* Position paper for GIScience workshop on the role of VGI in advancing science, Zurich, Switzerland.

Hale, K. (2010, August). *OSM used in a high school environment.* Presentation at State of the Map US Conference 2010, Atlanta, GA. http://www.vimeo.com/14592993.

Hennig, S., & Vogler, R. (2011). Participatory tool development for participatory spatial planning – The GEOKOM-PEP environment. In T. Jekel, A. Koller, K. Donert, & R. Vogler (Eds.), *Learning with geoinformation 2011*. Berlin: Wichmann Verlag.

Hirakawa, M., & Hewagamage, K. P. (2001). Situated computing: A paradigm for the mobile user-interaction with multimedia sources. *Annals of Software Engineering, 12*(1), 213–239.

Lasnia, D., Westermann, A., Tschorn, G., Weiss, P., & Ogundele, K. (2011). OpenFloorMap implementation. In A. Schwering, E. Pebesma, & K. Behncke (Eds.), *Geoinformatik 2011: Geochange* (Vol. 41 of *ifgiprints*. Münster, Germany). Heidelberg: AKA.

National Research Council Committee on Support for Thinking Spatially. (2006). *Learning to think spatially.* Washington, DC: National Academies Press.

OpenStreetMap. (2012). Wiki page on education. http://wiki.openstreetmap.org/wiki/Education. Accessed 26 Jan 2012.

Papert, S., & Harel, I. (1991). Situating constructionism. In *Constructionism* (p. 1). Norwood: Ablex Publishing Corporation.

Piaget, J. (1926). *The child's conception of the world.* New York: Harcourt Brace & World.

Rieffel, P. (2010, July). *Openstreetmap on OLPC XO-laptops in primary schools.* State of the Map Conference, Girona, Spain.

Rieffel, P. (2011, September). *Assessing learning aspects in digital maps – Using VGI to create a child suitable map.* State of the Map Conference, Denver, CO.

Rieffel, P., & Bartoschek, T. (2012, March). *Investigating cognitive aspects in digital maps – Using VGI to create a child suitable map.* GI Zeitgeist Conference Proceedings, Münster, Germany.

Schubert, J., & Bartoschek, T. (2010). Geoinformation im geographieunterricht – konzeption eines fachdidaktischen seminars an der universität Münster. In T. Jekel, A. Koller, K. Donert, & R. Vogler (Eds.), *Learning with geoinformation* (Vol. 5). Berlin: Wichmann Verlag.

Stark, H., & Bähler, L. (2009). Map your world – Schüler erfassen freie und offene geodaten. In T. Jekel, A. Koller, K. Donert, & R. Vogler (Eds.), *Learning with geoinformation* (Vol. 4). Berlin: Wichmann Verlag.

Trame, J., Rieffel, P., Tas, U., Baglatzi, A., & von Nathusius, V. (2011). LOSM – A lightweight approach to integrate OpenStreetMap into the web of data. In A. Schwering, E. Pebesma, & K. Behncke (Eds.), *Geoinformatik 2011: Geochange* (Vol. 41 of *ifgiprints*. Münster, Germany). Heidelberg: AKA.

Tschirner, S. (2009). GIS in deutschen klassenzimmern – von der Wandkarte zu Openstreetmap. In T. Jekel, A. Koller, K. Donert, & R. Vogler (Eds.), *Learning with Geoinformation* (Vol. 4). Berlin: Wichmann Verlag.

Wolff, P., & Wolff, V. (2009). Öffentliche Projekte für öffentliche karten. In T. Jekel, A. Koller, K. Donert, & R. Vogler (Eds.), *Learning with Geoinformation* (Vol. 4). Berlin: Wichmann Verlag.

Wolff, P., & Wolff, V. (2010). Über OpenStreetMap (OSM) stadt- und kulturgeschichte erleben. In T. Jekel, A. Koller, K. Donert, & R. Vogler (Eds.), *Learning with geoinformation* (Vol. 5). Berlin: Wichmann Verlag.

Chapter 20
Prospects for VGI Research and the Emerging Fourth Paradigm

Sarah Elwood, Michael F. Goodchild, and Daniel Sui

Abstract This concluding chapter reflects on some of the core themes that crosscut the contributed chapters, and further outlines some of the stimulating and significant relationships between volunteered geographic information (VGI) and the discipline of geography. We argue that future progress in VGI research depends in large part on building strong linkages with a diversity of geographic scholarship. We situate VGI research in geography's core concerns with space and place and offer several ways of addressing persistent challenges of quality assurance in VGI. We develop an argument for further research on the heterogeneous social relations through which VGI is produced and their implications for participation, power, and collective or civic action. The final two sections, closely related, position VGI as part of a shift toward hybrid epistemologies and potentially a fourth paradigm of data-intensive inquiry across the sciences.

20.1 Retrospective

Looking back over the chapters of this book, we are struck first and foremost by the diversity of work represented and the many research directions the authors have chosen to pursue. Contributions range from computational geography to education,

---

S. Elwood (✉)
Department of Geography, University of Washington, Seattle, WA, USA
e-mail: selwood@u.washington.edu

M.F. Goodchild
Department of Geography, University of California at Santa Barbara,
Santa Barbara, CA, USA
e-mail: good@geog.ucsb.edu

D. Sui
Department of Geography, The Ohio State University, Columbus, OH, USA
e-mail: sui.10@osu.edu

D. Sui et al. (eds.), *Crowdsourcing Geographic Knowledge: Volunteered Geographic Information (VGI) in Theory and Practice*, DOI 10.1007/978-94-007-4587-2_20,
© Springer Science+Business Media Dordrecht 2013

from Web demographics to public health, and from the social construction of place to the place of volunteered geographic information (VGI) in China. Many interesting and exciting issues emerge and prompt us to offer some thoughts in this final chapter on the prospects for future VGI research. While some might see VGI as a flash in the pan, a sideshow in the broader currents sweeping society, we believe nevertheless that it represents a very significant development for the discipline of geography and for the production of geographic information in general. In what follows, we explore several potential directions for VGI research. We have no desire to offer a prescription, and our discussion reflects no more than our personal perspectives; but by offering a discussion, we hope to stimulate further reflection and creative thinking about the future of the field.

Some, but by no means all, of the chapters of the book have been written by geographers, and all three of the editors hold appointments in academic departments of geography. To us, this makes very good sense: as with any area of research concerned with geographic information, VGI stands in the borderland between several disciplines, including computer science and several of the social sciences. Many of the issues surrounding VGI are common to other forms of user-generated content and are best studied in a broad context. But other issues are specific to the geographic nature of VGI and thus are best addressed within a framework of geographic knowledge and theory. In what follows, we explore some of the more stimulating relationships between VGI research and the discipline of geography, believing that progress depends in part on building stronger linkages with the accumulated body of geographic knowledge.

The next section follows such a strategy closely, by linking VGI research to the traditional dichotomy in geography between space and place (Tuan 1977), arguing that in contrast to much previous work in geographic information science (GIScience) and its concern with space, VGI research invites a more place-centric perspective and may even stimulate the development of a parallel, *platial* geographic information system. The second substantive section addresses the topic of VGI quality, a perennial topic in the field and one that surfaces in several of the chapters of this book. We argue in this section that quality assurance can benefit from a strong link to and formalization of the accumulated corpus of geographic knowledge, that is, knowledge of how the geographic world is constructed. The fourth section explores another persistent theme in VGI research – debates about the nature and implications of the individual and collective processes through which VGI is created. In the fifth section, we advance the concept of hybridity in relation to VGI and to future research and practice. Finally, we conclude this chapter by situating VGI, GIScience, and geography in the emerging fourth paradigm – data-intensive inquiries across the disciplines.

20.2 Space, Place, and VGI

Geographers and others have long drawn a distinction between space and place. While the distinction has many dimensions, by limiting our discussion to the digital world, we hope we can shed some light and more importantly address the implications for

VGI research. A spatial perspective emphasizes position on the surface of the Earth, the calculation of distances and directions, and indeed all of the apparatus of geographic information systems (GIS), based as it is on knowledge of the precise positions of features. A key GIS function is *topological overlay* or the *spatial join*, the ability to link together information based on common location. Thus, for example, one might link details of a road accident at some location to the attributes of the road on which the accident occurred, or the county containing its location, allowing various hypotheses to be investigated. From this perspective, features with indeterminate boundaries, such as "downtown Santa Barbara," present a problem (Montello et al. 2003). More broadly, GIS technology tends to favor features with precisely defined boundaries and locations (Burrough and Frank 1996), such as the counties and census tracts of the administrative hierarchy, or the rivers and lakes of surficial hydrology.

Things are very different in the informal world of human discourse. People tend to refer to locations by name, whether or not such names refer to precisely bounded areas. People tend not to know the precise latitude and longitude of features or to care very much about precise distances and directions. Instead, references to location occur largely through place-names, street addresses, and points of interest; and such references frequently allow something equivalent to a spatial join, as when a reference to Santa Barbara provides a link to other information about that place. Goodchild and Hill (2008) describe the gazetteer, an index of officially recognized place-names, as providing an essential link between the informal world of human discourse and the formal, spatial world of GIS. Nevertheless, a gazetteer provides only a limited form of link, being traditionally composed only of a triple (name, type, location) where location provides only a point coordinate even for a spatially extended object.

Because VGI is created by citizen volunteers, it tends to inherit the character of human discourse rather than scientific measurement and thus to have strong links to the world of place – in other words, VGI suggests a context that is more platial than traditional GIS. Indeed, it is remarkable that place-names have played a surprisingly minimal role in traditional GIS, being absent from the list of seven framework data types developed by the US Federal Geographic Data Committee (http://www.fgdc.gov/framework).

In a platial world, planimetric control is far less important than in a spatial world. Distorted maps, such as the Beck map of the London Underground and its many emulators around the world, sacrifice planimetric rigor for visual clarity and properties that are more important to users of mass-transit systems: the routes serviced, the order of stops along each route, and the interchanges between routes. Directions are typically simplified to a discrete set (8 for the Beck map), and distances distorted by as much as an order of magnitude. Yet such maps are clearly more useful to the traveler. Much the same could be said of the sketch maps that are often provided to visitors in many parts of the world: by depicting only essential detail, distorting distances and directions, and focusing on named places and points of interest, such maps provide a perfectly adequate service. Yet the lack of planimetric control means that they can never be used to effect a spatial join or to provide accurate estimates of distance or area, two of the motivations that drove the development of the earliest GIS (Foresman 1998).

It seems to us that VGI creates an opportunity for research on a very different, novel kind of GIS, in short a platial GIS, that would emphasize a platial rather than a spatial perspective. The sketch map would be its primary method of display, and functions would be needed to derive sketch maps from planimetric maps. It would support joining based on name rather than location; other key research questions might include:

- What is the appropriate functionality for a platial GIS? Which functions of spatial GIS would be relevant, and which new functions would be needed?
- What kinds of VGI could be solicited and used to populate a platial GIS?
- What research in cognitive issues would be needed to operationalize platial GIS?
- What would be the appropriate data model for a platial GIS? Would linked data provide an appropriate paradigm?
- How might uncertainty be characterized in a platial approach?

20.3 Assuring VGI Quality

The topic of data quality began to attract attention in GIScience in the late 1980s (Goodchild and Gopal 1988), building on efforts to develop quality standards for geographic information (Guptill and Morrison 1995) and on the application of statistical methods to error analysis in cartography (Maling 1989) and surveying. Since then, a large amount of work has been published, ranging from applications of geostatistics to the use of fuzzy sets. The growth of VGI clearly raises many issues of data quality, given the almost complete lack of quality control or any of the other mechanisms used to measure and assure quality in traditional geographic information.

There would seem to be three ways of addressing the problem of quality assurance in VGI. First, it is often argued that crowdsourcing converges on the truth if people have the opportunity to review and correct errors. Named in honor of Linus Torvalds (Raymond 1999) and formulated as a principle of software development, Linus' law asserts that, "given enough eyes, all bugs are shallow"; in other words, the probability of a bug in software being discovered rises with the number of people who have reviewed the software. Projects such as Wikipedia rely heavily on the principle, arguing that errors in the encyclopedia will eventually be discovered and corrected. The principle clearly works best when the number of reviewers is large and better therefore for errors in popular facts than for errors in obscure facts.

Unfortunately, Linus's law does not translate well to the geographic context. Errors in assertions about prominent features, such as Mt. Everest, will quickly attract attention, but errors concerning obscure features in areas of the world that are sparsely populated will likely escape attention. Sites such as Wikimapia may even attract malicious behaviors that exploit this principle, as when a contributor is tempted to find a minor feature in a remote part of the world and name it for himself or herself.

A more robust and reliable approach relies on the development of a social structure of review. Individuals with a track record of steady, reliable contributions can be invited to take on the role of moderator or gatekeeper, reviewing the contributions of others for quality. Moderators may specialize in certain feature types or certain areas of the world. In effect, social structures replicate the administrative structures of agencies, but in an entirely voluntary way. In Google's Map Maker, the highest levels of the hierarchy are reserved for Google employees (see Chap. 17 by Dobson and Chap. 14 by Coleman in this volume), and no asserted fact is accepted until it has been reviewed by an employee. This social approach to quality assurance appears to work well for OpenStreetMap (Haklay 2010), as it does for Wikipedia.

But a third approach appears to offer an even more effective way of assuring quality in VGI. As Tobler (1970) stated in his famous first law of geography – "All things are related but nearby things are more related than distant things." In short, the geographic world is far from an independent assemblage of phenomena and, instead, is subject to numerous relationships and constraints that determine what can exist at some location (and at some time, if the phenomenon is dynamic). One might term these the rules of geographic syntax, analogous to the syntactic rules that govern any language. In the case of a wildfire emergency, for example, the assertion "the fire is at **x**" might be checked against an array of rules: Has a fire been reported anywhere else near **x**? What is the current wind direction and is **x** downwind of any location already known to be burning? Is there suitable fire fuel at **x**, and is it sufficiently dry to ignite? A myriad of such rules can be formulated and if appropriately implemented could be used to examine a purported fact, give it a probability of being true, and recommend appropriate action: acceptance, rejection, query, etc. In effect, the set of such rules constitutes the corpus of knowledge about the geographic world and thus the core of the discipline of geography. Yet to date no coordinated effort has been made to assemble these rules into a formal set that can be implemented in a computing system. Such an effort, even in the comparatively narrow domain of wildfire, would be immensely valuable. Examples of such rule bases are reported to exist in the systems developed by some mapping companies to assess incoming volunteered corrections to their data, but only in an ad hoc form.

Before such an approach can be implemented, several issues will need to be researched:

- How might such rules of geographic syntax be formally structured and implemented in a computational system?
- How might rules be organized according to their degree of abstraction and general or specific nature?
- Can rules be used to assign quantitative measures of likelihood to purported geographic facts?
- What kinds of analysis of real-world geographies might yield suitable rules?
- How should rules be organized according to the various data types and domains of geographic knowledge?

20.4 VGI: Voluntary, Participatory, Individual, Collective?

A great deal of VGI research continues to focus on the nature of the social processes through which VGI is created, as a way of conceptualizing how it borrows on and departs from earlier paradigms for creating geographic information, and their implications for the resulting information. Among other things, this line of work has noted a blurring of boundaries of various roles associated with spatial data and mapping, such as expert/professional/amateur or producer/user (Goodchild 2007; Haklay et al. 2008; Budhathoki et al. 2010), and debated whether the notion of "volunteering" is sufficient to characterize the wide range of ways that VGI is generated (Sieber 2011; Chap. 3 by Harvey, this volume). The contributed chapters continue to detail the tremendous heterogeneity in the social processes and relations from which VGI emerges, as well as their implications for the content, quality, and social/political significance of the information. Across the initiatives detailed in this collection, we find data schemes and contribution protocols that range from highly structured and closely monitored national and local government efforts to fit VGI into existing conventional data infrastructures, to the radically open scheme of OpenStreetMap's user-generated wiki-ontologies (in which users are free to create their own spatial objects, feature classification types, and relationships among them). These examples show VGI being generated from practices that range from highly individualized to group-based efforts variably argued to be collective, civic, or participatory. And it is increasingly clear that any given VGI collection is the result of many different scopes/scales of contribution – from one-time contributions by some to sustained high-volume contributions by others.

With a rich body of evidence documenting these key dimensions in the heterogeneous social processes of VGI creation, a critical next step is deepening our understanding of what these factors mean for the character, content, and quality of VGI, as well as its societal implications. What dimensions of a contributor's experiential knowledge can we expect to be included or excluded in a VGI initiative that is highly structured through a preexisting data scheme versus one that emerges more organically through users' contributions and revisions? How do different kinds of social connectivities and interactivities in VGI initiatives shape the content and character of the information that emerges? The cases here provide examples of VGI emerging from social or collective encounters that include deeply individualized activities, initiatives by groups affiliated only through their involvement in a common VGI effort, initiatives in which individuals contribute information to an institutional or governmental project without necessarily joining it, and richly collaborative endeavors by preexisting groups already strongly connected through shared experiences, knowledge, and collective endeavors.

There is much yet to understand about how these differing social encounters around VGI creation and sharing shape the information that results. For example, the "long tail" effect in VGI (a few contributors will generate a great deal of information, while most others offer only a few contributions) is well documented, but as yet we know relatively little about what this means with respect to data content, quality, and reliability. In practice, the long tail effect means that any given collection of VGI is generated by contributors with very different commitments, familiarity with the initiative or its data schemes, or knowledge about the phenomena in question (Goodchild 2009). Yet we cannot assume the most prolific contributors are likely to be the most accurate or reliable. It may well be the case that low-volume contributors offer deep and direct "platial" knowledge, while high-volume contributors are offering more fleeting observations. In short, there remains much work to be done in understanding the forms of knowledge and social connection from which individuals and groups are generating VGI, and their implications for the data.

Within our efforts to better understand the social processes and relationships from which VGI emerges, the contributed chapters also point to a need to continue refining our theorizations of voluntarism, civic engagement, public participation, and collective action in relation to VGI. Much of the early conversation about the social significance of VGI has debated whether it stands to enhance these endeavors or undercut them by promoting "thin" forms of participation or individualism over social connectivity (Corbett 2011; Poore 2011; Sieber 2011). Clearly, actors who contribute a single data object to OpenStreetMap or report observations to local government via a VGI interface are participating or performing civic engagement of a sort, but in a very different manner than the dominant understandings of participation or engagement that have been part of, for example, prior work on participatory GIS or participatory planning (Sieber 2006; Dunn 2007). Beyond simply critiquing the sociopolitical processes of VGI exemplified in these kinds of activities as not sufficiently "participatory," we need context-specific accounts of how, when, and where individual and collective contributions of information may constitute participation or engagement. As well, we must consider whether the emergence and expansion of VGI as a phenomenon may be contributing to a shift in societal expectations about what it means to do these things. Sieber (2011), for example, asks whether VGI itself contributes to a transformation in what it means to volunteer. Finally, Lin's (Chap. 6, this volume) discussion of VGI activities in and about China reminds us of the need for careful contextualization of such claims. She shows that the meanings of participation and civic engagement are geographically contingent, as are the political and technological opportunity structures that govern the forms they can take in a particular place.

Circulating in these debates are very different assumptions about what the participatory, collective, or civic dimension of VGI entails and how it is constituted. For some, making and sharing collectively authored content is itself a collective action, one that contributes to what Hardy (Chap. 11, this volume) terms a "larger distributed civic participation." Along these lines, some readings of OpenStreetMap suggest that in a context in which access to some forms of spatial data has been deeply restricted and dominated by national mapping agencies, creating a crowdsourced, openly shared spatial data resource is deeply political (Haklay 2010). Other accounts, such as Corbett's (Chap. 13, this volume) case of long-distance "map-mediated dialogue," frame the participatory or collective dimensions of VGI as stemming not from simply contributing information but from involvement in deliberative processes that foster shared knowledge. Still others emphasize VGI as a site of political formation, arguing that making and sharing geographic information or representations contributes to the formation of self-aware civic or political actors, a critical precursor to political engagement, and a factor in shaping *how* these actors participate or engage (Lin, Chap. 6, this volume; Wilson 2011).

Drawing out these and other tacit assumptions about participation and the political context that underlie framings of VGI as participatory, collective, or civic (or not!) is an essential step in continuing to refine our theorizations of its social and political significance. In this effort, there is much we can draw from (and contribute to) research on isolation and connectivities rendered through social media (Turkle 2011), debates about shifts in the nature and scale of social and political ties that are created through new media (Shirky 2010), and resultant shifts in widely accepted meanings of activism or engagement (Lieverouw 2011). More broadly, some ongoing research questions may include:

- What kinds of state-civil society relationships are produced or transformed through the creation and use of VGI?
- What forms of deliberation, collaboration, or participation are being fostered (or constrained) around the creation and use of VGI, and how do these vary across cultural and political contexts?
- Does VGI imply transformations in the social construction and politics of "data," "science," or "geographic information"?

20.5 Hybrid VGI Futures: From Analysis to Synthesis

Throughout this collection, a persistent theme has been the ways that VGI straddles a host of boundaries. As a form of data, VGI may have a "spatial" dimension (in the precise Cartesian sense required in many digital environments) while simultaneously being richly platial. A particular data object or VGI artifact may have quantitative and qualitative dimensions (a photograph tagged with latitude/longitude coordinates) or may combine visual, textual, and numerical content. The production and use of VGI bridges the social roles that were part of preceding paradigms of geographic information creation and use: producer/user, expert/amateur/professional, citizen scientist, and "local" knowledge. Given the complex assemblages of hardware, software, organizations and institutions, and individual actors often involved in producing it, VGI cannot be situated squarely within the public, private, or nonprofit sectors. In terms of its use and applications, we see copresent trends toward integrating VGI into existing conventional spatial data initiatives (such as the work of national mapping agencies or census bureaus) and toward fostering entirely new modes of data creation and governance. With an eye toward all of these blendings and blurrings, Sui and DeLyser (2012) frame VGI as a hybrid, drawing on the notion of hybridity as referring "…to those things and processes that transgress and displace such boundaries and in so doing produce something that is ontologically new" (Rose 2000, p. 364). Here, we argue that a similar hybridity will be required in our engagements with VGI going forward, whether in terms of data handling, analytical tools, or epistemological and methodological frameworks.

Some of this needed hybridity is already in evidence in existing VGI practice, especially with respect to questions of data compilation, integration, quality, and sharing. In this volume, Dobson, Johnson and Sieber, Coleman, and Poore and Wolf all point to the emergence and continued need for blended metadata forms and data quality measures that integrate conventional approaches (often drawn from existing SDI models) with new forms specifically oriented toward VGI. While their examples focus especially on public- and private-sector spatial data initiatives, we can also see NGOs as needing and seeking such resources that support blendings of conventional spatial data and VGI, given the demonstrated tendency of these groups to flexibly mix a range of data and technology resources (Sieber 2000; Elwood 2006). User reviews of data quality, reliability ratings of contributors, and other social approaches to quality assurance are already in play in VGI services or initiatives such as Ushahidi or FracTracker. But as Poore and Wolf (Chap. 4, this volume), Dobson (Chap. 17, this volume), and Grira et al. (2010) suggest, an especially important next step lies in developing augmented forms of metadata appropriate for working with VGI or blended authoritative/VGI data sets, perhaps integrating structures and practices from conventional SDIs with those drawn from open-source mapping and other kinds of Web 2.0 initiatives. These efforts are necessary and important, but they will also be resource intensive. Some local and national government VGI initiatives are initially justifying their turn toward citizen-contributed information in terms of anticipated cost reductions (Esri 2010), but the initial research contributed here suggests that any realistic efforts to integrate VGI with existing conventionally curated data will require a host of new structures, techniques, and resources, augmenting rather than replacing existing tools and practices.

In working with VGI as a source of evidence in research, its spatial/platial hybridity introduces some challenges. Using VGI as a source of evidence almost inevitably requires methodologies prepared to handle both the precise locational information of a massive collection of VGI, as well as the context- and detail-rich knowledge that may also emerge from it. Of course no single methodological approach will suffice, given the diversity of VGI as a phenomenon and the range of questions researchers will inevitably use it to pursue. But it may be possible to identify different methodological trajectories best suited to particular kinds of VGI or research questions. For extremely large collections of textual VGI, quantitative techniques that discern semantic similarities or shifts in meaning have potential. And McKenzie's (Chap. 12, this volume) natural language processing techniques, for example, work toward being able to understand the content and fluidity of individuals' experiential knowledge, but do so via quantitative approaches to linguistic data that can be deployed on a scale previously impossible. Yet for smaller collections of VGI that span multiple representational forms (such as text, images, and maps), interpretive methods such as those emerging from qualitative GIS may be helpful. Such approaches include grounded visualization, computer-aided qualitative GIS, and geonarrative supported interpretive analysis of digital objects that have both quantitative and qualitative dimensions; mix visual, textual, and numerical representations; and span the "spatial/platial" divide (Knigge and Cope 2009; Jung and Elwood 2010).

Our notion of a platial GIS introduced in Sect. 20.2 flags the probable need for technologies that can handle Cartesian and non-Cartesian spatial relations or topologies together, or blend conventional modes of geovisual representation and spatial analysis with multimedia representation and modes of analysis. In spite of the many advances in mixed-method GIS, neither existing spatial technologies nor existing qualitative analysis software is fully prepared to do this, nor deal with dynamic content mediated over the Internet. Imagine, for example, a digital analysis and representation software that could support the kind of actor-network analysis that Palmer (Chap. 16, this volume) carries out in his analysis of severe-storm reports on Twitter and conventional spatial analysis of patterns in the contributed observations. In envisioning and beginning to implement such hybrid technologies for working with VGI, the recent push for spatial analysis and geovisual representation capabilities in computer-aided qualitative data analysis software may offer some fruitful possibilities (Fielding and Cisneros-Puebla 2009), as could the emergence of Web-based software oriented toward mixed-method analysis, such as Dedoose.[1]

As we try to build the analytical, methodological, and epistemological hybrids needed for ongoing research with and about VGI, there is much to be drawn from geography's long-standing tradition of mixed- and multi-method research (Elwood 2009; Sui and DeLyser 2012). Integrative frameworks such as critical quantitative methods (Sheppard 2001; Kwan and Schwannen 2009), qualitative and feminist GIS (Kwan 2002; Kwan and Knigge 2006; Cope and Elwood 2009), and, more recently, geohumanities or spatial humanities (Warf and Arias 2009; Bodenhamer et al. 2011; Dear et al. 2011) have paved paths to guide future VGI research and practices. In particular, geohumanities' approaches to integrating spatial-analytic and interpretive modes of analysis and working across visual, textual, numerical, archival, and ethnographic forms of evidence have much to offer. As well, given the position of VGI at the confluence of a number of constructed binaries – research/practice, scholarly/activist, science/society, and more – there is also much to be learned from other arenas in which geographers have grappled with them, including participatory action research or "public geographies" (Kindon et al. 2008; Fuller 2008; Ward 2007). At the broadest level, these approaches share an enacted commitment to working across multiple (and often very different) epistemologies, data types, modes of representation and analysis, and institutions. This is precisely what will be required of our work with and about VGI going forward, with some of the following questions potentially guiding these efforts:

- What advances in mixed-method epistemology and practice are needed to support researchers' efforts to work with VGI?
- How might hybrid methodological frameworks for working with VGI suggest productive avenues for "scaling up" qualitative methods or enriching efforts to discern meaning via quantitative techniques?
- What digital divides are produced or dismantled through VGI and blended VGI/authoritative data sets?

---

[1] We are grateful to Jin-Kyu Jung for bringing this example to our attention.

20.6 VGI, GIScience, and Geography: Toward the Fourth Paradigm?

We began this book by situating the VGI phenomenon in the emerging world of big data and the growing global digital divide. Throughout this book, contributors from multiple disciplines and backgrounds have covered a wide range of conceptual, computational, and social/political issues related to VGI and its applications. We would like to end by discussing the implications of VGI research for GIScience and geography in the context of the fourth paradigm – data-intensive inquiries across the sciences (both physical and social/behavioral) and the arts/humanities as a concomitant growth of big data (Hey et al. 2009; Bartscherer and Coover 2011; The U.S. National Science Foundation 2011; Berry 2012).

The fourth paradigm, also known as eScience in the literature, was advocated by Jim Gray (2007). According to Gray (2007), scientific discoveries until recently (the early days of the twenty-first century) were made according to three dominant paradigms – the empirical (by describing natural phenomena), the theoretical (by using and testing models and general laws), and the computational (by simulating complex phenomena using fictional/artificial or small real-world data sets) approaches. The new world of big data demands that researchers think beyond the three traditional methodological boxes. Gray (2007) argues that a new eScience aiming at seamlessly linking information technology with traditional domain inquiries can potentially serve as the new ark upon which we can survive the current big-data deluge. Unlike all the previous paradigms that deal with either artificial or relatively small data sets, the defining feature of the fourth paradigm eScience is data intensive, often dealing with data in peta- or even exabytes. Also quite distinct from the discovery processes in previous paradigms, data, methods, machines, and people are increasingly linked and networked in data-intensive inquiries (Nielsen 2012).

Evidently, VGI, or more generally geocoded data, is quickly becoming an important part of big data (Manyika et al. 2011). By all indications, GIScience, and to a limited extent geography, is moving rapidly to embrace the fourth paradigm – data-intensive inquiry. In fact, both geography and GIScience have deeper roots in data-intensive inquiries. Back in the early 1980s, geographer Peter Gould (1981) shouted that we should "let data speak for themselves" (p. 166), though with a limited following. Also despite its ups and downs, the International Symposium on Spatial Data Handling is still very much alive after almost 30 years since its commencement back in 1984 (www.sdh12.org). It is also interesting and gratifying for us to compare the content of Gray's (2007) vision of the fourth paradigm of eScience with that of GIScience laid out by one of us some 20 years ago (Goodchild 1992). With all the progress and advances that have been made during the past 20 years (Goodchild 2010), we strongly believe that GIScience and geography are poised for new breakthroughs if we can find creative ways to surf big data and the fourth-paradigm wave with VGI as our surfboard. Preliminary results following the data-intensive paradigm are exciting and promising, as demonstrated by the papers in a recent collection on data-intensive geospatial computing in the *International Journal of Geographic Information Science* (Jiang 2011, Vol. 25, No. 8). Furthermore, the emerging fourth paradigm in GIScience and geography will continue to push for more open access to both data and source code in GIScience/geographic research (Jiang 2011) – a practice long overdue within the geospatial community. We would like to encourage readers to think big in the age of big data. Here is a list of big questions for GIScientists and geographers (both paleo- and neogeographers); we invite you the reader to come up with your own questions (and answers) in the days ahead:

- Theoretical questions: What's the role of theories in the age of big data? What are the pros and cons of knowledge production through mining the big data? Do data have the same level of unreasonable effectiveness as mathematics (Wigner 1960) that will lead to new discoveries (Halevy et al. 2009)? Or has the current torrent of big data carried us to a new terra incognita? Is the rapidly expanding digital universe, similar to the physical one, ultimately too big to know (Weinberger 2012)?
- Methodological questions: Will synthesis be driving methodology in the age of big data? If so, how can it be done? Should we follow the model of big science to develop something along the lines of the Living Earth Simulator (http://www.futurict.ethz.ch/FuturICT), or, at a much smaller scale, a digital version of the microscope for science (Higginbotham 2011), or a plug-and-play macroscope envisioned by Börner (2011)?
- Technical questions: What kind of new cyber infrastructure is needed to deal the ever-expanding digital universe? Is there an ultimate physical limit of storing all the data humans create? Is cyber spatial infrastructure (Wright and Wang 2011) still needed or is it going to be submerged with the new cyber infrastructure increasingly dominated by cloud and mist computing?
- Social/political/ethical/legal issues: Are technologies related to big data truly liberating as articulated by Ghonim (2012) – "the power of people is bigger than the people in power"? What are the consequences and implications when big data is increasingly in the hands of big corporations such as Google and Facebook? What are the long-term implications of big data for privacy and personal liberty?

In 1900, German mathematician David Hilbert (1900) outlined a list of 23 fundamental problems in mathematics, which inspired new generations of mathematicians for making many exciting breakthroughs in the twentieth century. Three former AAG presidents (Cutter et al. 2002) have attempted a list of ten big questions in geography. We believe it is imperative, perhaps more urgent than ever, that we continue to ask even bigger questions in the age of big data. What are your big questions? Crowdsourcing them will shed a brighter light on the relevance of the questions themselves and perhaps even the answers.

References

Bartscherer, T., & Coover, R. (Eds.). (2011). *Switching codes: Thinking through digital technology in the humanities and the arts.* Chicago: University of Chicago.
Berry, D. M. (Ed.). (2012). *Understanding digital humanities.* Basingstoke: Palgrave Macmillan.

Bodenhamer, D., Corrigan, J., & Harris, T. (Eds.). (2011). *The spatial humanities: GIS and the future of humanities scholarship*. Bloomington: Indiana University Press.

Börner, K. (2011). Plug-and-play macroscopes. *Communications of the ACM, 54*(3), 60–69.

Budhathoki, N., Nedovic-Budic, Z., & Bruce, B. (2010). An interdisciplinary frame for understanding volunteered geographic information. *Geomatica, 64*(1), 11–26.

Burrough, P. A., & Frank, A. U. (Eds.). (1996). *Geographic objects with indeterminate boundaries*. Bristol: Taylor and Francis.

Cope, M., & Elwood, S. (2009). *Qualitative GIS: A mixed methods approach*. London: Sage.

Corbett, J. (2011, April 13). *The revolution will not be geotagged: Exploring the role of the participatory Geoweb in advocacy and supporting social change*. Paper presented at annual meeting of the Association of American Geographers, Seattle, WA.

Cutter, S., Golledge, R., & Graf, W. (2002). The big questions in geography. *The Professional Geographer, 54*, 305–317.

Dear, M., Ketchum, J., Luria, S., & Richardson, D. (2011). *GeoHumanities: Art, history, and text at the edge of place*. London/New York: Routledge.

Dunn, C. (2007). Participatory GIS: A people's GIS? *Progress in Human Geography, 31*(5), 617–638.

Elwood, S. (2006). Beyond cooptation or resistance: Urban spatial politics, community organizations, and GIS-based spatial narratives. *Annals of the Association of American Geographers, 96*(2), 323–341.

Elwood, S. (2009). Mixed methods: Thinking, doing, and asking in multiple ways. In D. DeLyser, M. Crang, L. McDowell, S. Aitken, & S. Herbert (Eds.), *The handbook of qualitative geography* (pp. 94–113). London: Sage.

ESRI. (2010). *The latest in citizen engagement*. ESRI advertising supplement. http://media2.govtech.com/documents/PCIO10_ESRI_V.pdf. Accessed 16 May 2011.

Fielding, N., & Cisneros-Puebla, C. (2009). CAQDAS-GIS convergence: Toward a new integrated mixed method research practice? *Journal of Mixed Methods Research, 3*(4), 349–370.

Foresman, T. W. (Ed.). (1998). *The history of geographic information systems: Perspectives from the pioneers*. Upper Saddle River: Prentice Hall.

Fuller, D. (2008). Public geographies I – Taking stock. *Progress in Human Geography, 32*(6), 834–844.

Ghonim, W. (2012). *Revolution 2.0: The power of the people is greater than the people in power*. New York: Houghton Mifflin Harcourt.

Goodchild, M. F. (1992). Geographical information science. *International Journal of Geographical Information Systems, 6*(1), 31–45.

Goodchild, M. (2007). Citizens as sensors: The world of volunteered geography. *GeoJournal, 69*(4), 211–221.

Goodchild, M. (2009). Neogeography and the nature of geographic expertise. *Journal of Location Based Services, 3*(2), 82–96.

Goodchild, M. F. (2010). Twenty years of progress: GIScience in 2010. *Journal of Spatial Information Science, 1*(1), 3–20.

Goodchild, M. F., & Gopal, S. (Eds.). (1988). *Accuracy of spatial databases*. New York: Taylor and Francis.

Goodchild, M. F., & Hill, L. L. (2008). Introduction to digital gazetteer research. *International Journal of Geographical Information Science, 22*(10), 1039–1044.

Gould, P. (1981). Letting data speaking for themselves. *Annals of the Association of American Geographers, 71*(2), 166–176.

Gray, J. (2007). eScience – A transformed scientific method. Presentation made to the NRC-CSTB. http://research.microsoft.com/en-us/um/people/gray/talks/NRC-CSTB_eScience.ppt. Accessed 19 Dec 2011.

Grira, J., Bedard, Y., & Roche, S. (2010). Spatial data uncertainty in the VGI world: Going from consumer to producer. *Geomatica, 64*(1), 61–71.

Guptill, S. C., & Morrison, J. L. (Eds.). (1995). *Elements of spatial data quality*. Oxford: Elsevier.

Haklay, M. (2010). How good is volunteered geographical information? A comparative study of OpenStreetMap and ordnance survey datasets. *Environment and Planning B, 37*(4), 682–703.

Haklay, M., Singleton, A., & Parker, C. (2008). Web mapping 2.0: The neogeography of the GeoWeb. *Geography Compass, 2*(6), 2011–2039.

Halevy, A., Norvig, P., & Pereira, P. (2009). The unreasonable effectiveness of data. *IEEE Intelligent Systems*, March/April, 8–12.

Hey, T., Tansley, S., & Tolle, K. (Eds.). (2009). *The fourth paradigm: Data-intensive scientific discovery*. Redmond: Microsoft Research.

Higginbotham, S. (2011). Big data: Science's microscope of the 21st century. http://www.businessweek.com/technology/big-data-sciences-microscope-of-the-21st-century-11092011.html. Accessed 19 Dec 2011.

Hilbert, D. (1900). Mathematical problems. *Göttinger Nachrichten*, 253–297 (original work in German; translated into English in 1902).

Jiang, B. (2011). Making GIScience research more open access. *International Journal of Geographical Information Science, 25*(8), 1217–1220.

Jung, J., & Elwood, S. (2010). Extending the qualitative capabilities of GIS: Computer-aided qualitative GIS. *Transactions in GIS, 14*(1), 63–87.

Kindon, S., Pain, R., & Kesby, M. (2008). *Participatory action research approaches and methods: Connecting people, participation and place*. London: Routledge.

Knigge, L., & Cope, M. (2009). Grounded visualization and scale: A recursive examination of community spaces. In M. Cope & S. Elwood (Eds.), *Qualitative GIS: A mixed methods approach* (pp. 95–114). London: Sage.

Kwan, M. (2002). Feminist visualization: Re-envisioning GIS as a method in feminist geography research. *Annals of the Association of American Geographers, 92*(4), 645–661.

Kwan, M., & Knigge, L. (2006). Doing qualitative research with GIS: An oxymoronic endeavor? *Environment and Planning A, 38*(11), 1999–2002.

Kwan, M., & Schwannen, T. (2009). Critical quantitative geographies. *Environment and Planning A, 41*(2), 261–264.

Lieverouw, L. (2011). *Alternative and activist new media*. Cambridge, MA: Polity Press.

Maling, D. H. (1989). *Measurements from maps: Principles and methods of cartometry*. Oxford: Pergamon.

Manyika, J., Chui, M., Brown, B., Bughin, J., Dobbs, R., Roxburgh, C., & Byers, A. H. (2011). Big data: The next frontier for innovation, competition, and productivity. http://www.mckinsey.com/Insights/MGI/Research/Technology_and_Innovation/Big_data_The_next_frontier_for_innovation. Accessed 19 Dec 2011.

Montello, D. R., Goodchild, M. F., Gottsegen, J., & Fohl, P. (2003). Where's downtown? Behavioral methods for determining referents of vague spatial queries. *Spatial Cognition and Computation, 3*(2–3), 185–204.

Nielsen, M. (2012). *Reinventing discovery: The new era of networked Science*. Princeton: Princeton University Press.

Poore, B. (2011, April 13). *VGI/PGI: Virtual community or bowling alone?* Paper presented at annual meeting of the Association of American Geographers, Seattle, WA.

Raymond, E. S. (1999). *The cathedral and the bazaar*. Sebastopol: O'Reilly.

Rose, G. (2000). Hybridity. In R. J. Johnston, D. Gregory, G. Pratt, & M. Watts (Eds.), *The dictionary of human geography* (pp. 364–365). Oxford: Blackwell.

Sheppard, E. (2001). Quantitative geography: Representations, practices, and possibilities. *Environment and Planning D: Society & Space, 19*(5), 535–554.

Shirky, C. (2010). *Cognitive surplus: How technology makes consumers into collaborators*. New York: Penguin.

Sieber, R. (2000). GIS implementation in the grassroots. *URISA Journal, 12*(1), 15–51.

Sieber, R. (2006). Public participation geographic information systems: A literature review and framework. *Annals of the Association of American Geographers, 96*(3), 491–507.

Sieber, R. (2011, April 13). *Volunteered geographic information: Motivation or empowerment?* Paper presented at annual meeting of the Association of American Geographers, Seattle, WA.

Sui, D., & DeLyser, D. (2012). Crossing the qualitative-quantitative chasm I: Hybrid geographies, the spatial turn, and volunteered geographic information (VGI). *Progress in Human Geography, 36*(1), 111–124.

The U.S. National Science Foundation. (2011). Rebuilding the mosaic: Fostering research in the social, behavioral, and economic sciences at the National Science Foundation in the next decade. www.nsf.gov/pubs/2011/nsf11086/nsf11086.pdf. Accessed 19 Dec 2011.

Tobler, W. R. (1970). A computer movie simulating urban growth in the Detroit region. *Economic Geography, 46*(2), 234–240.

Tuan, Y.-F. (1977). *Space and place: The perspective of experience*. Minneapolis: University of Minnesota Press.

Turkle, S. (2011). *Alone together: Why we expect more from technology and less from each other*. New York: Basic Books.

Ward, K. (2007). Geography and public policy: Activist, participatory and policy geographies. *Progress in Human Geography, 31*(5), 695–705.

Warf, B., & Arias, S. (Eds.). (2009). *The spatial turn: Interdisciplinary perspectives*. London/New York: Routledge.

Weinberger, D. (2012). *Too big to know: Rethinking knowledge now that the facts aren't the facts, experts are everywhere, and the smartest person in the room is the room*. New York: Basic Books.

Wigner, E. P. (1960). The unreasonable effectiveness of mathematics in the natural sciences. *Communications on Pure and Applied Mathematics, 13*, 1–14.

Wilson, M. (2011). 'Training the eye': Formation of the geocoding subject. *Social & Cultural Geography, 12*(4), 357–376.

Wright, D. J., & Wang, S. (2011). The emergence of spatial cyber infrastructure. *Proceedings of the National Academy of Sciences, 108*(14), 5488–5491.

Biographies of the Editors and Contributors

Editors

Daniel Sui is professor of geography and distinguished professor of social and behavioral sciences at the Ohio State University. He also serves as chair of geography (since July 2011) and director of the Center for Urban and Regional Analysis (CURA) (since July 2009). Sui holds adjunct professorship at the John Glenn School of Public Affairs, Knowlton School of Architecture (the City and Regional Planning program), and College of Public Health at OSU. Prior to assuming his position at OSU in July 2009, Sui was professor of geography (1993–2009) and holder of the Reta A. Haynes Endowed Chair (2001–2009) at Texas A&M University. He holds a BS (1986) and MS (1989) from Peking University and PhD from University of Georgia (1993). His current research interests include geographic information science, urban geography, and geographic thought. Sui has authored/coauthored 4 books and over 100 articles in these areas. Sui was a 2009 Guggenheim Fellow. He is also a current member of the US National Mapping Science Committee and serves as editor in chief for *GeoJournal*.

Sarah Elwood is professor of geography at the University of Washington. She received a BA in Geography from Macalester College in 1994 and an MA and PhD in Geography from the University of Minnesota in 1996 and 2000. Her recent research bridges critical GIS, urban political geography, qualitative methods, and participatory action research, including a long-term collaborative project on the use and impacts of geographic information systems and GIS-based spatial knowledge in neighborhood revitalization, and a coedited volume on qualitative GIS. She is currently concluding a 3-year project on interactive mapping technologies in collaborative learning and civic engagement with young teens, and beginning research on the spatial politics of poverty and class identities in economic crisis and recovery.

Michael F. Goodchild is Jack and Laura Dangermond professor of geography at the University of California, Santa Barbara, and director of UCSB's Center for Spatial Studies. He received his BA degree in Physics from Cambridge University in 1965 and his PhD in Geography from McMaster University in 1969 and has received four honorary doctorates. He was elected to be a member of the National Academy of Sciences and Foreign Member of the Royal Society of Canada (2002), member of the American Academy of Arts and Sciences (2006), and Foreign Member of the Royal Society and Corresponding Fellow of the British Academy (2010), and in 2007, he received the Prix Vautrin Lud. He serves on the editorial boards of ten journals and book series and has published more than 15 books and 400 articles. His current research interests center on geographic information science, spatial analysis, and uncertainty in geographic data.

Contributors (By Alphabetical Order of Last Names)

Benjamin Adams is a PhD candidate in the Department of Computer Science at the University of California, Santa Barbara, and a member of the Spatial Cognitive Engineering (SpaCE) Lab. His research is concerned with developing new computational methods to organize and represent geographic knowledge drawn from heterogeneous data sources in order to aid both humanistic and scientific inquiry. This research touches on both the practical development of tools and methods for exploring and synthesizing these data as well as theoretical issues regarding the representation of geospatial data semantics. Benjamin is the program manager for the Cognitive Science program at the University of California, Santa Barbara. He received his MA in Computer Science from the University of California, Santa Barbara (2011), and BA (*summa cum laude*) in Social Science and Computer Science from Eastern Michigan University (2006). He was an NSF Integrative Graduate Education and Research Trainee in Interactive Digital Multimedia from 2007 to 2009.

Thomas Bartoschek is a doctoral student at Institute for Geoinformatics, University of Münster, Germany, where he leads the GI@School initiative (http://www.gi-at-school.de) in its 6th year. In his PhD studies, he is doing research on geospatial learning and thinking with geotechnologies and on GIScience education. Volunteered geographic information plays a key role in his scientific and professional work. Several prototypes of applications for geospatial learning, which he has developed or supervised, were based on VGI or made for the production of VGI in educational contexts. In teaching, especially in his K-12 education efforts and teacher trainings, he "evangelizes" the use and collection of VGI and has reached thousands of students and teachers over the last years. Thomas is also involved in the organization of the Vespucci Summer Institute on Geographic Information Science (http://vespucci.org).

T. Edwin Chow is an assistant professor in the Department of Geography at Texas State University–San Marcos. He holds a PhD in Geography from the University of South Carolina. His research interests focus in Internet GIS, volunteered geographic information (VGI), and GIS-based modeling. His recent passion in research related to spatial demography is to investigate the potential of Web demographics to unearth spatial patterns of population dynamics. His other work involves a broad spectrum of GIS-based modeling, including site assessment, quality of life, risk assessment, hydrology, and wildland fire. He has been working on several collaborative research projects to model raccoon habitats and conduct ecological risk assessment, associate precipitation and surface runoff with ENSO periods, and simulate wildfire behavior in grassland ecosystems. Edwin teaches GIS at all levels (i.e., undergraduate, masters, and doctoral) as well as a graduate course in quantitative methods. He has the pleasure to mentor, and learn a lot more from, several masters and PhD students. He has also been serving as an ad hoc reviewer for many GIScience journals (e.g., IJGIS) and the NSF grant proposals.

David J. Coleman is dean of engineering and professor of geomatics engineering at the University of New Brunswick in Canada. Prior to obtaining his PhD, he spent 15 years in the Canadian geomatics industry – first as a project engineer, then as an executive with one of Canada's largest digital aerial mapping companies, and later as an owner and partner in a GIS and land information management consulting firm. The former chair of UNB's Department of Geodesy and Geomatics Engineering, Dr. Coleman has authored over 150 publications and reports dealing with land information policy development, geomatics operations management, geographic information standards, and spatial data infrastructure. He is a fellow of the Canadian Academy of Engineering, a past member of the GEOIDE Research Network Board of Directors, a former president of the Canadian Institute of Geomatics, and president-elect of the Global Spatial Data Infrastructure Association.

Jon Corbett is an assistant professor in the Community, Culture and Global Studies Department at UBC Okanagan and the codirector of the Centre for Social, Spatial and Economic Justice. He has three primary research interests: firstly, to explore how digital multimedia technologies can be combined with maps and used by people to document, store, and communicate their spatial knowledge; secondly, to examine how geographic representation of this knowledge using these technologies can strengthen community internally as well as externally by increasing people's influence over decision-making processes; and, thirdly, to address the process and implementation of sustainable development, particularly with respect to community-based resource management. All aspects of his research incorporate a core community element. Within the context of his research program, this means that the research is of tangible benefit for the communities with which he works, that those communities feel a strong sense of ownership over the research process, and that community members are engaged by and engage in the research endeavor.

Nicolás di Tada spends most of his time designing and managing software projects. Before starting his company, Manas Technology Solutions, Nicolás spent 10 years as a software architect and project leader for many organizations, including start-ups and large corporations, acquiring a background in information retrieval, machine learning, information visualization, and Web development. During the last 7 years, he founded two other companies in the fields of e-learning and consumer-end social applications and guided several development teams through a wide variety of projects ranging from digital photogrammetry and biomedical signal processing to enterprise applications. Both for small start-ups and Fortune 500 companies, his teams have always proudly delivered usable and effective software on time. Passionate about the convergence between technology, science, and art to make a better world,

Nicolas currently leads the design and development of InSTEDD's software platform, coordinating the distributed development team, open-source contributors, interns, and volunteers.

Michael W. Dobson is president and principal consultant for TeleMapics LLC. He provides strategic and technical consulting to national and international clients involved in mapping, local search, location-based services, navigation, and telematics. His blog "Exploring Local" is widely read by members of the community interested in maps and mapping. He was previously employed as the CTO and executive vice president of technology for go2 Systems where he served as the corporate officer responsible for technology, software engineering, IT, and product development. go2 supplied location services, including mapping and routing to major wireless carriers and services, such as MSN Wireless, AT&T Wireless, Sprint, Nextel, Verizon Wireless, and others. Earlier, Dobson was the chief technologist and chief cartographer for Rand McNally, where he managed the technology and software development organizations supporting the company's map-based commercial and consumer products and services. He also served as the company's vice president of business development and public relations spokesperson. Dobson began his professional career with the Geography Department at the State University of New York at Albany where he was an assistant and then associate professor.

Rob Feick is an associate professor in the School of Planning at the University of Waterloo. His research focuses broadly on the application of spatial information technology to assist decision making and public participation in land management and planning. His current research focuses on the development and evaluation of PPGIS, VGI, and Web 2.0 tools that facilitate citizen involvement in community planning contexts, spatial-decision-aiding methods such as GIS-based multicriteria analysis, and Web-based spatial data visualization.

Marcus Goetz is a research assistant at the GIScience group (Geoinformatics) at Heidelberg University since early 2010. He has a background in mathematics and computer science and holds a diploma degree in Computer Science from the Karlsruhe Institute of Technology (KIT). He is currently working towards his PhD at the University of Heidelberg. Among other things, his current research interests include 3D city models, CityGML, (3D) VGI, and OSM, as well as (3D) indoor routing and indoor LBS.

Christopher Goranson is the director of the Parsons Institute for Information Mapping at The New School. PIIM is a research, development, and professional services facility that specializes in data and knowledge visualization. Previous to working at PIIM, Chris was the director of the GIS Center within the Bureau of Epidemiology Services, NYC DOHMH. The GIS Center is tasked with providing centralized resources including training, consulting, map production, and geographic information analysis support to the department. Prior to joining the department in 2004, Chris worked for a professional services firm supporting various federal client GIS projects including those for the EPA, FHWA, GSA, and USGS.

Muki Haklay is professor of geographic information science in the Department of Civil, Environmental and Geomatic Engineering, University College London. He is the director of the UCL Extreme Citizen Science research group, which aims to allow any community, regardless of their literacy, to use scientific methods and tools to collect, analyze, interpret, and use information about their area and activities. His research interests include (1) public access and use of environmental information, (2) human-computer interaction (HCI) and usability engineering aspects of GIS, and (3) societal aspects of GIS use – in particular, participatory mapping and citizen science. He received his PhD in Geography from UCL and holds a BS in Computer Science and Geography and MA in Geography from the Hebrew University of Jerusalem.

Darren Hardy is a senior analyst for the Ocean Health Index project at the National Center for Ecological Analysis and Synthesis, UC Santa Barbara. His research interests include spatial analysis, open scientific computing, distributed systems, and technology and society. He earned his PhD (2010) and Masters (2005) in Environmental Science & Management at the Bren School, UC Santa Barbara. His interdisciplinary dissertation examines the production and use of volunteered geographic information in Wikipedia. He has over 20 years of professional experience in roles for academia and industry, including the Harvest information discovery and access system (1995), an early Web search engine and proxy software (now Squid), and software engineering positions in Silicon Valley at Netscape Communications Corp., Affinia, Inc., and Napster, Inc. He also earned MS (1993) and BS (1991) degrees in Computer Science from the University of Colorado, Boulder.

Francis Harvey is an associate professor at the University of Minnesota. His research interests include location privacy, spatial data infrastructures, geographic information and sharing, semantic interoperability, and critical GIS. He serves on the editorial boards of the *International Journal for Geographical Information System*, *Cartographica*, *GeoJournal*, and the journal of the URISA. He published *A GIS Primer* with Guildford Press in 2008. He is currently finishing work on a long-term research project in Poland considering discrepancies between the cadastre and land use. He continues to work on SDI research, currently through an FGDC-supported project examining the return on investment for parcel data in regional data sharing (MetroGIS). He also contributed to the development of a model curriculum and resources for GIS ethics teaching (gisprofessionalethics.org).

Bin Jiang is professor in geoinformatics and computational geography at University of Gävle, Sweden. He is also affiliated to Royal Institute of Technology (KTH) at Stockholm via KTH Research School. He worked in the past with the Hong Kong Polytechnic University and the Centre for Advanced Spatial Analysis at University College London. He is the founder and chair of the International Cartographic Association Commission on Geospatial Analysis and Modeling. He has been coordinating the NordForsk-funded Nordic Network in Geographic Information Science. His research interest is geospatial analysis and modeling, in particular topological analysis of urban street networks in the context of geographic information systems.

He is currently an associate editor of the international journal *Computers, Environment and Urban Systems*.

Peter A. Johnson is a postdoctoral researcher and lecturer in the Department of Geography at McGill University, Montreal, Canada. In 2010, he completed his PhD in Geography, also from McGill University. For his dissertation, he developed an agent-based model to support the development of tourism planning scenarios. His research interests include the Geoweb, participatory GIS, and the use of geospatial technology in a community development context. In 2012, he will begin an appointment as assistant professor in the Department of Geography and Environmental Management at the University of Waterloo in Waterloo, Ontario.

Carsten Keßler (http://carsten.io) is a postdoc researcher at Institute for Geoinformatics, University of Münster, Germany, where he has been working in the Semantic Interoperability Lab (MUSIL) for several years. He has done research on context-aware information retrieval on the Semantic Web during his PhD studies. Volunteered geographic information has been one of his focus topics in teaching and research over the last years, with a strong focus on the semantics of VGI. His recent work focuses on the development of provenance-based measures of trustworthiness for VGI. His other research interests are in the areas of geosemantics, linked data and semantic technologies, context modeling, and collaborative and participatory geographic information systems. Carsten has co-organized a number of workshops and conferences, and he currently coordinates the University of Münster's Linked Open Data initiative (http://lodum.de).

Scott Kraushaar graduated from the University of Missouri–Columbia with an MA in 2011. Scott's thesis research focuses on volunteered geographic information and storm chasing including motivations for reporting, observing, and documenting storm reports. Scott has an interest in meteorology and recreational storm chasing and has been observing/spotting storms since the late 1990s. He is currently employed with the United States Department of Defense. In his spare time, Scott has been known to travel great distances in pursuit of rotating supercell thunderstorms and tornadoes.

Wen Lin has been a lecturer at the School of Geography, Politics and Sociology at Newcastle University in Newcastle upon Tyne, UK, since spring 2012. Prior to taking her current position, Wen was an assistant professor in the Department of Geography and Earth Science at University of Wisconsin–La Crosse (2009–2012). Her research interests include critical GIS, public participation GIS, and urban geography. Her main research centers on examining the intersection between the development and usage of geospatial technologies and the sociopolitical conditions in which these practices are situated. In particular, she has worked on three related themes: investigating the sociopolitical implications of recent mapping practices combined with Web 2.0 technologies, examining GIS-related practices in China's urban planning agencies, and examining public participation GIS practices in urban governance.

Grant McKenzie is a second year PhD student in the Department of Geography at the University of California, Santa Barbara. He holds a Master of Applied Science degree from the University of Melbourne (2008) and an Advanced Diploma in Geographic Information Science from the British Columbia Institute of Technology (2004). During his time in Melbourne, Grant was the recipient of the J H Mirams Memorial Research Scholarship and was awarded a Google Doctoral Colloquium Award and Scholarship for promising research proposal at the 2007 Conference on Spatial Information Theory (COSIT 2007). Prior to starting his PhD, Grant was a founding member of the Seattle-based start-up Spatial Development International and worked as a geospatial software developer for the engineering consulting firm CH2M HILL. Grant completed his BA in Geography at the University of British Columbia in 2002.

Mark H. Palmer is an assistant professor of geography at the University of Missouri–Columbia. His current research interests focus upon the social aspects of geographic information systems including the uneven development of geographic information networks within government agencies and their connections/disconnections within local and indigenous communities in North America. Palmer is also interested in understanding the dynamic interfaces between local knowledge systems, the geosciences, and digital technologies like GIS, in hopes of determining how elements of culture like language, storytelling, education, and performance influence the use and understanding of geographic information. Since moving to the forested hills of Missouri, Palmer has officially retired from amateur storm chasing.

Barbara S. Poore is a research geographer in the Center of Excellence in GIScience at the US Geological Survey specializing in volunteered geographic information and the use of social media in crowdsourced mapping. She has an AB in Art History from Wellesley College, an AM in Art History from Brown University, and a PhD in Geography from the University of Washington. Before embarking on a research career, Barbara worked for the Federal Geographic Data Committee during the establishment of the National Spatial Data Infrastructure. She lives and works by the bay in Saint Petersburg, FL.

Stéphane Roche is a professor of geographical information sciences in the Département des sciences géomatiques at the Université Laval, Québec, Canada. Stéphane is a surveying engineer (ESGT, France). He did a Masters in Planning and a PhD in Geography in the University of Angers (France). He is mainly interested in the analysis of the relationships between space and society with regard to the development process of the GeoWeb 2.0 and in the design of geospatial collaborative solutions (participatory GIS, WikiGIS) to address participatory geodesign practices. Stéphane coordinated with Claude Caron (University of Sherbrooke) for the book "Organizational Facets of GIS," published in 2009 by John Wiley and Sons. He was also (with Rob Feick, University of Waterloo) guest editor of the *Geomatica* special issue (vol. 64, n. 1) on volunteered geographic information. He is currently guest editor (with Mike Goodchild, University of California, Santa Barbara) of the special issue on geodesign of the *International Journal of Geomatics and Spatial Analysis*.

Renee E. Sieber received her PhD from Rutgers University and is currently an associate professor at McGill University in Montreal, Canada. Her prime research focus is public participation geographic information systems (PPGIS), the methods by which those who are marginalized from public policy can use computational mapping and spatial databases to better participate in policy making. These individuals may be inner city or indigenous peoples. She brings to this a background as a community organizer and activist as well as a computer programmer. She increasingly researches PPGIS on the geospatial Web 2.0 (Geoweb). Her research areas are diverse. She leads a team of 10 researchers in the use of the participatory Geoweb for global environmental and climate change. She also conducts research in the digital humanities. She organized the first public participation GIS conference. She cofounded the GIS specialty group of the Canadian Association of Geographers and co-organized Spatial Knowledge and Information Canada, the first academic GIS conference in Canada.

Jim Thatcher is a PhD candidate in Geography at Clark University. His research focuses on the intersection of global capitalist systems and mobile geospatial technology. As the use of mobile geospatial technologies rises, what is known, what can be known, and what can be done all shift for both the individual and the state. Jim's research focuses on how programmatic decisions can delimit the episteme of the end user. He can be followed on twitter @alogicalfallacy.

Sayone Thihalolipavan worked with Christopher Goranson on alcohol-related mapping projects in the past. He currently serves as the director of the Cessation Unit at the Bureau of Chronic Disease Prevention and Tobacco Control at NYC Department of Health and Mental Hygiene. He advises on BTC's policy, clinical, and educational interventions, ensuring conformity with current evidence and best practices as well as helping to devise innovative strategies to deliver population cessation interventions through direct mail, the internet, email, text messaging, and other media including the annual nicotine and gum program which gives and serves about 40,000 New Yorkers annually. He also oversees the employee smoking cessation clinic for any NYC government employee.

Eric B. Wolf has worked for over two decades in the software industry and is a geographer in the Center of Excellence in GIScience at the US Geological Survey. He does research on integrating citizen contributions into *The National Map*. Eric has a BS in Applied Mathematics from the University of Tennessee at Chattanooga and an MS in GIScience from Northwest Missouri State University. He is currently a doctoral candidate at the University of Colorado at Boulder under the advisement of Dr. Barbara P. Buttenfield. His dissertation is focused on the structure of metadata for spatial data infrastructures that include volunteered geographic information. Eric and his wife, Asha W. Wolf, live in Longmont, Colorado, with their dog and two cats.

Alexander Zipf is chair of GIScience (geoinformatics) at Heidelberg University since late 2009 and is a member of the Interdisciplinary Center for Scientific Computing (IWR), the Heidelberg Center for the Environment (HCE), and the

Department of Geography. He has previously been chair of cartography at University of Bonn and, earlier, was professor for applied computer science and geoinformatics at the University of Applied Sciences in Mainz, Germany. He has a background in mathematics and geography and finished his PhD at the EML European Media Laboratory in Heidelberg. Current research interests include among others 3D GIS, Spatial Data Infrastructures 2.0, as well as volunteered and crowdsourced geoinformation or location-based services (http://giscience.unihd.de).

Index

A
Aboriginal knowledge, 233–236
Aboriginal people, 224, 229
Absolute space, 135, 136
Accuracy, 8, 20, 22, 33, 39, 40, 58, 67, 70, 73, 74, 79, 98, 111–116, 161, 162, 189, 233, 236, 251, 258, 259, 266, 267, 271, 273, 275–277, 279, 281, 282, 297, 302, 307, 310–315, 317, 325, 332, 335, 348
Action, 34, 69, 71, 74–78, 109, 161, 162, 164, 165, 168–171, 194, 226, 289, 290, 293, 294, 296, 298, 308, 365, 367, 370
Active community input, 309–311, 314
Active contribution, 112, 258, 268, 269, 311–314
Activity topic, 208
Actor, 133, 165, 179, 288, 290, 292, 296
Actor-network theory (ANT), 8, 287–304, 370
Address, 2, 9, 26, 32, 66, 69, 73, 76, 78, 85–87, 89, 92, 94, 95, 97, 98, 128, 132, 136, 148, 162–165, 167, 170, 171, 178, 180, 183, 184, 189, 191, 192, 194, 195, 232, 251, 256, 258, 266–268, 270–276, 278, 289, 310, 313, 325, 330, 332, 334–336, 338, 343, 344, 362–364
Adoption, 7, 39, 52, 65–79, 170, 258, 330, 339, 375
Aerial surveys, 248
Aeronautical charts, 247
Aggregate, 22, 33, 34, 57, 134, 161, 204, 207, 209, 212, 228, 232, 280, 282, 331, 332, 336
Alcohol, 331, 332, 335

Alcoholics Anonymous meeting, 335
Amazon.com, 48, 56
Android, 168–170, 348
Anonymity, 193
Anonymizing, 193
Anonymous users, 184
ANT. *See* Actor-network theory (ANT)
Artificial intelligence, 127
Attributes, 37, 39, 46, 47, 58, 60, 141, 143–148, 156, 226, 245, 246, 249, 252, 257–259, 266, 271, 272, 276, 280, 281, 304, 310–312, 317, 318, 321, 325, 330, 332, 334, 363
Authoritative data, 72, 79, 247–248, 252, 314, 321, 370
Authoritative GI, 16, 18–25
Authoritative knowledge, 229, 232
Authority, 61, 68, 252, 301, 319
Authorship, 8, 175–195, 204

B
Bacon numbers, 133
BADUPCT. *See* Bureau of Alcohol and Drug Use, Prevention, Care and Treatment (BADUPCT)
Behavior, 48, 61, 132, 135, 177, 187, 189, 193–195, 224, 278, 280, 309, 312, 331, 332, 335, 337, 352, 364, 370
Big data, 2–9, 333, 371, 372
Bing, 4, 141, 147, 153, 163, 228, 258, 259, 330
 aerial imagery, 4, 143, 147, 258, 311
 maps, 4, 56, 141, 147, 153, 163, 228, 330
Blog. *See* weblog
Border numbers, 133

Bots, 177, 178, 180, 184, 191, 192
Bottlenecks, 44, 78, 164, 166, 171
Bottom-up ontology, 55, 58
Boundaries, 16, 75, 86, 88, 97, 98, 100, 106, 107, 134, 190, 191, 207, 225, 292, 310, 324, 332, 333, 336, 363, 366, 368
Boundary, 75, 190, 324, 332, 336
British Columbia, 223, 230, 232
British Columbia Treaty Process, 234
Bruno Latour, 289
Building footprints, 148–154, 249
Building generation, 150
Built environment, 331
Bureau of Alcohol and Drug Use, Prevention, Care and Treatment (BADUPCT), 331

C
Cadastral, 17, 247
Callon, Michel, 289, 290
Canada, 19, 226, 228, 229, 234, 247, 250, 313
Carrier IQ, 335
Cartographer, 10, 226, 227, 293, 309
Casual geographers, 309
Cathedral and bazaar, 25
Center of calculation, 288, 292–294, 300–303
Changeset, 56, 57
Charting, 246
Chief Privacy Officer, 337
China, 7, 83–100, 216, 362, 367
Citizen participation, 7, 65–68, 73, 78, 84, 85, 91, 92, 100
Citizens, 2, 19, 35, 46, 65, 84, 109, 165, 228, 245, 267, 288, 363
  science, 2, 7, 84, 86, 105–120
  scientist, 66, 115, 118, 119, 297–299, 368
  weather observers, 295
Citizenship, 7, 85, 90–92, 98, 100, 281
Civic engagement, 85, 90, 99, 100, 367
Clearinghouse, 44, 47, 537
Cloud, 2, 3, 20, 56, 69, 128, 177, 208, 232, 259, 298, 302, 330, 372
Cloud computing, 2, 3, 128, 259
Clustering algorithm, 134
Collaboration, 19, 60, 76, 77, 94, 118–120, 126, 178, 195, 230, 234, 248, 308, 330, 343, 368
Collective action, 194, 308, 367
Collective authorship, 175–179, 194
Colonization, 229

Communication, 6, 9, 21, 26, 44, 45, 58, 65–68, 78, 85, 89–91, 93–98, 100, 106, 109, 111, 118, 171, 176, 178, 180, 193, 195, 224, 227–229, 234, 237, 238, 257, 266, 295–299, 301, 302
Communities of users, 44, 46, 49
Community, 3, 17, 45, 68, 84, 107, 127, 139, 177, 223, 248, 267, 309, 331, 344, 372
Community-based research (CBR), 230
Computational geography, 7, 125–136, 361
Computational social science, 127
Confidential, 18, 24, 273, 280, 310, 333–335
Connection to the land, 237
Constraints, 7, 47, 67–70, 72, 74–76, 78, 136, 155, 166, 193, 251, 325, 365
Constructionism, 259
Constructivism, 359
Content Standard for Geospatial Metadata (CSDGM), 46, 47
Contestation, 85, 91, 92, 98
Contour, 53, 211
Contributors, 19, 23, 24, 26, 70–73, 76, 107, 112, 126, 141, 176–178, 182, 186, 193, 194, 228, 246, 250–252, 255–258, 268, 281, 309, 310, 312, 315–318, 323–326, 364, 366, 367, 369, 371
Conventional datasets, 248
Co-production of data, 44, 46
Corporate Spatial Data Library, 253
Credibility, 73, 236, 251, 252, 256, 258, 297, 300
Crenulation, 53, 54
Crisis
  camps, 55
  mapping, 20, 92, 94, 161
  response, 8, 162, 164–167, 170, 171
Critical GIS, 67, 85, 87, 99
Crowdsourced, 2, 7, 16, 31–40, 45, 58, 66, 76, 135, 204, 310–324, 367
  compilation systems, 310, 315, 316, 319
  data, 2, 7, 16, 31–40, 204, 312, 314, 315, 317–319, 324, 325
  geodata, 139–158
CSDGM. *See* Content Standard for Geospatial Metadata (CSDGM)
CUNY School of Public Health-Hunter College, 331
Curators, 47
Curriculum, 343–350, 357
Cyber infrastructure, 2, 9, 52, 278, 372
Cyborg, 87, 88, 288
Cycle of accumulation, 293

Index

D
3D. *See* Three-dimensional (3D)
Data extraction, 183, 184, 269
Data infrastructure, 366
Data integration, 47, 191, 192
Data-intensive computing, 126, 127, 133
Data mining, 127, 178, 180, 181, 184, 192–194, 202, 204, 270, 280, 314, 321, 322
Data quality, 22, 23, 32, 38–40, 47, 52, 60, 73, 74, 76, 106, 113, 235, 247, 251, 307–310, 312, 313, 315–321, 325, 339, 358, 364, 369
Data sharing, 45, 191
Data spam, 310, 325
DE. *See* Digital earth (DE)
Death of distance, 9–10
Decision-making, 18, 19, 65–69, 72–79, 87, 109, 115, 126, 162, 165, 171, 224, 227, 230, 278, 298
De-identification, 336–338
Densification, 53
Dérives, 7, 25, 27
Dialogue, 77, 224, 232, 235, 268, 290, 303, 367
Diaspora, 193, 223, 235
Diffusion, 77, 90, 187
DigiPlace, 87–89, 92, 98–100
Digital commons, 176, 195
Digital divide, 1–10, 99, 100, 269, 280, 288, 370, 371
Digital earth (DE), 4
DigitalGlobe, 55
Digital line graphs (DLGs), 53
Digital map maintenance, 248
Digital terrain model (DTM), 143, 146, 148
Digital traces, 193, 280
Direct observation, 311
Discourse, 7, 16, 85, 90, 91, 266, 278, 280, 301, 363
Displacement, 225, 226, 278
Distance decay, 8, 182, 184, 187, 188
Distribution of volunteers, 316
Division of labor, 317
DLGs. *See* Digital line graphs (DLGs)
Domestic violence, 335
3D-SLD. *See* 3D styled layer descriptor (3D-SLD)
DTM. *See* Digital terrain model (DTM)
Dublin Core, 179

E
Edit wars, 192, 194
Education, 3, 9, 25, 111, 112, 119, 280, 281, 297, 317, 334, 341–359, 361

E-governance, 66, 68, 91
Embedded metadata, 48–49
Entropy, 205, 209–211, 215
Erosion, 68, 71, 75, 238, 266
Error, 20, 22, 56, 61, 72–74, 95, 235, 251, 266, 268, 275, 276, 307, 310–316, 318, 319, 321, 323, 364
eScience, 2, 37, 126, 127, 371
Esri, 54
Ethics, ethical problems, 17, 33, 35, 334, 337
Exaflood, 1–10
Exploitation, 88, 333, 337
Exploratory analysis, 206, 211
Exponential, 61, 126, 130, 182, 187, 188, 209, 335
Exponential decay, 182, 187, 209
eXtensible Markup Language (XML), 49, 50, 56, 111, 163, 348, 349

F
Facebook, 3, 5, 24, 71, 136, 261, 271, 299, 301, 372
Feature topic, 208
Federal Geographic Data Committee, 44, 46, 363
Federal Privacy Rule, 336
Field, 7, 8, 21, 23, 24, 27, 47, 52, 54, 59, 61, 72, 106, 111, 113, 125–128, 132, 161, 163, 187, 207, 211, 228, 249, 253, 256, 273, 275, 277, 288, 294–299, 301–303, 312, 314, 331, 332, 345, 346, 349, 350, 359, 362
Findability, 44, 46, 49, 51, 52, 54, 55, 58, 59
First law of geography, 135, 209, 278, 365
Fitness-for-use, 6, 7, 16, 18, 22, 26, 33, 38–40, 52, 58, 315
Flickr, 3, 39, 136, 141, 180, 192, 304, 346, 350
Folksonomies, 52
Food Retail Expansion to Support Health (FRESH), 333
Fourth paradigm, 9, 126, 361–372
FRESH. *See* Food Retail Expansion to Support Health (FRESH)

G
Gazetteer, 58, 180, 191, 363
Geoblogosphere, 45
GeoChat, 331, 345
Geo-coding, 9, 207, 276, 334, 335
Geocomputation, 127

Geo crowd sourcing
GeoEye, 55, 56
Geographic
    data, 1, 3, 18, 44, 60, 125, 140, 143, 180, 234, 273, 304, 313, 330, 332, 334, 335, 337, 345, 363
    proximity, 334
    effects, 182, 194
    footprint, 334, 335, 338, 339
    forms and processes, 127, 128, 135, 136
    imagination, 133
Geographic information (GI), 2, 15, 31, 45, 68, 106, 125, 161, 178, 202, 227, 246, 288, 308, 330, 341, 362
Geographic information science, 114, 125, 202, 334, 362,371
Geographic information systems (GIS), 68, 106, 126, 204, 228, 303, 341, 363
Geographic Names Information System, 58, 180
Geoinformatics, 126–128, 136, 139, 345–348, 351, 357
Geolive, 232, 234, 235
Geomatics, 127
Geometric center, 133
Geometry, 128, 129, 140, 141, 143, 148–154, 156, 179, 309, 310, 322, 324
Georeferencing, 179, 191, 192, 288, 300–302, 304
(Geo)slavery, 24, 87, 269
Geospatial, 3, 21, 44, 66, 85, 126, 162, 191, 228, 246, 269, 287, 330, 345, 371
Geospatial metadata, 7, 44–46, 49, 58
Geotag, 3, 4, 9, 20, 22, 34, 113, 141, 176, 179–184, 186, 189–192, 258, 288, 300, 302, 303, 346, 349–354, 356, 358
Geotagging, 9, 179, 180, 190–192, 288, 300, 302, 346, 349, 352, 354, 356, 358
GeoWeb, 2, 7, 16, 20, 24, 43–61, 66–71, 75, 84, 163, 223–238
Gibbs sampling, 205, 208
GIScience, 2, 9, 17, 45, 52, 125, 127, 128, 191, 192, 289, 343, 345, 351, 357, 362, 364, 371–372
GIS professional, 44, 45, 50, 51, 337, 338
Global positioning system (GPS), 3, 107, 109–112, 115, 116, 126, 140, 141, 143, 148, 163, 169, 192, 193, 228, 253, 258, 269, 288, 297, 299, 300, 302, 309, 310, 312, 314, 317, 330, 332, 334, 343, 344, 346–349
Global scale, 4, 191, 194

GMM. *See* Google MapMaker (GMM)
Google, 3, 4, 10, 20, 33, 36, 37, 47, 51, 52, 54, 59, 74, 87, 88, 90, 92, 93, 97, 176, 228, 246, 256–259, 279, 308, 311–314, 321–324, 330, 333–335, 350, 352–354, 356, 358, 365, 372
Google Earth, 4, 87, 88, 90, 92, 228, 313, 330
Google MapMaker (GMM), 10, 36, 37, 59, 107, 245, 313, 322, 323, 365
Google Maps, 3, 10, 20, 69–71, 74, 86, 88, 92, 93, 95–97, 126, 163, 169, 207, 228, 232, 245, 276, 312, 313, 321–323, 330, 345, 347, 350, 352
Governance, 19, 66–70, 73, 74, 77–79, 90, 91, 368
Government, 2, 15, 33, 51, 65, 93, 118, 136, 165, 223, 246, 266, 288, 311, 330, 366
Government mapping agencies, 246, 248, 250
GPS. *See* Global positioning system (GPS)
GPS traces, 317
Graduate studies, 9, 341–359
Gravity model, 184–189
Great circle distance, 189

H
Haiti, 6, 20, 21, 46, 55, 56, 58, 59, 94, 164, 165
Have2p, 6
Head/tail division rule, 132
Health Information Technology for Clinical and Economic Health (HITECH), 337
Health Insurance Portability and Accountability Act (HIPAA), 336, 337
Health, public health, 8, 126, 329–339, 345, 362
Heavy-tailed distribution, 128, 130–133, 135
High-performance computing, 127, 128
HIPAA. *See* Health Insurance Portability and Accountability Act (HIPAA)
HITECH. *See* Health Information Technology for Clinical and Economic Health (HITECH)
HIV, 334, 335
Homophile, 6
HON Code of Conduct, 336
Human geography topics, 204
Human mobility patterns, 132

Index 391

Humans as sensors, 287, 288, 298, 301
Human-sensors, 288, 291, 294, 295, 297, 298, 300, 301, 303, 304
Human subjects, 114, 337, 338
Hybrid map compilation, 315–319
Hydrography data, 53, 54

I

Identity, 73, 89, 100, 223, 224, 226, 233, 236, 237, 252, 279, 310, 335
Incentives, 22, 73, 334, 342, 349, 350, 354, 355, 358
Indirect observation, 311
Information field, 187
Information production, 176–178, 249
Informed consent, 337
Innovation, 9, 20, 24–26, 78, 187, 205, 294
InSTEDD, 331
Institutional review board (IRB), 336–339
Intellectual property, 229, 249
Intelligent crowd, 317
Interactive, 7, 44–46, 48–49, 59, 89, 98, 99, 176, 191, 192, 228, 256, 313
Intermediary, 59, 74, 237, 288, 290
Internet, 2, 21, 38, 44, 65, 84, 105, 176, 228, 266, 288, 309, 370
Internet censorship, 6, 85, 91, 92, 97
IP geolocation, 184, 185, 193
IRB. *See* Institutional review board (IRB)

J

Java, 347
Jensen-Shannon divergence, 205
Jumping scale, 74
Jurisdiction, 66, 68, 70, 74–76, 230, 245, 247, 248, 258, 259, 324

K

K-12, 341–359
Kernel density estimation (KDE), 211, 212
Keyhole Markup Language (KML), 78, 330, 347
Keyword, 47, 51, 52, 190, 270, 301
KML. *See* Keyhole Markup Language (KML)
Knowledge production, 2, 7, 10, 85, 86, 88, 90, 92, 99, 100, 106, 118, 119, 176, 233, 372
Knowledge transfer, 229, 330
Kullback-Liebler divergence, 205

L

Ladder of participation, 115
Land claims, 224, 228, 231
Land management, 66, 71
Latent Dirichlet allocation (LDA), 8, 203–209, 215, 216, 219
Latitude, 25, 56, 179, 191, 207, 302, 335, 336, 363, 368
Latitude (Google), 207, 335
Law, John, 289
LDA. *See* Latent Dirichlet allocation (LDA)
Learning, 24–26, 53, 69–71, 117, 205, 228, 237, 270, 297, 342–347, 349, 357, 359
Lego block, 7, 24–27
Line, 38, 53, 86, 87, 119, 129, 188, 216, 252, 273, 299, 366
Linus's Law, 126, 364
Locality topic, 208, 216
Localization, 215–216, 252
Local knowledge, 66, 67, 72, 74, 86, 293, 300, 310, 311, 315–317, 321, 323, 325, 368
Local users, 310, 324
Location, 4, 20, 31, 46, 66, 84, 106, 129, 148, 162, 178, 202, 224, 251, 267, 297, 308, 330, 341, 363
Location-based services; location based information, 2, 36, 46, 72, 163, 189, 316, 330–332, 335, 338, 341
Location entropy (of topic distribution), 211
Lognormal, 130, 132
London underground map, 129
Longitude, 56, 179, 191, 207, 302, 335, 336, 363, 368

M

Maintenance, 8, 20, 70, 119, 224, 236, 248, 260
Mandate, 52, 69, 75, 77, 247, 248, 279, 307
Map compilation, 249, 307, 309–311, 313–326
Map-mediated dialogue, 232, 367
Mapping, 3, 16, 37, 45, 83, 107, 125, 141, 161, 176, 207, 224, 245, 266, 288, 307, 344, 365
  party, 38, 39
  topics, 214
Maps, 3, 20, 48, 68, 86, 112, 126, 140, 163, 176, 207, 226, 245, 276, 288, 307, 330, 344, 363

MapShare, 253, 255–256
Maptivism, 86, 87
Markov chain Monte Carlo (MCMC), 205
Markup language, 47, 158, 179, 183, 192
Mashup, 26, 70, 86, 92, 95, 96, 126, 163, 192, 232, 266
MCMC. *See* Markov chain Monte Carlo (MCMC)
Measuring work (edit count), 193
Media, 5, 20, 34, 45, 71, 89, 112, 128, 177, 228, 246, 270, 288, 313, 330, 368
Medial axis, 133
MediaWikimetadata, 182
Memories, 224–226, 230, 231, 235–238
Metadata, 7, 38, 43, 168, 179, 257, 267, 325, 369
Metadata squared, 43–61
Metadata standards, 44, 47, 50, 52, 60, 191
Meteorologist, 288, 294, 296, 298–304
Microformat, 179
Microsoft, 4, 33, 228, 259, 308, 330, 334, 336
MINUSTAH. *See* United Nations Stabilization Mission in Haiti (MINUSTAH)
Mobile application, 278, 343
Mobile Device Privacy Act, 335
Mode of information, 85, 89–90, 98, 100
Motivations, 9, 19, 23, 24, 67, 68, 113, 114, 117, 251, 268, 277, 281, 304, 342, 349–351, 353, 357, 358, 363
Multimedia, 84, 232, 237, 270, 279, 370
Municipal government, 74, 75, 77
Murphy's law, 6

N
Named streets, 131
National Hydrography Dataset, 54
National Institutes of Health, 337
National Map, 45, 46, 58, 125, 247, 253, 367, 368
National Map Corps, 253–255, 257
National Oceanic and Atmospheric Administration (NOAA), 296–298
National Science Foundation, 334, 342, 371
National Severe Storms Laboratory (NSSL), 295, 297
National Weather Service (NWS), 288
Natural cities, 133–135
Natural language, 201–219
Natural language processing (NLP), 202, 272, 369
Natural Resources Canada, 19
Natural streets, 131
Nautical charts, 247

Navigation, 25, 34, 112, 129, 176, 256, 270, 307–326
NAVTEQ, 245, 246, 250, 256, 308, 311, 313, 314, 319, 320
NCGIA, 2
Nearness, 194
Neighborhood, 39, 167, 257, 278, 330–333, 344
Neocartography, 330
Neogeographer, 45, 46, 49–52, 60, 72, 372
Neogeographic datasets, 246
Neogeography, 2, 49, 50, 52, 59, 84, 92, 194, 246, 330
Neoliberalization, 66
NES. *See* Notification for Edit Service (NES)
Netizen, 91, 92, 99
Network, 8, 18, 25, 53, 55, 109, 111, 131, 132, 163, 193–195, 250, 253, 268, 273, 288, 290–292, 294–296, 298–300, 302, 304, 319, 326, 339, 346, 358, 359
Neutral point of view, 177
New media, 46, 47, 49, 78, 89, 90, 100, 194, 246, 368
New York City Department of Health and Mental Hygiene (NYC DOHMH), 331, 332
NLP. *See* Natural language processing (NLP)
NOAA. *See* National Oceanic and Atmospheric Administration (NOAA)
Non-expert, 45, 70, 72–74
Normal distribution, 130, 132, 135
Notification for Edit Service (NES), 253
NSERC. *See* Natural Sciences and Engineering Research Council of Canada (NSERC)
NSSL. *See* National Severe Storms Laboratory (NSSL)
NWS. *See* National Weather Service (NWS)
NYC Department of City Planning, 332, 333
NYC DOHMH. *See* New York City Department of Health and Mental Hygiene (NYC DOHMH)
NYC Economic Development Corporation, 333
NYC Projection Area, 332

O
Object-level metadata, 55, 56, 281
Occupy wall street, 23
Office of the National Coordinator for Health Information Technology (ONCHIT), 336, 337

Index 393

OGC. *See* Open Geospatial
    Consortium (OGC)
OLPC. *See* One Laptop Per Child (OLPC)
ONCHIT. *See* Office of the National
    Coordinator for Health Information
    Technology (ONCHIT)
One Laptop Per Child (OLPC), 347
Online mapping, 45, 49, 84, 86, 87, 92, 99,
    176, 180, 224, 312
Ontology, 27, 52, 55, 58, 61, 168, 270
OpenDataKit, 331, 335
Open Geospatial Consortium (OGC),
    143, 147, 148, 158, 330
Open source, 20, 25, 45, 46, 51, 56, 60, 61, 77,
    87, 93, 106, 162, 167, 170, 182,
    232, 246, 257, 331, 344, 369
OpenStreetMap (OSM), 7, 20–22, 25, 27, 32,
    34, 37–39, 45, 52, 55, 59–61, 88,
    100, 107, 112, 126, 128, 131, 134,
    136, 140, 156–158, 161, 165, 245,
    251, 255, 267, 269, 312, 315–317,
    319, 323, 330, 342–343, 358–346,
    351, 354, 357, 358, 365–367
OpenTopography, 4
Opportunity costs, 71
Ordnance Survey, 248
Organization, 18, 19, 21, 44, 46, 47, 52–55,
    67–71, 75, 77, 78, 84–86, 91, 93,
    98–100, 167, 171, 178, 202, 207,
    227, 235, 245–251, 253, 255–260,
    288, 289, 294, 296, 301, 307, 324,
    337–339, 347, 368
OSM. *See* OpenStreetmap (OSM)
OSM-3D, 140, 143, 144, 147–150, 152, 154,
    156–158

P

Participation, 7, 16, 32, 56, 65, 83, 105,
    140, 194, 225, 267, 288, 308,
    338, 342, 367
Participatory culture, 325–326
Participatory geoweb, 226
Participatory GIS, 106, 115, 119, 367
Participatory mapping, 8, 83–100, 107, 224,
    226–228, 232, 235, 236
Participatory VGI, 84–86, 92, 100, 119, 358
Passive community input, 309–310, 321, 325
Passive contribution, 258, 269, 310, 321
Peer production, 178
People-finder sites, 266–268, 270–272, 275,
    277, 280
Personal health record (PHR), 336
Personal location data, 4

Personally identifiable information,
    333–335, 338
Personal travel history; travel paths,
    334–336
Photographs, 38, 84, 97, 98, 105, 126, 176,
    230, 232, 233, 237, 304, 368
PHR. *See* Personal health record (PHR)
Physical geography topics, 215
Place, 7, 16, 54, 66, 87, 112, 132, 142, 162,
    176, 201, 223, 251, 266, 289, 311,
    332, 349, 362
Place-names, 215, 302, 363
Platial, 362–364, 367–370
POI. *See* Point-of-Interests (POI)
Point, 7, 19, 21, 35, 36, 51, 56, 88, 96, 111,
    112, 141, 152, 176, 177, 179, 201,
    207, 211, 212, 216, 225, 227, 234,
    235, 251, 252, 258, 276, 318, 332,
    334, 335, 342, 356, 363, 367, 369
Point-of-Interests (POI), 142, 143, 148, 307,
    343, 346
Polygon, 141, 148, 179, 211, 212, 215, 300,
    303, 330, 332
Population dynamics, 266, 273, 279
Portal, 3, 47, 237, 278, 343
Positional accuracy, 22, 58, 189, 251, 252,
    276, 307, 310, 315, 317
Positioning, 251, 252, 343
PostGIS, 2
Power law, 6, 128, 130, 134, 135, 177
Power law distribution, 128, 130,
    134, 177
Power relations, 7, 84–86, 100, 106,
    115, 177
PPGIS. *See* Public participation
    GIS (PPGIS)
Precision, 20, 73, 113, 114, 192, 271, 272
Privacy, 8, 17, 21, 24, 33–36, 39, 40, 67, 84,
    86, 87, 193, 229, 232, 234, 247,
    252, 266, 268, 269, 273, 279–282,
    333–338, 354, 372
  policy, 36
  rule, 336–338
Private good, 17, 18
Private sector, 19, 87, 250, 258, 268, 369
Privatization, 301
Procedural extrusion, 153
Produsage, 249
Produsers, 22, 25, 164, 249
Public good, 17, 18
Public health, 8, 126, 329–339, 345, 362
Public health researcher, 331, 333, 334, 339
Public participation GIS (PPGIS), 7, 16, 23,
    67, 68, 78, 83–100, 140, 194

Public sector, 4, 19, 21, 22, 246, 258–259
Public-sector mapping, 246, 258–259
Public sphere, 90, 92, 100, 118

Q

Qualitative, 17–19, 45, 61, 73, 113, 128, 143, 145, 156, 164, 228, 232, 233, 235, 237, 276, 331, 356, 368–370
Qualitative information, 113, 143, 145, 231
Quality assurance, 247, 249–256, 272, 312, 362, 364, 365, 369
Quebec, 19, 67, 68, 75, 79

R

Radar, 294, 295, 298, 300–303
Realistic 3D models, 143, 150
Region, 75, 140, 154, 190, 202, 207, 215, 231, 275, 295, 297
Regional geography, 184, 202
Registered users, 141, 232, 313
Relative entropy. *See* Kullback-Liebler divergence
Relative space, 135
Reliability, 236, 252, 256, 258, 259, 276, 295, 314, 367, 369
Representation, 8, 16, 84, 88, 89, 96, 98, 100, 130, 135, 162, 164, 165, 191, 207, 211, 218, 219, 227, 249, 251, 252, 257, 280, 288, 302, 303, 332, 344, 368–370
Research, 2, 17, 33, 44, 67, 84, 109, 126, 141, 162, 176, 202, 224, 251, 265, 287, 312, 331, 343, 361
Resistance, 68, 89, 91, 92, 97, 98, 226
Resource cost, 70, 71
Return on investment (ROI), 18–20
Revenge of geography, 9–10
Reverse geocoding, 334, 335
Revisions, 182–184, 188, 192, 249, 252, 253, 281, 312, 321, 322, 366
RFID, 3, 32
Risk, 25, 73, 88, 118, 250–257, 294, 334, 335, 337, 338
ROI. *See* Return on investment (ROI)
Routing, 130, 148, 258, 312

S

Scale free, 130, 177
Scaling, 128–133, 135, 136, 279, 370
Scaling of geographic space, 128, 130, 132, 135

Science and technology studies (STS), 289
SDI. *See* Spatial data infrastructures (SDI)
Secondary school, 341–343
Self-synchronization, 309, 318
Semantic similarity. *See* Similarity
Semantic Web, 2, 170, 270
Sense of place, 8, 176, 203, 204, 209, 223–238
Sensor network, 3, 46, 295, 298
Sensors, 3, 34, 46, 66, 289, 295, 298, 301, 304, 311
Sentinel, 333, 335
Serendipity, 25, 27
SES. *See* Socioeconomic Status (SES)
Severe weather, 288, 292–304
Short message service (SMS), 20, 39, 331, 347
Signature distance, 183, 188–190, 193
Similarity, 8, 115, 193, 205, 209, 210, 217, 218, 272, 273
Simulation, 89, 127, 128, 136, 208, 212, 216, 290, 292–294
Situated knowledge, 53
Situated learning, 53, 359
Six degrees of separation, 5
Six Rivers National Forest, 53, 54
SKYWARN, 294, 297
Smart phones, 2, 3, 7, 32, 33, 163, 193, 297, 330, 331, 335, 336, 343, 348, 350
SMS. *See* Short Message Service (SMS)
SNI. *See* Supermarket need index (SNI)
Snow, John, 334
Social capital, 233
Social economy, 79
Social media, 5, 71, 128, 135, 136, 281, 298, 368
Social networking, 34, 55, 67, 92, 93, 96, 98, 228, 233, 252, 267, 268, 277, 279, 290, 301, 304, 308
Social networking sites, 34, 92, 93, 96, 98, 252, 268, 279, 301, 304
Social networks, 23, 34, 52, 55, 67, 92, 93, 96, 98, 193, 195, 228, 233, 234, 250, 252, 266–268, 277–279, 290, 301, 304, 308, 326, 350, 358, 359
Social security number, 339
Socioeconomic Status (SES), 193, 331
Software-as-a-service (SaaS), 69
Space, 7, 8, 21, 25, 86, 89, 126, 128, 130, 132, 135, 136, 140, 147, 162–166, 169, 176, 189, 202, 203, 219, 278, 289,

Index

292, 295, 298, 302, 309, 344, 362–364
Space-time convergence, 5
Spatial analysis, 70, 71, 126, 193, 207, 276, 370
Spatial autocorrelation, 135, 203
Spatial century, 9
Spatial data infrastructures (SDI), 6, 16, 18, 19, 44–46, 51, 52, 61, 66, 158, 178, 191–193, 369
Spatial data mining, 193
Spatial data quality, 22, 106, 315
Spatial footprint, 184–188
Spatial heterogeneity, 130, 135, 202
Spatial information theory, 127, 187
Spatial interaction model, 184, 187
Spatial pattern, 182, 184, 194, 277
Spatial resolution, 300–303
Spatial Thinking, 135, 342, 343
SPC. *See* Storm Prediction Center (SPC)
SPOT map, 334
Spotter Network, 294, 299, 301
State control, 85, 91, 92, 100
State of Victoria, Australia, 253
STD, 335
Storm Prediction Center (SPC), 297
Straight skeleton, 153
Strategic planning, 69
Street block, 132–133
Street nodes, 134
Streets, 23, 45, 55, 71, 126, 129–135, 141, 142, 147, 155, 193, 211, 218, 267, 271, 272, 275, 279, 310–312, 322, 324, 325, 330, 334, 336, 363
Strengths, Weaknesses, Opportunities, and Threats (SWOT), 6
Subjectivity, 85, 88–90, 252
Supermarket need index (SNI), 333
Survey, 4, 9, 97, 105, 107, 108, 193, 248, 253, 268, 270, 273, 275–279, 293, 312, 313, 331, 332, 335, 342, 346, 350–358, 364
SWOT. *See* Strengths, Weaknesses, Opportunities, and Threats (SWOT)
Symbology, 93, 96, 344
Synthesis hybirity, 368–370

T
Tag, 56–58, 177, 288, 348, 349
TeleAtlas, 256, 257, 308, 313
Templates, 90, 179, 180, 183
Temporal
 data, 335, 338
 proximity, 334
Temporal analysis, 203, 216–217
TerraServer, 4
Text analysis, 206
Thematic geography, 202
Third dimension, 139–158
Three-dimensional (3D), 7, 140, 141, 143, 144, 146–150, 152–158, 176
 building models, 143, 148–150, 153, 156
 city models, 141, 156
 navigation, 176
 routing, 148
 VGI, 156
3D styled layer descriptor (3D-SLD), 148
TIGER. *See* Topologically integrated geographic encoding and referencing (TIGER)
Time-space compression, 4, 5
Time-space distanciation, 4
TIP. *See* Tornado Intercept Project (TIP)
Tlowitsis Nation, 223, 224, 226, 230, 232
Tobler's first law, 203, 209, 278
Tobler's First Law of Geography, 203, 209, 278
TomTom, 20, 245, 246, 250, 253, 255–256, 258, 308, 310, 313, 314
Top-down schema, 58
Topic model, 8, 202–206, 208, 217
Topic trend, 217
Topographic, 8, 40, 53, 135, 230, 245–260, 308, 312
Topological center, 133
Topologically integrated geographic encoding and referencing (TIGER), 129, 269
Topology, 128–131, 136
Toponym, 180, 191, 278
Tornado, 294–296, 298–303
Tornado alley, 295
Tornado Intercept Project (TIP), 295
Totable Tornado Observatory (TOTO), 295
TOTO. *See* Totable Tornado Observatory (TOTO)
Translation, 53, 78, 97, 288, 290–292, 294, 298–300, 302–304
Trust, 38, 56, 61, 79, 97, 108, 112, 114, 115, 170, 234, 236, 249–252, 256, 275, 279, 311, 312, 314, 319, 322, 323
Twitter, 3, 37, 70, 71, 73, 95–97, 136, 202, 299, 301, 302, 370

U

Uncertainty, 22–24, 26, 27, 114, 120, 273, 364, 378
Undergraduate students, 296, 357
United Nations Stabilization Mission in Haiti (MINUSTAH), 59
United States Geological Survey (USGS), 2, 45, 53, 58, 61, 253–255, 293
Updating, 8, 47, 58, 95, 106, 165, 246, 247, 249–251, 253, 256, 257, 260, 281, 311, 313, 320
Usability, 43–61, 155, 227, 330, 350, 354, 356, 357, 359
User-created content, 246, 270
User-generated content, 2, 23, 125, 126, 161, 164, 178, 246, 362
User interface, 74, 142, 206, 330, 350
USGS. *See* United States Geological Survey (USGS)

V

Value chain, 19, 20, 24–26
Value, definitions
  economic perspective, 16–20
  philosophical perspective, 17
Value, type
  economic value, 18–20
  qualitative value, 17–19
  social value, 16, 20, 21
Vancouver Island, 223
Vector, 4, 207, 258, 330, 346
Verifiable, 58, 107
Verification of the Origins of Rotation in Tornadoes Experiment (VORTEX), 295, 296
VGS. *See* Volunteered geographic services (VGS)
Virtual Earth, 4, 330
Virtual globe, 140, 143, 144, 343
Virtual personas, 235
Virtual spaces, 128, 135
Visualization, 89, 99, 127, 128, 136, 140, 143, 148, 154, 161, 163, 194, 207, 208, 211, 369
Volunteer, 21, 24, 31–40, 46, 55, 66, 91, 94, 112–114, 116, 117, 245, 250–251, 253, 255, 257, 267, 269, 295, 296, 317, 331, 367

Volunteered geographic information (VGI)
  heterogeneity, 22, 26, 27, 366
  quality, 362, 364–365
  value, 15–27
Volunteered geographic services (VGS), 7, 8, 161–171
VORTEX. *See* Verification of the Origins of Rotation in Tornadoes Experiment (VORTEX)
VRML, 147, 154

W

WARNGEN, 303
Water Hackathon, 10
W3DS. *See* Web 3D Service (W3DS)
Web 2.0, 2, 3, 7, 9, 16, 25, 26, 46, 65–67, 83–100, 125, 140, 161–164, 228, 233, 236, 246, 249, 266, 267, 277, 288, 300, 369
Web crawling, 176
Web demographics, 8, 265–282, 362
Web 3D Service (W3DS), 143, 147, 148
Weblog, 45, 50, 95, 97, 202, 206, 209, 211, 215, 216, 288
Web-scraping, 266, 270
Web 2.0 technologies, 100, 125, 277
Wiki, 55–59, 126, 142, 175, 177, 180, 182, 183, 191, 194, 195, 344, 348, 366
Wikification, 85, 178, 234
Wikipedia, 3, 8, 55, 56, 113, 126, 163, 175–195, 202, 252, 257, 281, 364, 365
Wikipedian, 177, 178, 183, 187, 189, 193, 194
Wikipedia-World, 180, 184
Wikiscanner, 178, 258
Wikitext markup language, 179, 180, 183
Wise crowds, 314
Workflow, 52, 71, 181, 189, 246, 249–257
WorldWind, 4

X

XML. *See* eXtensible Markup Language (XML)
XNavigator, 143, 144, 148

Z

Zettabyete, 3
ZIP Code, 271, 332, 333, 335
Zipf's Law, 133–135, 177, 271

Printed by Printforce, the Netherlands